DIMENSION AND EXTENSIONS

North-Holland Mathematical Library

VOLUME 48

NORTH-HOLLAND
AMSTERDAM • LONDON • NEW YORK • TOKYO

Dimension and Extensions

J.M. AARTS
Faculty of Technical Mathematics and Informatics
Delft University of Technology
Delft, The Netherlands

T. NISHIURA
Department of Mathematics
Wayne State University
Detroit, MI, USA

1993
NORTH-HOLLAND
AMSTERDAM • LONDON • NEW YORK • TOKYO

ELSEVIER SCIENCE PUBLISHERS B.V.
Sara Burgerhartstraat 25
P.O. Box 211, 1000 AE Amsterdam, The Netherlands

Library of Congress Cataloging-in-Publication Data

Aarts, J. M.
Dimension and extensions / J.M. Aarts, T. Nishiura.
p. cm. -- (North-Holland mathematical library ; v. 48)
Includes bibliographical references and index.
ISBN 0-444-89740-2 (alk. paper)
1. Dimension theory (Topology) 2. Mappings (Mathematics)
3. Compactifications. I. Nishiura, T. II. Title. III. Series.
QA611.3.A27 1993
514'.32--dc20 92-44402
CIP

ISBN: 0 444 89740 2

This book is printed on acid-free paper

Printed in The Netherlands

Dedicated to the memory of
Johannes de Groot

PREFACE

Two types of seemingly unrelated extension problems are discussed in this book. Their common focus is a long standing problem proposed by Johannes de Groot, the main conjecture of which was recently resolved. As is true of many important conjectures, a wide range of mathematical investigations had developed. These investigations have been grouped into the two extension problems under discussion.

The problem of de Groot concerned compactifications of spaces by means of an adjunction of a set of minimal dimension. This minimal dimension was called the compactness deficiency of a space. Early successes in 1942 lead de Groot to invent a generalization of the dimension function, called the compactness degree of a space, with the hope that this function would internally characterize the compactness deficiency which is a topological invariant of a space that is externally defined by means of compact extensions of a space. From this, two extension problems were spawned.

The first extension problem concerns the extending of spaces. Among the various extensions studied here, the important ones are compactifications and metrizable completions. Also, σ-compact extensions are discussed. The natural problems are the construction of dimension preserving extensions satisfying various extra conditions and the construction of extensions possessing adjoined subsets satisfying certain restrictions.

The second extension problem concerns extending the theory of dimension by replacing the empty space with other spaces. The compactness degree of de Groot was defined by the replacement of the empty space with compact spaces in the initial step of the definition of the small inductive dimension. Such replacements lead to two kinds of investigations. The first is the development of a generalized dimension theory. The other kind is the search for a dimension-like invariant which would internally characterize the first extension problem.

It took almost fifty years to settle the original problem of de Groot. Surprisingly, the analogous problem for metric completions turned out to be much more manageable. It also became apparent in many other analogous problems that "excision" was a more natural concept than that of extension. In 1980 Pol produced an example to show that the compactness degree was not the right candidate for characterizing the compactness deficiency. Then Kimura showed in 1988 that the compactness dimension function that was introduced by Sklyarenko in 1964 characterized the compactness deficiency. Also, from 1942 to the present, a substantial theory of the extension of dimension theory to dimension-like functions evolved. This book is a presentation of the current status of the two extension problems.

The organization of the book

The material has been arranged in six chapters with the following themes: history, mappings into spheres, inductive invariants, covering dimension, basic dimension and compactifications. The first chapter is a historical introduction, dealing mainly with separable metrizable spaces. In it we discuss prototypes of the results which will be obtained in the subsequent chapters. We have also included a short introduction into dimension theory, making the book essentially self-contained. In Chapter II the theory of dimension and mappings into spheres is generalized to dimension-like functions. A wealth of examples is presented, including those related to the Borel classes. Throughout the book, use is made of the material in Chapters I and II. Chapters III and IV deal with functions of inductive dimensional type and with functions of covering dimensional type. In Chapter V we return to the class of metrizable spaces to discuss the basic dimension functions, that is, functions that depend upon the existence of special bases for the open sets of a space. These functions are used to relate the many dimension functions that were developed in the earlier chapters. In the final chapter the various compactifications are discussed in a unified way. Included in this discussion are the compactification of de Groot [1942] which was the origin of the theory presented in the book and the compactification of Kimura [1988] which finally resolved the compactification problem.

What is new?

Much of the material is taken from the literature. Some of the new material appearing in the book is summarized below.

- The discussion of the basic and the order dimension leading to the equality of the generalized covering and inductive dimensions in many cases.
- The characterization of the generalized covering dimension by means of mappings into spheres, resulting in the proof of the equality of strong inductive compactness dimension and covering compactness dimension.
- The introduction of the class of spaces called the Dowker universe which is also of interest in dimension theory proper.
- The line-up of the various compactness dimension functions.
- The discussion of recent results in the axiomatics of the dimension functions, with emphasis on the class of metrizable spaces, including the result of Hayashi.
- The integrated discussion of compactifications constructed by Zippin, Freudenthal, de Groot, de Vries and Kimura as Wallman compactifications.

What is next?

The future is not ours to see, of course, but surely there are many possible investigations yet to be pursued. Some unsolved problems have been posed at the end of each chapter. The listing is by no means exhaustive. We have tried to indicate what we think are the more interesting open problems. This book has not touched upon the transfinite dimension and its possible generalizations. The 1987 paper by Pol may serve as an introduction into this problem area.

Acknowledgements

We are indebted to K. P. Hart for valuable advice on the use of TeX and AMS-TeX. To Eva Coplakova go our thanks for reading various versions of the book and for providing us with the English translations of the Russian papers cited in the Bibliography whose translations were not available in the literature. Our thanks is due to Jan van Mill for helpful comments and continuous encouragement.

Finally, we would like to thank both the Faculty of Technical Mathematics and Informatics of the Delft University of Technology and the Department of Mathematics of Wayne State University. Their generous support enabled us to work together on the book.

The book was typeset by $\mathcal{A}\mathcal{M}\mathcal{S}$-TEX.

J. A. and T. N.

CONTENTS

Sculpture "Needle Tower II" by Kenneth Snelson
Photo: Jan Aarts
(reprinted with permission of K. Snelson)

CHAPTER I

THE SEPARABLE CASE IN HISTORICAL PERSPECTIVE

This introductory chapter presents an exposition of the history of a compactification problem in dimension theory and the dimension-like functions that have grown out of it. At the same time, it presents the plan of the book. By no means is the chapter just a collection of historical facts, rather it is a mathematical text in which some emphasis has been placed on the history of the compactification problem. For the sake of simplicity, many of the results will not be presented in full generality; the later chapters will make up for this.

1. A compactification problem

An appropriate subtitle for the book would be "A compactification problem in dimension theory." The theory presented in this book originated from a compactification problem posed by Johannes de Groot [1942] in his thesis. This problem was posed in a form that made it appear to be a natural generalization of the small inductive dimension. It was finally solved in the negative by Pol [1982]. Meanwhile a substantial theory of dimension-like functions had been developed in the various attempts to solve the problem. It is this theory that is the core of our book.

Since the emphasis in this chapter is being placed on the historical perspective of the compactification problem and the theory of dimension-like functions, it is proper that the chapter begin with a few remarks about the origins of dimension theory itself and the most important results of dimension theory. The dimension-like functions and the compactification problem will then appear in their natural historical settings.

Though the chapter has been made as self-contained as possible, there are details of two major points which have been delayed to later chapters. The first of these points is the sum and decomposition theorems whose details will be discussed in Chapter III, and the second is the constructions of various compactifications whose details will be given in Chapter VI.

2. Dimensionsgrad

The need for a dimension theory came at the end of the 19th century when examples of "dimension-raising" maps had been discovered. In 1878 Cantor showed that the line and the plane have the same number of points by constructing a bijective function from the real line $\mathbb{R}$ onto the plane $\mathbb{R}^2$. Then in 1890 Peano constructed a continuous mapping of the interval $[0,1]$ onto its square and thereby exhibited a continuous parametrization of the square by means of an interval. It was the first space-filling curve. Apparently a naive concept of dimension was no longer adequate.

In the paper [1912], published in a philosophical journal, Poincaré pointed out that dimension can be defined in an inductive way by using the notion of separation of a space. In the next year Brouwer [1913] gave the first definition of a topological dimension function, the *Dimensionsgrad*, and showed that the Dimensionsgrad of the n-cube $\mathbb{I}^n$ is equal to n. The definition of the Dimensionsgrad is an inductive one based on a notion of separation. But Brouwer's notion of separation is somewhat different from the one which is used nowadays in dimension theory. Because of the inductive nature of the definition of dimension, the value of the dimension function depends in a sensitive way on the start of the definition, that is, on the definition of the zero-dimensional sets. In this respect, Brouwer's definition was not the best possible choice. In Chapter III we shall study the inductive dimensions and their generalizations and reveal the dependence of the value of the dimension-like functions upon the start of the definitions.

After the structure of zero-dimensional spaces had been explored by various mathematicians, Urysohn in [1922] and Menger in [1923] finally and independently defined the dimension function which eventually became known as the *small inductive dimension*. The fine point of the definition is the step before the first step, namely the assigning of the dimension number minus one to the empty space. In this way the correct choice of the collection of zero-dimensional spaces was made. But one should not be deceived, the step from 0 to 1 still contains many delicate points as the general theory of dimension-like functions will show.

3. The small inductive dimension ind

We shall now give a precise formulation of the definition of the small inductive dimension and state the most important theorems.

The following notion of separation will be used.

3.1. Definition. Let X be a topological space and let F and G be disjoint sets in X. A subset S of X is called *a partition between F and G* if $X \setminus S$ can be written as a union of disjoint open sets U and V with one containing F and the other containing G, that is,

$$X \setminus S = U \cup V, \quad U \cap V = \emptyset, \quad F \subset U, \quad G \subset V.$$

If F consists of only one point, $F = \{p\}$, we say that S is *a partition between p and G*.

Notice that a partition is always a closed set.

3.2. Definition. To every topological space X one assigns the *small inductive dimension*, denoted by $\operatorname{ind} X$, as follows.

(i) $\operatorname{ind} X = -1$ if and only if $X = \emptyset$.

(ij) For each natural number n, $\operatorname{ind} X \leq n$ if for each point p in X and for each closed set G of X with $p \notin G$ there is a partition S between p and G such that $\operatorname{ind} S \leq n-1$.

(iij) $\operatorname{ind} X = n$ if $\operatorname{ind} X \leq n$ and $\operatorname{ind} X \not\leq n-1$.

(iv) $\operatorname{ind} X = \infty$ if the inequality $\operatorname{ind} X \leq n$ does not hold for any natural number n.

The reader may have noticed that we have adopted the convention that *the natural numbers start with* 0. We shall denote the set of natural numbers by $\mathbb{N}$.

It is immediately clear that ind is a topological invariant, i.e., $\operatorname{ind} X = \operatorname{ind} Y$ whenever the topological spaces X and Y are homeomorphic. A somewhat closer look at the definition of ind reveals that $\operatorname{ind} X = \infty$ for each nonregular topological space X. Here are a few examples to illustrate the definition.

3.3. Examples. The sets of real, rational and irrational numbers will be denoted by $\mathbb{R}$, $\mathbb{Q}$ and $\mathbb{P}$ respectively.

a. For a and b in $\mathbb{P}$ with $a < b$ the set $(a, b) \cap \mathbb{Q}$ is closed as well as open in $\mathbb{Q}$. It follows that for p in $(a, b) \cap \mathbb{Q}$ the empty set is a partition between p and $\mathbb{Q} \setminus (a, b)$ in $\mathbb{Q}$. Consequently, $\operatorname{ind} \mathbb{Q} = 0$. A similar argument will show that $\operatorname{ind} \mathbb{P} = 0$.

b. As $\mathbb{R}$ is connected, $\operatorname{ind} \mathbb{R} > 0$. For $\varepsilon > 0$ and for $p \in \mathbb{R}$ the two-point set $\{ p - \varepsilon, p + \varepsilon \}$ is a partition between p and the closed set $(-\infty, p - 2\varepsilon] \cup [p + 2\varepsilon, +\infty)$. It easily follows that $\operatorname{ind} \mathbb{R} = 1$. Notice that the 1-dimensional space $\mathbb{R}$ can be covered by two 0-dimensional spaces since $\mathbb{R} = \mathbb{P} \cup \mathbb{Q}$.

c. The n-dimensional sphere $\mathbb{S}^n$ is the subset of $\mathbb{R}^{n+1}$ defined by

$$\mathbb{S}^n = \{ x : x \in \mathbb{R}^{n+1}, \ \|x\| = 1 \}$$

where n is in $\mathbb{N}$ and $\|\cdot\|$ denotes the Euclidean norm. Denoting the spherical neighborhood of the point p in $\mathbb{S}^n$ with radius ε by $S_\varepsilon(p)$, we have that its boundary $B(S_\varepsilon(p))$ is homeomorphic with $\mathbb{S}^{n-1}$ when $0 < \varepsilon < 2$ and $n \geq 1$. It readily follows that $\operatorname{ind} \mathbb{S}^n \leq n$. The inequality $\operatorname{ind} \mathbb{S}^n \geq n$ (whence $\operatorname{ind} \mathbb{S}^n = n$) will be established in Theorem 4.8.

d. With the aid of **c**, namely $\operatorname{ind} \mathbb{S}^n = n$, we shall prove $\operatorname{ind} \mathbb{R}^n = n$. Recall that $\mathbb{S}^n$ is homeomorphic with the one-point compactification of $\mathbb{R}^n$. Note that $\mathbb{S}^n$ is homogeneous, that is, for any two points x and y of $\mathbb{S}^n$ there is a homeomorphism of $\mathbb{S}^n$ onto itself sending x to y. Thus

$$\operatorname{ind} \mathbb{R}^n = n.$$

Therefore the small inductive dimension distinguishes $\mathbb{R}^n$ from $\mathbb{R}^m$ for $n \neq m$ and thus satisfies a basic requirement of a dimension function.

The theory of the small inductive dimension was developed between 1920 and 1940. A beautiful account can be found in the book by Hurewicz and Wallman [1941]. See also the more recent book by Engelking [1978], in particular Chapter 1. This theory of ind has become the paradigm of the theories about inductive invariants. The *subspace theorem*, the *sum theorem* and the *decomposition theorem*, to be discussed below, are essential features of it. As these types of theorems make their appearances in the book they will be so identified.

3.4. Theorem (Subspace theorem). *For every subspace Y of a space X,*

$$\operatorname{ind} Y \leq \operatorname{ind} X.$$

Proof. The proof is a very nice example of the type of inductive proofs found in dimension theory. The statement in the theorem holds trivially for $\operatorname{ind} X = \infty$. So, without loss of generality, we may assume that $\operatorname{ind} X = n < \infty$. The statement is obvious if $X = \emptyset$ (equivalently, $\operatorname{ind} X = -1$, the first step of the inductive proof). Let us assume for a natural number n that the statement has been proved for all spaces Z with $\operatorname{ind} Z \leq n - 1$ and let X be a space with $\operatorname{ind} X = n$. Suppose that p is a point in a subspace Y of X and that G is a closed subset of Y with $p \notin G$. Then $\operatorname{cl}_X(G)$ is a closed subset of X with $p \notin \operatorname{cl}_X(G)$. Because $\operatorname{ind} X = n$, there is a partition S in X between p and $\operatorname{cl}_X(G)$ such that $\operatorname{ind} S \leq n - 1$. It follows directly from the definition of a partition that $S \cap Y$ is a partition in Y between p and G. By the induction hypothesis we have $\operatorname{ind}(S \cap Y) \leq \operatorname{ind} S \leq n - 1$. Thus we have $\operatorname{ind} Y \leq n$. Thereby the theorem is proved.

The following two propositions suggest other ways of defining the small inductive dimension. For regular spaces these alternative definitions result in the same dimension function. The straightforward proofs will be left to the reader.

3.5. Proposition. *Let X be a regular space. For each natural number n, $\operatorname{ind} X \leq n$ if and only if each point p of X has a neighborhood base $\mathcal{B}(p)$ such that $\operatorname{ind} \mathrm{B}(U) \leq n - 1$ for every U in $\mathcal{B}(p)$.*

In the statement of the proposition, the notation $\mathrm{B}(U)$ has been used to denote the boundary of the set U. The content of the proposition is sometimes stated as follows:

> $\operatorname{ind} X \leq n$ *if and only if each point p of X has arbitrarily small neighborhoods U with* $\operatorname{ind} \mathrm{B}(U) \leq n - 1$.

3.6. Proposition. *Suppose that X is a regular space. For each natural number n, $\operatorname{ind} X \leq n$ if and only if there exists a base $\mathcal{B}$ for the open sets of X such that $\operatorname{ind} \mathrm{B}(U) \leq n - 1$ for every U in $\mathcal{B}$.*

We shall now formulate the countable sum theorem and the decomposition theorem. The proofs of both of them will be presented in Chapter III in a much more general setting, see Theorem III.2.10 in particular. Up to the discussion in Section III.2 we shall freely use both theorems. The reader should not worry too much about this because the sum and decomposition theorems can be regarded as axioms of small inductive dimension until Chapter III. Though this arrangement of material is not the most logical one, it will enable us to present some motivating examples first. Indeed, the arrangement is quite natural for the thematic approach in this chapter.

3.7. Theorem (Countable sum theorem). *Let* $\{ F_i : i \in \mathbb{N} \}$ *be a countable, closed cover of a separable metrizable space* X. *Then*

$$\operatorname{ind} X = \sup \{ \operatorname{ind} F_i : i \in \mathbb{N} \}.$$

Recall that a subset of a space X is said to be an *F_σ-set* if it can be represented as a countable union of closed subsets of X. Note that an open subset of a metrizable space is an F_σ-set of that space.

3.8. Theorem (Decomposition theorem). *Let* X *be a separable metrizable space. If* $\operatorname{ind} X \leq n$, *then* X *can be partitioned into* $n+1$ *disjoint subsets* X_i *with* $\operatorname{ind} X_i \leq 0$, $i = 0, 1, \ldots, n$. *Moreover, one of the* X_i*'s may be assumed to be an* F_σ*-set. Conversely, if* $X = \bigcup \{ X_i : i = 0, 1, \ldots, n \}$ *is such that* $\operatorname{ind} X_i \leq 0$ *for every* i, *then* $\operatorname{ind} X \leq n$.

The sum theorem and the decomposition theorem are very powerful. Their power will be best illustrated by presenting some applications. Just as in the statements of the previous two theorems we shall sometimes assign a type to a theorem for easy reference.

3.9. Theorem (Addition theorem). *Let* X *be a separable metrizable space. If* $X = Y \cup Z$, *then*

$$\operatorname{ind} X \leq \operatorname{ind} Y + \operatorname{ind} Z + 1.$$

Proof. We may assume that both Y and Z are not empty. Let $\operatorname{ind} Y = n$ and $\operatorname{ind} Z = m$. Then we have the two decompositions

$$Y = \bigcup \{ Y_i : i = 0, 1, \ldots, n \}, \quad \operatorname{ind} Y_i = 0, \quad i = 0, 1, \ldots, n,$$
$$Z = \bigcup \{ Z_j : j = 0, 1, \ldots, m \}, \quad \operatorname{ind} Z_j = 0, \quad j = 0, 1, \ldots, m.$$

Because of the equality

$$X = \left(\bigcup\{ Y_i : i = 0, 1, \ldots, n \}\right) \cup \left(\bigcup\{ Z_j : j = 0, 1, \ldots, m \}\right),$$

we have by the decomposition theorem that

$$\operatorname{ind} X \leq (n+1) + (m+1) - 1.$$

The theorem now follows.

By applying the decomposition theorem to $\mathbb{P}$, $\mathbb{Q}$ and $\mathbb{R}$ of Examples 3.3.a and b, we can prove $\operatorname{ind} \mathbb{R} \leq 1$ in a sophisticated way.

In the sequel we shall frequently use the product theorem for the small inductive dimension.

3.10. Theorem (Product theorem). *Let $X \times Y$ be the topological product of two separable metrizable spaces X and Y at least one of which is not empty. Then*

$$\operatorname{ind}(X \times Y) \leq \operatorname{ind} X + \operatorname{ind} Y.$$

Proof. The proof is by way of a double induction on $\operatorname{ind} X = n$ and $\operatorname{ind} Y = m$. The statement of the theorem is evident if either $\operatorname{ind} X = -1$ or $\operatorname{ind} Y = -1$. Let $\operatorname{ind} X = n$ and $\operatorname{ind} Y = m$ and suppose that the statement has been proved correct for the cases

(1) $\operatorname{ind} X \leq n$ and $\operatorname{ind} Y \leq m - 1$,
(2) $\operatorname{ind} X \leq n - 1$ and $\operatorname{ind} Y \leq m$.

Each point (x, y) in $X \times Y$ has arbitrarily small neighborhoods of the form $U \times V$ where U and V are open neighborhoods of x and y in X and Y respectively with $\operatorname{ind} \mathrm{B}_X(U) \leq n - 1$ and $\operatorname{ind} \mathrm{B}_Y(V) \leq m - 1$. We have

$$\mathrm{B}_{X \times Y}(U \times V) = \left(\mathrm{cl}_X(U) \times \mathrm{B}_Y(V)\right) \cup \left(\mathrm{B}_X(U) \times \mathrm{cl}_Y(V)\right).$$

As both sets in the union of the right-hand side are closed, we have

$$\operatorname{ind} \mathrm{B}_{X \times Y}(U \times V) \leq n + m - 1$$

by the subspace and sum theorems and the induction hypotheses (1) and (2); thereby the induction is completed.

Another consequence of the sum theorem is the point addition theorem: *The dimension of a nonempty separable metrizable space cannot be raised by the adjunction of a single point.*

3.11. Theorem (Point addition theorem). *Suppose that X is a separable metrizable space consisting of more than one point. For each point p in X,*

$$\operatorname{ind} X = \operatorname{ind}(X \setminus \{p\}).$$

Proof. By the subspace theorem we have $\operatorname{ind}(X \setminus \{p\}) \leq \operatorname{ind} X$. For each n in $\mathbb{N}$ define $F_n = \{x \in X : \frac{1}{n+1} \leq d(x,p)\}$. As X is the union $\{p\} \cup (\bigcup\{F_n : n \in \mathbb{N}\})$, the countable sum theorem gives $\operatorname{ind} X = \max\{0, \sup\{\operatorname{ind} F_n : n \in \mathbb{N}\}\}$. Because F_n is not empty for some n, we have $\operatorname{ind} X \leq \sup\{\operatorname{ind} F_n : n \in \mathbb{N}\} = \operatorname{ind}(X \setminus \{p\})$.

Henceforth in our discussions involving the small inductive dimension we shall always assume that our spaces are regular. This is done so as to avoid some pathological situations. As an illustration, consider the following example.

3.12. Example. Let the space Y be the set of real numbers $\mathbb{R}$ with the topology determined by the subbase consisting of the usual open sets of $\mathbb{R}$ and the set $\mathbb{Q}$ of the rational numbers. We note that Y is the standard example of a Hausdorff space which is not regular; there are no partitions between the closed set $\mathbb{P}$ of irrational numbers and any rational point q.

By the observation made after Definition 3.2 we have $\operatorname{ind} Y = \infty$. The collection $\mathcal{B}$ of all sets of the forms (a,b) and $(a,b) \cap \mathbb{Q}$, $a < b$, is a base for the open sets of Y. Also, a set of the form $(a,b) \cap \mathbb{Q}$ has $([a,b] \cap \mathbb{P}) \cup \{a, b\}$ as its boundary. Now note that the subspace topology of $\mathbb{P}$ induced by Y coincides with the one induced by the usual topology of $\mathbb{R}$. From this we find $\operatorname{ind} \mathrm{B}(U) = 0$ for each member U of $\mathcal{B}$. Let us define a dimension function ind^* for (not necessarily regular) topological spaces in the way suggested by Proposition 3.6, namely $\operatorname{ind}^* \emptyset = -1$ and $\operatorname{ind}^* X \leq n$ if there exists a base $\mathcal{B}$ for the open sets of X such that $\operatorname{ind}^* \mathrm{B}(U) \leq n - 1$ for every U in $\mathcal{B}$. For this dimension function we find $\operatorname{ind}^* Y = 1$ as opposed to $\operatorname{ind} Y = \infty$. In this way one can define a dimension function for which some nonregular spaces will have finite dimensional values. However, such a function will have very limited consequences for the theory that will evolve. As we have seen in Proposition 3.6, the gap between the dimension functions ind and ind^* will disappear in the realm of regular spaces.

4. The large inductive dimension Ind

In [1931] Čech introduced the large inductive dimension Ind which is closely related to the Dimensionsgrad of Brouwer. We shall present the definition of Ind and deduce some basic results.

The definition of Ind is similar to that of ind. It is based on the separation of disjoint closed sets rather than the separation of points and closed sets.

4.1. Definition. To every topological space X one assigns the *large inductive dimension*, denoted by $\operatorname{Ind} X$, as follows.

(i) $\operatorname{Ind} X = -1$ if and only if $X = \emptyset$.
(ij) For each natural number n, $\operatorname{Ind} X \leq n$ if for each pair of disjoint closed sets F and G there is a partition S between F and G such that $\operatorname{Ind} S \leq n - 1$.
(iij) $\operatorname{Ind} X = n$ if $\operatorname{Ind} X \leq n$ and $\operatorname{Ind} X \not\leq n - 1$.
(iv) $\operatorname{Ind} X = \infty$ if the inequality $\operatorname{Ind} X \leq n$ does not hold for any natural number n.

It is immediately clear that Ind is a topological invariant. If the topological space X is not a normal space, then $\operatorname{Ind} X = \infty$. The following results can be established by easy inductive arguments.

4.2. Proposition. *For every T_1-space X,*

$$\operatorname{ind} X \leq \operatorname{Ind} X.$$

4.3. Proposition. *For every closed subspace F of a space X,*

$$\operatorname{Ind} F \leq \operatorname{Ind} X.$$

For many spaces X the equality $\operatorname{ind} X = \operatorname{Ind} X$ holds. More specifically, ind and Ind coincide for every separable metrizable space as we shall prove shortly. But, as a matter of fact, it was not until [1962] that an example was constructed by Roy of a metrizable space Δ with $\operatorname{ind} \Delta = 0$ and $\operatorname{Ind} \Delta = 1$. Other examples of such metrizable spaces have been presented in Kulesza [1990] and Ostaszewski [1990]. The space Δ will appear later in various examples. A detailed exposition of the space Δ can be found in Pears [1975]. The coincidence of ind and Ind will be discussed further in Chapters III and V.

4.4. Theorem (Coincidence theorem). *Suppose that X is a separable metrizable space. Then*

$$\operatorname{ind} X = \operatorname{Ind} X.$$

Proof. In view of Proposition 4.2, only $\operatorname{Ind} X \le \operatorname{ind} X$ needs to be proved. We may assume that $\operatorname{ind} X < \infty$. The proof is by induction on the value of $\operatorname{ind} X$. The inequality obviously holds for $X = \emptyset$. We assume that the inequality has been proved for all separable metrizable spaces Z with $\operatorname{ind} Z \le n-1$. Let X be a separable metrizable space with $\operatorname{ind} X = n$. Suppose that F and G are disjoint closed sets of X. By proposition 3.6 there exists a base $\mathcal{B}$ for the open sets such that $\operatorname{ind} \mathrm{B}(U) \le n-1$ for every U in $\mathcal{B}$. As X is a second countable space, we may assume that $\mathcal{B}$ is countable as well (see Engelking [1977], Theorem 1.1.15). Consider the collection $\{(C_j, D_j) : j \in \mathbb{N}\}$ of all pairs of elements of $\mathcal{B}$ with $\operatorname{cl}(C_j) \subset D_j$ such that $\operatorname{cl}(D_j) \cap G = \emptyset$ or $D_j \cap F = \emptyset$. For each k in $\mathbb{N}$ define

$$V_k = D_k \setminus \bigcup\{\operatorname{cl}(C_j) : j = 0, 1, \ldots, k-1\}.$$

The open collection $\mathbf{V} = \{V_k : k \in \mathbb{N}\}$ is a locally finite cover of X, and

$$\mathrm{B}(V_k) \subset \mathrm{B}(C_0) \cup \cdots \cup \mathrm{B}(C_{k-1}) \cup \mathrm{B}(D_k)$$

for each k. Now let W be the open set $\bigcup\{V_k : \operatorname{cl}(D_k) \cap G = \emptyset\}$. Clearly F is a subset of W. As $\mathbf{V}$ is locally finite, $\operatorname{cl}(W) \cap G = \emptyset$ and $\mathrm{B}(W) \subset \bigcup\{\mathrm{B}(V_k) : \operatorname{cl}(D_k) \cap G = \emptyset\}$ hold. Subsequently we have $\operatorname{ind}\big(\bigcup\{\mathrm{B}(V_k) : \operatorname{cl}(D_k) \cap G = \emptyset\}\big) \le n-1$ from the countable sum theorem and $\operatorname{ind} \mathrm{B}(W) \le n-1$ from the subspace theorem. The induction hypothesis yields $\operatorname{Ind} \mathrm{B}(W) \le n-1$. As $\mathrm{B}(W)$ is a partition between F and G, we have $\operatorname{Ind} X \le n$.

Consider a space Y with $\operatorname{Ind} Y = n$ that is placed in a hereditarily normal space X. Suppose that F and G are disjoint closed subsets of Y. It follows immediately from the definition of Ind that there is a partition S between F and G in Y with $\operatorname{Ind} S \le n-1$. We shall prove now that such partitions may be extended to partitions between F and G in X.

4.5. Lemma. *Suppose that Y is a subspace of a hereditarily normal space X. Let F and G be disjoint closed subsets of Y. Then for each partition S between F and G in Y there exists a partition T between F and G in X such that $S = T \cap Y$.*

Proof. For a partition S between F and G in Y there are disjoint open subsets U and V of Y such that $Y \setminus S = U \cup V$, $F \subset U$ and $G \subset V$. The sets U and V have disjoint closures in the open subspace $Y' = X \setminus (\mathrm{cl}_X(S) \cup (\mathrm{cl}_X(U) \cap \mathrm{cl}_X(V)))$ of X. So there are disjoint open subsets U' and V' of Y' (whence of X) with $U \subset U'$ and $V \subset V'$. The required set T is $X \setminus (U' \cup V')$.

4.6. Proposition. *Suppose that X is a hereditarily normal space. Let Y be a subspace of X with $\mathrm{Ind}\, Y \leq n$. Then for every pair of disjoint closed subsets F and G of X there is a partition S between F and G in X with $\mathrm{Ind}\,(S \cap Y) \leq n-1$.*

Proof. Let F' and G' be disjoint closed neighborhoods of F and G respectively. There is a partition T in the subspace Y between $F' \cap Y$ and $G' \cap Y$ with $\mathrm{Ind}\, T \leq n-1$. Note that T is also a partition between F and G in the subspace $F \cup Y \cup G$. By Lemma 4.5 there is a partition S in X between F and G such that $T = S \cap Y$.

By repeatedly applying the previous proposition we get the following corollary.

4.7. Corollary. *Suppose that X is a hereditarily normal space and Y is a subspace of X with $\mathrm{Ind}\, Y \leq n$. For each collection of $n+1$ pairs (F_i, G_i) of disjoint closed subsets of X, $i = 0, 1, \ldots, n$, there are partitions S_i between F_i and G_i in X for every i such that $Y \cap S_0 \cap S_1 \cap \cdots \cap S_n = \emptyset$.*

We are now in a position to prove the results announced in the Example 3.3.c.

4.8. Theorem. *For every positive integer n,*

$$\mathrm{ind}\, \mathbb{R}^n = \mathrm{ind}\, \mathbb{S}^n = \mathrm{Ind}\, \mathbb{R}^n = \mathrm{Ind}\, \mathbb{S}^n = n.$$

Proof. Denote the interval $[-1, 1]$ by $\mathbb{I}$. In view of the inequalities that were established in Example 3.3 and also of the subspace and coincidence theorems, it will be sufficient to establish $\mathrm{Ind}\, \mathbb{I}^n \geq n$.

By Corollary 4.7 we need to exhibit a collection of n pairs (F_i, G_i) of disjoint closed subsets of $\mathbb{I}^n$, $i = 1, \ldots, n$, such that the intersection $S_1 \cap \cdots \cap S_n$ is not empty for all partitions S_i between F_i and G_i in $\mathbb{I}^n$. This property of $\mathbb{I}^n$ is the content of the following theorem which is of interest in itself.

Henceforth we shall adopt the notation $\mathbb{I}$ of the previous proof to denote the interval $[-1, 1]$.

4.9. Theorem. *Let $\{(F_i, G_i) : i = 1, \ldots, n\}$ denote the collection of the pairs of opposite faces of $\mathbb{I}^n$; that is, for $i = 1, \ldots, n$,*

$$F_i = \{(x_1, \ldots, x_n) \in \mathbb{I}^n : x_i = -1\},$$
$$G_i = \{(x_1, \ldots, x_n) \in \mathbb{I}^n : x_i = 1\}.$$

If S_i is a partition between F_i and G_i for $i = 1, \ldots, n$, then

$$S_1 \cap \cdots \cap S_n \neq \emptyset.$$

Proof. The proof is based on the famous Brouwer fixed-point theorem which reads as follows.

> *For every continuous map $g \colon \mathbb{I}^n \to \mathbb{I}^n$ there is an x in $\mathbb{I}^n$ such that $g(x) = x$, that is, x is a fixed-point of g.*

We shall follow the ingenious proof found in Engelking [1978]. The proof is by way of contradiction. Assume that there are partitions S_i between F_i and G_i in $\mathbb{I}^n$, $i = 1, \ldots, n$, such that $S_1 \cap \cdots \cap S_n = \emptyset$. For each i let $\mathbb{I}^n \setminus S_i = U_i \cup V_i$ where U_i and V_i are disjoint open sets with $F_i \subset U_i$ and $G_i \subset V_i$. Consider the continuous map $f \colon \mathbb{I}^n \to \mathbb{I}^n$ whose n component functions f_i are given by

$$f_i(x) = \begin{cases} \dfrac{d(x, S_i)}{d(x, S_i) + d(x, F_i)}, & \text{for } x \in U_i \cup S_i \\[2ex] \dfrac{-d(x, S_i)}{d(x, S_i) + d(x, G_i)}, & \text{for } x \in V_i \cup S_i. \end{cases}$$

It is readily verified that $f_i^{-1}[0] = S_i$, $f_i^{-1}[1] = F_i$ and $f_i^{-1}[-1] = G_i$. So $f[F_i] \subset G_i$ and $f[G_i] \subset F_i$ for $i = 1, \ldots, n$. As $S_1 \cap \cdots \cap S_n = \emptyset$, we have $0 \notin f[\mathbb{I}^n]$. Now the composition of the map f and the projection of $\mathbb{I}^n \setminus \{0\}$ from 0 onto the combinatorial boundary of $\mathbb{I}^n$ is a fixed-point free map, a contradiction.

Let us elaborate on the idea of Proposition 4.6 to obtain a technical lemma which will be useful later on. First we need the following variant of the decomposition theorem.

4.10. Theorem. *Let $\{Y_k : k \in \mathbb{N}\}$ be a collection of F_σ-sets of a separable metrizable space X. Suppose that $\operatorname{ind} Y_k = n_k \geq 0$ for k in $\mathbb{N}$. Then there exists an F_σ-set Z of X such that $\operatorname{ind} Z = 0$ and $\operatorname{ind}(Y_k \setminus Z) = n_k - 1$ for k in $\mathbb{N}$.*

Proof. Using the decomposition theorem, we put $Y_k = Y_k^0 \cup Y_k^1$ for each k in $\mathbb{N}$, where Y_k^0 is an F_σ-set of Y_k with $\operatorname{ind} Y_k^0 = 0$ and where $\operatorname{ind} Y_k^1 = n_k - 1$. Because Y_k is an F_σ-set of X, the set Y_k^0 is an F_σ-set of X. Let $Z = \bigcup\{Y_k^0 : k \in \mathbb{N}\}$. By the subspace theorem and the sum theorem we have $\operatorname{ind} Z = 0$. It follows from the addition theorem that $\operatorname{ind}(Y_k \setminus Z) = n_k - 1$ for k in $\mathbb{N}$.

4.11. Lemma. *Let Y be a subspace of a separable metrizable space X with $\operatorname{ind} Y \leq n$. Suppose that $\boldsymbol{F} = \{S_i : i \in \mathbb{N}\}$ is a family of closed subsets of X such that*

$$\operatorname{ind}(S_{i_1} \cap \cdots \cap S_{i_k} \cap Y) \leq n - k$$

whenever $i_1 < \cdots < i_k$ and $1 \leq k \leq n+1$. Then for each pair of disjoint closed subsets F and G of X there is a partition S between F and G in X such that

$$\operatorname{ind}(S_{i_1} \cap \cdots \cap S_{i_{k-1}} \cap S \cap Y) \leq n - k$$

whenever $i_1 < \cdots < i_{k-1}$ and $1 \leq k \leq n+1$.

The crux of the lemma is that if $\boldsymbol{F}$ is a countable collection of closed sets such that the intersection of each of its k-tuples with some n-dimensional set Y has dimension $n - k$, then the collection $\boldsymbol{F}$ can be enlarged by adding a partition so as to obtain a new collection with the same dimensional property as the original collection $\boldsymbol{F}$.

Proof. Write $X_0 = X$ and $X_k = \bigcup\{S_{i_1} \cap \cdots \cap S_{i_k} : i_1 < \cdots < i_k\}$ for $k = 1, \ldots, n$. Let $Y_k = Y \cap X_k$, $k = 0, 1, \ldots, n$, and note that Y_k is an F_σ-set of Y. By the sum theorem we have $\operatorname{ind} Y_k \leq n - k$. The previous theorem provides a subset Z of Y with $\operatorname{ind} Z = 0$ and $\operatorname{ind}(Y_k \setminus Z) \leq n - k - 1$ for $k = 0, 1, \ldots, n$. In particular, we have $\operatorname{ind}(Y_0 \setminus Z) = \operatorname{ind}(Y \setminus Z) \leq n - 1$. Now recall Theorem 4.4 to get $\operatorname{Ind} Z = 0$. Then by Proposition 4.6 there is a partition S between F and G in X such that $\operatorname{Ind}(S \cap Z) \leq -1$, whence $S \cap Z = \emptyset$. So we have $(S_{i_1} \cap \cdots \cap S_{i_{k-1}} \cap S \cap Y) \subset X_{k-1} \cap S \cap Y \subset Y_{k-1} \setminus Z$ whenever $i_1 < \cdots < i_{k-1}$ and $1 \leq k \leq n+1$ hold, and the lemma follows.

The point addition theorem for Ind is next.

4.12. Theorem (Point addition theorem). *Suppose that X is a regular space consisting of more than one point. For each point p in X,*

$$\operatorname{Ind}(X \setminus \{p\}) \geq \operatorname{Ind} X.$$

Proof. We may assume $\operatorname{Ind}(X \setminus \{p\}) < \infty$. Let F and G be disjoint closed subsets of X. Assume $p \notin F$. Let V be any closed neighborhood of p with $V \cap F = \emptyset$. Define D to be $(G \cup V) \setminus \{p\}$. Then D and F are disjoint closed subsets of $X \setminus \{p\}$. Observe that $X \setminus \{p\}$ must be a normal space because we have assumed $\operatorname{Ind}(X \setminus \{p\}) < \infty$. As any set partitioning D and F in $X \setminus \{p\}$ also partitions G and F in X, the theorem follows.

Let us remark here that the example M of Dowker [1955] shows that the inequality of the above theorem cannot be sharpened to an equality. See Isbell [1964] for a detailed discussion.

5. The compactness degree; de Groot's problem

Now that our introduction of inductive dimension functions has been completed we shall turn our discussion towards the compactification problem of de Groot.

It was mentioned earlier that the compactification problem was posed as a generalization of dimension. The small inductive dimension was the first example of a topological property that was defined by imposing conditions on the boundaries of elements of a suitable base for the open sets (Proposition 3.6). In [1935] Zippin introduced another such property, namely that of a rim-compact space. The terms "semi-compact" or "peripherally compact" have also been used in place of "rim-compact".

5.1. Definition. A regular space X is called *rim-compact* if there exists a base $\mathcal{B}$ for the open sets of X such that $\mathrm{B}(U)$ is compact for each U in $\mathcal{B}$.

Equivalently, a regular space X is rim-compact if for each point p in X and for each closed set G of X with $p \notin G$ there is a compact partition S between p and G.

Zippin [1935] discovered the following link between compactification and dimension.

5.2. Theorem. *Let X be a separable completely metrizable space. Then X is rim-compact if and only if there exists a metrizable compactification Y of X such that the set $Y \setminus X$ is countable.*

It is to be observed that the subspace $Y \setminus X$, like any countable metrizable space, is zero-dimensional. This will follow from the countable sum theorem because the small inductive dimension of a singleton point set is zero. The observation also can be proved directly. Indeed, the set $\{\mathrm{d}(x,p) : x \in (Y \setminus X)\}$ of distances is countable for any point p in Y. It follows that, with at most countably many exceptional ε's, the ε-neighborhood $\mathrm{S}_\varepsilon(p)$ has empty boundary. So there are arbitrarily small neighborhoods $\mathrm{S}_\varepsilon(p)$ with empty boundaries.

This result of Zippin was modified by de Groot in his thesis [1942].

5.3. Theorem (de Groot [1942]). *Suppose that X is a separable metrizable space. Then X is rim-compact if and only if there exists a metrizable compactification Y of X such that* $\operatorname{ind}(Y \setminus X) \leq 0$.

(The proofs of Theorems 5.2 and 5.3 will be presented in Section VI.3.) What makes de Groot's theorem so interesting is that an external property of a space X, namely the existence of a compactification with special properties, is characterized by the internal property of rim-compactness. There is a strong similarity between the definition of rim-compactness and the characterization of zero-dimensionality included in Proposition 3.6. In fact, one obtains the definition of rim-compactness by replacing the empty set by a compact set. In passing, we note that rim-compactness was not the first such generalization of the definition of small inductive dimension. In curve theory, the notions of regular and rational curves already had been introduced. A curve X is called *regular* (*rational*) if there exists a base $\mathcal{B}$ for the open sets of X such that $\mathrm{B}(U)$ is finite (countable) for each U in $\mathcal{B}$. For more details see Whyburn [1942].

In his thesis [1942] de Groot suggested the following possible generalization of Theorem 5.3 which will be called the *compactification problem.* For its formulation we need two more definitions. The first one is an extension of Definition 5.1.

5.4. Definition. To every regular space X one assigns the *small inductive compactness degree* $\operatorname{cmp} X$ as follows.

(i) $\operatorname{cmp} X = -1$ if and only if X is compact.

(ij) For each natural number n, $\operatorname{cmp} X \leq n$ if for each point p in X and for each closed set G of X with $p \notin G$ there is a partition S between p and G such that $\operatorname{cmp} S \leq n-1$.
(iij) $\operatorname{cmp} X = n$ if $\operatorname{cmp} X \leq n$ and $\operatorname{cmp} X \not\leq n-1$.
(iv) $\operatorname{cmp} X = \infty$ if the inequality $\operatorname{cmp} X \leq n$ does not hold for any natural number n.

5.5. Definition. For a separable metrizable space X let $\mathcal{K}(X)$ denote the set of all metrizable compactifications of X. The *compactness deficiency* of X is defined by

$$\operatorname{def} X = \min \{ \operatorname{ind} (Y \setminus X) : Y \in \mathcal{K}(X) \}.$$

At this moment there is no point in defining def in a more general setting. Consideration of only separable metrizable compactifications has the advantage that all possible dimension functions will coincide. Thus def is unambiguously defined. Because every separable metrizable space can be embedded in the Hilbert cube, there will be at least one compactification of X. The announced possible generalization of Theorem 5.3 is as follows.

5.6. Compactification problem (de Groot [1942]). For a separable metrizable space X, does the equality $\operatorname{cmp} X = \operatorname{def} X$ hold?

De Groot conjectured the answer to be yes. Indeed, he noted that Theorem 5.3 can be restated as follows.

5.7. Theorem. *Let $n = 0$ or -1. For every separable metrizable space X, $\operatorname{cmp} X = n$ if and only if $\operatorname{def} X = n$.*

Related to the conjecture of de Groot are the following questions that have motivated this book.

1. What are necessary and sufficient internal conditions on separable metrizable spaces X so that $\operatorname{def} X \leq n$?
2. Is it possible to obtain a fruitful generalization of dimension theory by replacing the empty space in the definition with other spaces?
3. What is the special role of the empty space in the theory of dimension?

Let us disclose the end of the story right away. De Groot's conjecture is not correct. In [1982] Pol gave an example of a separable metrizable space X with $\operatorname{cmp} X = 1$ and $\operatorname{def} X = 2$. (We shall discuss the example in Section 11.) But was this really the end of the story? In the early efforts to resolve de Groot's conjecture or to answer Question 1, various functions like cmp were introduced. One of these functions, namely the one introduced by Sklyarenko [1960] (see Section 6), finally turned out to be successful in answering Question 1. This is the result of Kimura [1988] (see Section 12 and Section VI.5).

Now let us return to the conjecture and the related questions. The remainder of the section contains the very earliest results.

5.8. Theorem. *For every separable metrizable space* X,

$$\operatorname{cmp} X \leq \operatorname{def} X.$$

Proof. We may assume, of course, that $\operatorname{def} X < \infty$. The proof is by induction on $\operatorname{def} X$. If $\operatorname{def} X = -1$, then X is compact and hence $\operatorname{cmp} X = -1$. Suppose that $\operatorname{def} X = n$. Let Y be any metrizable compactification of X with $\operatorname{ind}(Y \setminus X) = n$. Assume that p is a point of X and that G is a closed subset of X with $p \notin G$. Then $\operatorname{cl}_Y(G)$ is a closed subset of Y with $p \notin \operatorname{cl}_Y(G)$. The coincidence theorem gives $\operatorname{Ind}(Y \setminus X) = n$. From Proposition 4.6 there is a partition S between p and $\operatorname{cl}_Y(G)$ in Y with $\operatorname{Ind}\big(S \cap (Y \setminus X)\big) \leq n - 1$. The set $S \cap X$ is a partition between p and G in X. As the set S is compact and $S = \big(S \cap X\big) \cup \big(S \cap (Y \setminus X)\big)$, we have $\operatorname{def}(S \cap X) \leq n - 1$ by the subspace theorem. Then $\operatorname{cmp}(S \cap X) \leq n - 1$ by the induction hypothesis. The proof is completed.

5.9. Propositions. Before discussing several examples let us list a few propositions. The first one is obvious.

A. (Topological invariance). *If X and Y are homeomorphic, then $\operatorname{cmp} X = \operatorname{cmp} Y$ and $\operatorname{def} X = \operatorname{def} Y$ whenever the functions are defined.*

A simple inductive proof will yield the following result.

B. *For every regular space X,* $\operatorname{cmp} X \leq \operatorname{ind} X$.

A proof of the next result is obtained by copying the proof of the subspace theorem.

C. *For each closed subspace Y of a regular space X,* $\operatorname{cmp} Y \leq \operatorname{cmp} X$.

An analogue of Proposition 3.6 is next.

D. *For every regular space X,* $\operatorname{cmp} X \leq n$ *if and only if there exists a base $\mathcal{B}$ for the open sets of X such that* $\operatorname{cmp} \mathrm{B}(U) \leq n-1$ *for every U in $\mathcal{B}$.*

E. *For every open subspace Y of a regular space X with* $\operatorname{cmp} X \geq 0$, $\operatorname{cmp} Y \leq \operatorname{cmp} X$.

The easy proof will be omitted. The following proposition has a straightforward proof.

F. *For every closed subspace Y of a separable metrizable space X,* $\operatorname{def} Y \leq \operatorname{def} X$.

5.10. Examples. A common feature of many of the examples in dimension theory is that the examples themselves are easy to describe but the values are hard to compute. The same is true for some of the examples given here.

a. Define X to be $\{(x,y) \in \mathbb{R}^2 : -1 \leq x < 1,\ -1 \leq y \leq 1\}$. The set X is a closed square with a closed edge removed. As X is locally compact, we find that $\operatorname{cmp} X = 0$ and $\operatorname{def} X = 0$.

b. Form the subspace Y of $\mathbb{I}^2$ by adjoining the points $p = (1,1)$ and $q = (1,-1)$ to the set X in Example **a**, i.e., $Y = X \cup \{p, q\}$. The space Y is then a closed square with an open edge removed. Let us prove $\operatorname{cmp} Y = \operatorname{def} Y = 1$. As Y can be compactified by adding the open edge $\{(1,y) : -1 < y < 1\}$, we have $\operatorname{def} Y \leq 1$. By Theorem 5.8, it remains to be shown that $\operatorname{cmp} Y > 0$. To this end, consider the sets $V_t = \{(t,y) : -1 \leq y \leq 1\}$, $t \in [-1,1)$. Let U be any neighborhood of p that is contained in $\mathrm{S}_{\frac{1}{2}}(p)$. Then $\mathrm{B}(U) \cap V_t$ is not empty for any t for which $(t,1) \in U$ because V_t is a connected interval. It follows that $\mathrm{B}(U)$ cannot be compact. Thus $\operatorname{cmp} Y > 0$ has been shown.

c. $\operatorname{cmp} \mathbb{Q} = \operatorname{def} \mathbb{Q} = 0$. Observe that $\mathbb{Q}$ is also not locally compact.

d. Recall that $\mathbb{I} = [-1,1]$ and let $X = \mathbb{Q}' \times \mathbb{I}^n$ where $\mathbb{Q}' = \mathbb{Q} \cap \mathbb{I}$ and $n \geq 1$. As X can be compactified to $\mathbb{I}^{n+1}$ by adding $(\mathbb{I} \setminus \mathbb{Q}') \times \mathbb{I}^n$ and as the product theorem yields $\operatorname{ind}\big((\mathbb{I} \setminus \mathbb{Q}') \times \mathbb{I}^n\big) \leq n$, it follows that $\operatorname{def} X \leq n$. In Example 7.12 we shall prove that $\operatorname{cmp} X \geq n$. It will then follow that $\operatorname{cmp} X = n = \operatorname{def} X$.

e. The subspace Z of the n-dimensional cube $\mathbb{I}^n$, where $n \geq 2$, is defined as follows. We begin with the set

$$E^{n-1} = \{ (x_1, \ldots, x_{n-1}, 1) : -1 < x_i < 1,\ i = 1, \ldots, n-1 \}.$$

Clearly the set E^{n-1} is one of the open faces of $\mathbb{I}^n$. Then define Z to be $\mathbb{I}^n \setminus E^{n-1}$, the n-cube with an open face removed. We shall show $\operatorname{def} Z = n - 1$. Surprisingly the exact value of $\operatorname{cmp} Z$ is still unknown when $n \geq 4$. Obviously $\operatorname{def} Z \leq n - 1$; just add E^{n-1} to Z. The inequality $\operatorname{def} Z \geq n - 1$ will be established by way of contradiction. Assume $\operatorname{def} Z \leq n - 2$ and let Z^* be a compactification of Z with $\operatorname{ind}(Z^* \setminus Z) \leq n - 2$. Adopting the notation of Theorem 4.9, we consider the pairs $(F_1, G_1), \ldots, (F_{n-1}, G_{n-1})$ of opposite faces of $\mathbb{I}^n$. These faces are also subsets of Z. By Corollary 4.7, there are partitions S_i^* between F_i and G_i in Z^* for $i = 1, \ldots, n-1$ with $(Z^* \setminus Z) \cap S_1^* \cap \cdots \cap S_{n-1}^* = \emptyset$. It follows that $S_1^* \cap \cdots \cap S_{n-1}^*$ is a compact subset of Z. For each S_i^* there is a partition S_i between F_i and G_i in $\mathbb{I}^n$ with $S_i \cap Z = S_i^* \cap Z$ (Lemma 4.5). For t in the open interval $(-1, 1)$ we let

$$V_t = \{ (x_1, \ldots, x_{n-1}, t) : -1 \leq x_i \leq 1,\ i = 1, \ldots, n-1 \}.$$

The set V_t is a partition between F_n and G_n in $\mathbb{I}^n$ for each t. With the aid of Theorem 4.9 it will follow that $S_1^* \cap \cdots \cap S_{n-1}^* \cap V_t \neq \emptyset$ for each t. From this we can readily deduce that $S_1^* \cap \cdots \cap S_{n-1}^*$ is not compact, a contradiction.

f. Let X be the subset of $\mathbb{R}^2$ defined by $X = \mathrm{S}_1((0,0)) \cup \{ (1,0) \}$, an open disc together with one point from its boundary. It can be shown in almost the same way as in Example **b** that $\operatorname{cmp} X = 1 = \operatorname{def} X$. Denote the point $(1,0)$ by p and let A and B be defined by

$$A = \{ x \in X : \tfrac{2}{n+2} \leq d(x,p) \leq \tfrac{2}{n+1} \text{ for an odd } n \text{ in } \mathbb{N} \} \cup \{ p \},$$
$$B = \{ x \in X : \tfrac{2}{n+2} \leq d(x,p) \leq \tfrac{2}{n+1} \text{ for an even } n \text{ in } \mathbb{N} \} \cup \{ p \}.$$

Both A and B are closed and rim-compact. As $X = A \cup B$, this example shows that the (finite) sum theorem does not hold for the function cmp. The example also shows the failure of the (point) addition theorem for cmp.

A useful relationship between def and ind can be established by using dimension preserving compactifications. The existence of such compactifications will be established in Section VI.2. We have the following theorem.

5.11. Theorem. *For every separable metrizable space X,*

$$\operatorname{def} X \leq \operatorname{ind} X.$$

Proof. Let Y be a metrizable dimension preserving compactification for X (Theorem VI.2.7). Then $\operatorname{def} X \leq \operatorname{ind}(Y \setminus X) \leq \operatorname{ind} Y = \operatorname{ind} X$.

6. Splitting the compactification problem

Early on, when the original conjecture turned out to be too difficult to resolve, splittings of the compactification problem were considered. We shall discuss several of these splittings.

The first attempt at splitting used a function which is a large inductive variation of cmp.

6.1. Definition. To every normal space X one assigns the *large inductive compactness degree* $\mathcal{K}\text{-Ind}\, X$ as follows.

(i) $\mathcal{K}\text{-Ind}\, X = -1$ if and only if X is compact.
(ij) For each natural number n, $\mathcal{K}\text{-Ind}\, X \leq n$ if for each pair of disjoint closed sets F and G of X there is a partition S between F and G such that $\mathcal{K}\text{-Ind}\, S \leq n-1$.
(iij) $\mathcal{K}\text{-Ind}\, X = n$ if $\mathcal{K}\text{-Ind}\, X \leq n$ and $\mathcal{K}\text{-Ind}\, X \not\leq n-1$.
(iv) $\mathcal{K}\text{-Ind}\, X = \infty$ if the inequality $\mathcal{K}\text{-Ind} \leq n$ does not hold for any natural number n.

The following propositions can be proved by easy inductions.

6.2. Proposition. *For every normal space X,*

$$\operatorname{cmp} X \leq \mathcal{K}\text{-Ind}\, X \leq \operatorname{Ind} X.$$

6.3. Proposition. *For every closed subspace F of a space X,*

$$\mathcal{K}\text{-Ind}\, F \leq \mathcal{K}\text{-Ind}\, X.$$

That $\mathcal{K}$-Ind is not suitable for the compactification problem will be clear from the next example.

6.4. Example. Let $X = (0,1]$. Then $\operatorname{cmp} X = 0 = \operatorname{def} X$ and $\mathcal{K}\text{-}\operatorname{Ind} X = 1 = \operatorname{Ind} X$. The first equalities are obvious as X is locally compact but not compact. To prove the second equalities, we observe that the disjoint closed sets

$$F = \{ \tfrac{1}{n+1} : n \in \mathbb{N},\ n \text{ odd} \} \quad \text{and} \quad G = \{ \tfrac{1}{n+1} : n \in \mathbb{N},\ n \text{ even} \}$$

cannot be separated by a compact set. Hence $\mathcal{K}\text{-}\operatorname{Ind} X > 0$. Obviously, $\mathcal{K}\text{-}\operatorname{Ind} X \leq \operatorname{Ind} X = \operatorname{ind} X = 1$.

In order to obtain a useful splitting of the compactification problem, we should define a large inductive variation of cmp which agrees with cmp in the values -1 and 0.

6.5. Definition. To every normal space X one assigns the *large compactness degree* $\operatorname{Cmp} X$ as follows.

(i) For $n = -1$ or 0, $\operatorname{Cmp} X = n$ if and only if $\operatorname{cmp} X = n$.
(ij) For each positive natural number n, $\operatorname{Cmp} X \leq n$ if for each pair of disjoint closed sets F and G there is a partition S between F and G such that $\operatorname{Cmp} S \leq n-1$.

$\operatorname{Cmp} X = n$ and $\operatorname{Cmp} X = \infty$ are defined as usual. (The corresponding equalities in the definitions appearing in the remainder of the chapter will no longer be explicitly stated.)

Obviously Cmp is a topological invariant. It also provides a splitting of the compactification problem as the next theorem will show.

6.6. Theorem. *For every separable metrizable space X,*

$$\operatorname{cmp} X \leq \operatorname{Cmp} X \leq \operatorname{def} X \leq \mathcal{K}\text{-}\operatorname{Ind} X.$$

Proof. The first inequality is easily proved by induction. We shall prove the second by induction on $\operatorname{def} X$. We may assume, of course, that $\operatorname{def} X < \infty$. If $\operatorname{def} X < 1$, then $\operatorname{cmp} X = \operatorname{def} X$ by de Groot's theorem and therefore $\operatorname{cmp} X = \operatorname{Cmp} X$ by definition. Let $n \geq 1$ and assume that the inequality holds for all values of $\operatorname{def} X$ less than n. Consider a space X with $\operatorname{def} X = n$ and let Y be any metrizable compactification of X with $\operatorname{ind}(Y \setminus X) = \operatorname{def} X = n$. For disjoint closed sets F and G of X, let us construct a partition S in X between them with $\operatorname{Cmp} S \leq n-1$. Denote $\operatorname{cl}_Y(F) \cap \operatorname{cl}_Y(G)$ by D. There are two cases to consider, namely $D = \emptyset$ and $D \neq \emptyset$.

When $D = \emptyset$, by Proposition 4.6 and the coincidence theorem, there exists a partition S' between $\mathrm{cl}_Y(F)$ and $\mathrm{cl}_Y(G)$ in Y such that $\mathrm{ind}\,\big(S' \cap (Y \setminus X)\big) \leq n-1$. For the resulting partition $S = S' \cap X$ between F and G in X we have $\mathrm{def}\, S \leq \mathrm{ind}\,\big(S' \cap (Y \setminus X)\big) \leq n-1$ because S' is a compactification of S. By the induction hypothesis, $\mathrm{Cmp}\, S \leq n-1$. To contruct S in the contrary case of $D \neq \emptyset$, we let $Z = Y \setminus D$. Clearly we have $X \subset Z$, and by the subspace theorem we have $\mathrm{ind}\,(Z \setminus X) \leq n$. In the same way as in the previous case, there is a partition S' between $\mathrm{cl}_Z(F)$ and $\mathrm{cl}_Z(G)$ in Z such that $\mathrm{ind}\,\big(S' \cap (Z \setminus X)\big) \leq n-1$. We shall show that the partition $S = S' \cap X$ between F and G in X satisfies $\mathrm{Cmp}\, S \leq n-1$. The space S' is locally compact. So let $\alpha(S')$ denote its one-point compactification obtained by adjoining a new point ω. Note that $\alpha(S')$ contains a metrizable compactification of S. Consequently we have $\mathrm{def}\, S \leq \mathrm{ind}\,(\alpha(S') \setminus X)$. Using the point addition theorem and the identity $\alpha(S') \setminus X = \big(S' \cap (Z \setminus X)\big) \cup \{\omega\}$, we get $\mathrm{ind}\,(\alpha(S') \setminus X) \leq n-1$ and thereby $\mathrm{def}\, S \leq n-1$. We have $\mathrm{Cmp}\, S \leq \mathrm{def}\, S \leq n-1$ by the induction hypothesis. So $\mathrm{Cmp}\, X \leq n$ and the proof of the second inequality is completed.

The third inequality is a theorem of de Vries [1962] and will be proved in Section VI.4. In passing, we remark that an easy inductive proof of $\mathrm{Cmp}\, X \leq \mathcal{K}\text{-}\mathrm{Ind}\, X$ can be made.

6.7. Example. As in Example 5.10.e we let $Z = \mathbb{I}^n \setminus E^{n-1}$ with $n \geq 2$. We have already computed $\mathrm{def}\, Z = n-1$. We shall prove $\mathrm{Cmp}\, Z = n-1$. In view of Theorem 6.6, it will suffice to show that $\mathrm{Cmp}\, Z \geq n-1$ holds. To this end, we shall introduce in the next paragraph another invariant that is related to compactness.

We write $\mathrm{Comp}\, X = -1$ if and only if X is compact. And for a natural number n we write $\mathrm{Comp}\, X \leq n$ if for any $n+1$ pairs $(F_0, G_0), \ldots, (F_n, G_n)$ of disjoint compact subsets of X there are partitions S_i between F_i and G_i in X, $i = 0, \ldots, n$, such that the intersection $S_0 \cap \cdots \cap S_n$ is compact. It is easily verified that every rim-compact space X has $\mathrm{Comp}\, X \leq 0$. And $\mathrm{Comp}\, X \leq \mathrm{Cmp}\, X$ is readily established by induction for separable metrizable spaces X.

Returning now to the proof of $\mathrm{Cmp}\, Z \geq n-1$, we observe that in the discussion of Example 5.10.e it was shown that the assumption $\mathrm{Comp}\, Z < n-1$ leads to a contradiction. So, $\mathrm{Comp}\, Z \geq n-1$. Consequently $\mathrm{Cmp}\, Z = n-1$ holds as promised.

It is to be observed that $\operatorname{Comp} X$ can be strictly less than $\operatorname{Cmp} X$. Indeed, for the space X defined in Example 5.10.f (the open disc together with a point on its boundary) we find $\operatorname{Comp} X = 0$ and $\operatorname{cmp} X = \operatorname{Cmp} X = \operatorname{def} X = 1$.

Another splitting of the compactification problem of a totally different nature was introduced by Sklyarenko in [1960]. He defined a new invariant which is connected to a characterization of dimension by means of special bases for the open sets of a space.

6.8. Definition. Let $n = -1$ or $n \in \mathbb{N}$. A separable metrizable space X is said to have $\operatorname{Skl} X \leq n$ if X has a base $\mathcal{B} = \{ U_i : i \in \mathbb{N} \}$ for the open sets such that for any $n+1$ different indices $i_0, \ldots, i_n$ the intersection $\mathrm{B}(U_{i_0}) \cap \cdots \cap \mathrm{B}(U_{i_n})$ is compact.

Skl is a topological invariant. There is the following splitting of the compactification problem.

6.9. Theorem. *For every separable metrizable space X,*

$$\operatorname{cmp} X \leq \operatorname{Skl} X \leq \operatorname{def} X.$$

The proof is divided into two parts.

6.10. Lemma. *For every separable metrizable space X,*

$$\operatorname{cmp} X \leq \operatorname{Skl} X.$$

Proof. $\operatorname{Skl} X = -1$ means that the intersection of the empty collection of subsets of X, that is X itself, is compact. According to Definitions 5.1 and 6.8, we have $\operatorname{Skl} X = 0$ if and only if $\operatorname{cmp} X = 0$. The proof of the lemma will be by induction on $\operatorname{Skl} X$. The first step of the proof has just been completed. The inductive step is as follows. Suppose that $\operatorname{Skl} X \leq n$ with $n \geq 1$. Let $\mathcal{B} = \{ V_i : i \in \mathbb{N} \}$ be a base for the open sets of X witnessing the fact that $\operatorname{Skl} X \leq n$. For each j in $\mathbb{N}$ define

$$\mathcal{B}_j = \{ V_i \cap \mathrm{B}(V_j) : i \neq j \}.$$

The collection $\mathcal{B}_j$ is a base for the open sets of $\mathrm{B}(V_j)$. For clarity of notation let ∂ denote the boundary operator in the subspace $\mathrm{B}(V_j)$. Then we have $\partial\left(V \cap \mathrm{B}(V_j)\right) \subset \mathrm{B}(V) \cap \mathrm{B}(V_j)$ for any subset V of X.

Thus for any n different indices $i_0, \ldots, i_{n-1}$ that are all different from j it follows that

$$\begin{aligned} \bigcap\{\, \partial\,(V_{i_k} \cap \mathrm{B}\,(V_j)) : k = 0, \ldots, n-1 \,\} \\ \subset \bigcap\{\, \mathrm{B}\,(V_{i_k}) \cap \mathrm{B}\,(V_j) : k = 0, \ldots, n-1 \,\} \\ = \mathrm{B}\,(V_{i_0}) \cap \cdots \cap \mathrm{B}\,(V_{i_{n-1}}) \cap \mathrm{B}\,(V_j). \end{aligned}$$

But the last intersection is compact because $\operatorname{Skl} X \leq n$. We have established $\operatorname{Skl} \mathrm{B}\,(V_j) \leq n-1$. By the induction hypothesis we have $\operatorname{cmp} \mathrm{B}\,(V_j) \leq n-1$. Since this holds true for any V_j in $\mathcal{B}$, it follows that $\operatorname{cmp} X \leq n$.

Thereby the first part of Theorem 6.9 has been proved. Here is the second part.

6.11. Lemma. *For every separable metrizable space X,*

$$\operatorname{Skl} X \leq \operatorname{def} X.$$

Proof. We may assume $\operatorname{def} X < \infty$. The proof is by induction on $\operatorname{def} X$. As $\operatorname{def} X = -1$ if and only if $\operatorname{Skl} X = -1$, we shall focus on the inductive step. Let $\operatorname{def} X = n$ and let Y be a metrizable compactification of X with $\operatorname{ind}(Y \setminus X) = n$. Let $\mathcal{B}_1$ be any countable base for the open sets of Y and consider the countable collection $\boldsymbol{D} = \{\, (V_i, W_i) : i \in \mathrm{N} \,\}$ of pairs of elements of $\mathcal{B}_1$ such that $\operatorname{cl}_Y(V_i) \subset W_i$. We shall construct inductively a collection $\boldsymbol{S} = \{\, S_i : i \in \mathrm{N} \,\}$ with the properties

(1) $\quad S_i$ is a partition between $\operatorname{cl}_Y(V_i)$ and $Y \setminus W_i$ for i in N

and, when $i_1 < \cdots < i_k$ and $1 \leq k \leq n+1$,

(2) $$\operatorname{ind}\big(S_{i_1} \cap \cdots \cap S_{i_k} \cap (Y \setminus X)\big) \leq n-k.$$

This will be done by repeatedly applying Lemma 4.11. Let $S_i' = \emptyset$ for all i in N. By induction on m we first construct families

$$\boldsymbol{S}_m = \{\, S_i : i < m \,\} \cup \{\, S_i' : i \geq m \,\}$$

satisfying (2) and such that the set S_i is a partition between $\operatorname{cl}_Y(V_i)$ and $Y \setminus W_i$ in Y for $i < m$. Let $\boldsymbol{S}_0 = \{\, S_i' : i \in \mathrm{N} \,\}$. Suppose that the collection $\boldsymbol{S}_m$ has already been constructed. By Lemma 4.11 there is a partition S_m between $\mathrm{cl}_Y(V_m)$ and $Y \setminus W_m$ in Y such that

$$\boldsymbol{S}_{m+1} = \{ S_i : i \leq m \} \cup \{ S'_i : i > m \}$$

satisfies (2). The resulting collection $\boldsymbol{S} = \{ S_i : i \in \mathrm{N} \}$ obviously satisfies both conditions (1) and (2).

Because of condition (1), for each S_i in $\boldsymbol{S}$ there is an open set U_i of Y such that $\mathrm{cl}_Y(V_i) \subset U_i \subset W_i$ and $\mathrm{B}_Y(U_i) \subset S_i$. Now we observe that the collection $\mathcal{B}_2 = \{ U_i : i \in \mathrm{N} \}$ is a base for the open sets of Y. This can be seen as follows. For any open set U of Y and any point p in U there is a pair (V_i, W_i) in $\boldsymbol{D}$ such that $p \in V_i$ and $W_i \subset U$; thus, $p \in U_i \subset U$ for this index i. In view of condition (2), this base has the property that the equality

$$\mathrm{ind}\,\big(\mathrm{B}_Y(U_{i_0}) \cap \cdots \cap \mathrm{B}_Y(U_{i_n}) \cap (Y \setminus X)\big) = -1$$

holds for $i_0 < \cdots < i_n$. Consequently $\mathrm{B}_Y(U_{i_0}) \cap \cdots \cap \mathrm{B}_Y(U_{i_n})$ is a compact set contained in X. Let $\mathcal{B} = \{ U_i \cap X : U_i \in \mathcal{B}_2 \}$, a base for the open sets of X. Note that for each i we have

$$\begin{aligned} \mathrm{B}_X(U_i \cap X) &= \mathrm{cl}_X(U_i \cap X) \setminus U_i \\ &= (\mathrm{cl}_Y(U_i) \cap X) \setminus U_i = \mathrm{B}_Y(U_i) \cap X. \end{aligned}$$

In deducing this formula, we have used the fact that X is dense in Y. It now easily follows that $\mathcal{B}$ is a base for X which witnesses the fact that $\mathrm{Skl}\, X \leq n$.

We have mentioned already that the invariant Skl is related to a characterization of dimension. The following is this characterization.

6.12. Corollary. *Let X be a separable metrizable space. Then $\mathrm{ind}\, X \leq n$ if and only if X has a base $\mathcal{B} = \{ U_i : i \in \mathrm{N} \}$ for the open sets such that for any $n+1$ different indices $i_0, \ldots, i_n$ the intersection $\mathrm{B}\,(U_{i_0}) \cap \cdots \cap \mathrm{B}\,(U_{i_n})$ is empty.*

Proof. The proof is quite easy. We shall give the last condition of the corollary the name *order dimension* and denote it by Odim. Thus $\mathrm{Odim}\, X \leq n$ if X has a base $\mathcal{B} = \{ U_i : i \in \mathrm{N} \}$ for the open sets such that for any $n+1$ different indices $i_0, \ldots, i_n$ we have $\mathrm{B}\,(U_{i_0}) \cap \cdots \cap \mathrm{B}\,(U_{i_n}) = \emptyset$. Copying almost verbatim the proof of

Lemma 6.10, we get $\operatorname{ind} X \leq \operatorname{Odim} X$ for every separable metrizable space X. We also can redo a large part of the proof of Lemma 6.11 to get $\operatorname{Odim} X \leq \operatorname{ind} X$ by making the following modifications. In that proof we replace the superset Y by the space X (the compactness of the space Y is not relevant for the present proof) and the subset $Y \setminus X$ by the empty set. To obtain the required base $\mathcal{B}$ for the open sets of X we take a countable base $\mathcal{B}_1$ for the open sets of X and the collection $\boldsymbol{D}$ of pairs of elements (V_i, W_i) of $\mathcal{B}_1$ such that $\operatorname{cl}_X(V_i) \subset W_i$. Applying Lemma 4.11 (with the set Y as X, resulting in $\operatorname{ind} Y = \operatorname{ind} X = n$), we then inductively construct a collection $\boldsymbol{S} = \{ S_i : i \in \mathbb{N} \}$ with the properties

$$S_i \text{ is a partition between } \operatorname{cl}_X(V_i) \text{ and } X \setminus W_i \text{ for } i \text{ in } \mathbb{N}$$

and, when $i_1 < \cdots < i_k$ and $1 \leq k \leq n+1$,

$$\operatorname{ind}(S_{i_1} \cap \cdots \cap S_{i_k}) \leq n - k.$$

Continuing along these lines, we obtain $\operatorname{Odim} X \leq \operatorname{ind} X$. This completes the proof.

At this point we want to come again to Definition 6.8 of $\operatorname{Skl} X$. The definition slightly differs from the original one of Sklyarenko. The condition used in Sklyarenko's [1960] paper and also in Isbell's [1964] book is that the intersection of $n+1$ different boundaries is compact instead of the condition that the intersection is compact for any $n+1$ different indices that is used in Definition 6.8. We have chosen our definition for two reasons. First, in Kimura's [1988] paper solving Sklyarenko's compactification problem—the paper we have alluded to in the comments on Theorem 5.7—Definition 6.8 is used. And second, Definition 6.8 avoids a pitfall that arises from the other definition. The pitfall is best illustrated by analyzing the corresponding one for the function Odim of Corollary 6.12. Consider the following example.

6.13. Example. Let X be the subset of $\mathbb{R}^2$ defined by

$$X = \{ (x,y) : x \neq 0 \} \cup \{ (0,0) \}.$$

We begin with the following three open sets:

$$U_0 = \{ (x,y) : (x,y) \in X \text{ and } y > 0 \},$$
$$U_1 = \mathrm{S}_1((-1,0)) = \{ (x,y) : (x+1)^2 + y^2 < 1 \},$$
$$U_2 = \mathrm{S}_1((2,0)) = \{ (x,y) : (x-2)^2 + y^2 < 1 \}.$$

Note that for any k distinct *indices* out of the collection $\{0, 1, 2\}$ the intersection of the corresponding sets in $\{\mathrm{B}(U_0), \mathrm{B}(U_1), \mathrm{B}(U_2)\}$ has small inductive dimension at most $2-k$ when $k \leq 3$. As in the proof of Corollary 6.12 we can enlarge the collection $\{U_0, U_1, U_2\}$ to a base $\mathcal{B}' = \{U_i : i \in \mathrm{N},\ i \neq 3\}$ for the open sets of X such that for any 3 distinct indices i_1, i_2, i_3 from $\mathrm{N} \setminus \{3\}$ we have

$$\mathrm{B}(U_{i_1}) \cap \mathrm{B}(U_{i_2}) \cap \mathrm{B}(U_{i_3}) = \emptyset.$$

(The condition $i \neq 3$ is introduced here to facilitate our later calculations.) In the same way as in the proof of Lemma 6.11 it can be shown that for any index j different from 3 the collection $\mathcal{B}'_j = \{U_i \cap \mathrm{B}(U_j) : i \notin \{j, 3\}\}$ is a base for the open sets of $\mathrm{B}(U_j)$ such that for any k and l that are different from 3 and j and from each other the intersection $\mathrm{B}_{U_j}(U_k \cap \mathrm{B}(U_j)) \cap \mathrm{B}_{U_j}(U_l \cap \mathrm{B}(U_j))$ is also empty, thus witnessing the fact that $\operatorname{ind} \mathrm{B}(U_i) = 1$ according to Corollary 6.12. Now, to indicate where the pitfall lies in the unindexed version of Odim analogous to the original definition of Sklyarenko, we shall introduce the fourth set

$$U_3 = \{(x, y) : (x+1)^2 + y^2 > 1 \text{ and } x < 0\}.$$

From the observation that $\mathrm{B}(U_1) = \mathrm{B}(U_3)$ it follows that for any k distinct *sets* out of the collection $\{\mathrm{B}(U_0), \mathrm{B}(U_1), \mathrm{B}(U_2), \mathrm{B}(U_3)\}$ the intersection has small inductive dimension $2-k$ when $k \leq 3$. (It is to be noted that one set has been labeled twice.) From the base $\mathcal{B}'$ we form the base $\mathcal{B} = \{U_i : i \in \mathrm{N}\}$ by adjoining the open set U_3 to it. Then $\mathcal{B}$ has the property that any three different boundaries of members of $\mathcal{B}$ have empty intersection. Now let us concentrate on $\mathrm{B}(U_0)$ and its base $\mathcal{B}_0 = \{U_i \cap \mathrm{B}(U_0) : i \neq 0\}$. One would expect $\mathcal{B}_0$ to be a base for $\mathrm{B}(U_0)$ that witnesses the fact that $\operatorname{ind} \mathrm{B}(U_0) = 1$ when one uses Odim. But we shall see that this is not the case. Denoting the boundary operator in $\mathrm{B}(U_0)$ by ∂, we have

$$\begin{aligned}
U_1 \cap \mathrm{B}(U_0) &= \{(x, 0) : -2 < x < 0\} \\
&\qquad \text{and} \quad \partial\big(U_1 \cap \mathrm{B}(U_0)\big) = \{(-2, 0), (0, 0)\}, \\
U_2 \cap \mathrm{B}(U_0) &= \{(x, 0) : 1 < x < 3\} \\
&\qquad \text{and} \quad \partial\big(U_2 \cap \mathrm{B}(U_0)\big) = \{(1, 0), (3, 0)\}, \\
U_3 \cap \mathrm{B}(U_0) &= \{(x, 0) : x < -2\} \\
&\qquad \text{and} \quad \partial\big(U_3 \cap \mathrm{B}(U_0)\big) = \{(-2, 0)\}.
\end{aligned}$$

It follows that

$$\partial\left(U_1 \cap \mathrm{B}\left(U_0\right)\right) \neq \partial\left(U_3 \cap \mathrm{B}\left(U_0\right)\right).$$

And we find

$$\partial\left(U_1 \cap \mathrm{B}\left(U_0\right)\right) \cap \partial\left(U_3 \cap \mathrm{B}\left(U_0\right)\right) = \{(-2,0)\}.$$

Thus it is not true in general that any two different boundaries of $\mathcal{B}_0$ have empty intersection. In other words, the base $\mathcal{B}_0$ does not witness the fact that $\operatorname{ind} \mathrm{B}\left(U_0\right) = 1$. Thereby the pitfall has been exhibited.

The approach to dimension via order dimension will be studied in detail in Chapter V.

7. The completeness degree

In the discussion of de Groot's compactification problem 5.6 and Theorem 5.7 we have posed the question whether it is possible to obtain a fruitful generalization of dimension theory by replacing the empty space in the definition of dimension with compact spaces. On examining Example 5.10.f, one will get the impression that the analogy between the functions ind and cmp is rather poor. The example shows the failures of the finite sum theorem and the (point) addition theorem for the function cmp. In this section we shall discuss a dimension function that will show a better analogy with the small inductive dimension. This function, called the small inductive completeness degree, was introduced in [1968] by Aarts. It was the first generalized dimension function for which a substantial theory similar to dimension theory was developed.

Before presenting the definition of completeness degree, we shall discuss some results about topological completeness that will play an important role in our development.

In this book the term "complete" means "topologically complete". That is to say, a metrizable topological space is called *complete* if it is homeomorphic to a space with a complete metric. Such spaces are often called "completely metrizable". A space Y is called an *extension* of a space X if X is a dense subspace of Y. Recall that every metrizable space X has a metrizable extension that is complete. As we shall see, complete extensions are by no means unique. Our investigation will begin by determining which subspaces of a complete space are complete. It is well known that a closed subspace of a complete space is complete. The full determination of the complete subspaces will be settled in Theorem 7.4.

Recall that a subset of a space X is called a G_δ*-set* if it can be represented as a countable intersection of open subsets of X. It is to be noted that each closed subset F of a metrizable space X is a G_δ-set of X because $F = \bigcap\{\{x : d(x, F) < \frac{1}{n+1}\} : n \in \mathbb{N}\}$ for any metric d on X.

7.1. Theorem. *Any* G_δ*-set of a complete metrizable space is complete.*

Proof. Let X be a complete space. Assume that a complete metric ρ has been selected. Suppose that A is a G_δ-set of X, that is, $A = \bigcap\{U_i : i = 1, 2, \ldots\}$ where U_i is open in X for each i. Our tactic will be to find a topological embedding of A which is a closed subset of $X \times \mathbb{R}^{\omega_0}$, where $\mathbb{R}^{\omega_0}$ is the countable product of real lines. (It is to be observed that $X \times \mathbb{R}^{\omega_0}$ with the usual product metric is complete; thus this copy of A will be complete.) We shall denote points of $X \times \mathbb{R}^{\omega_0}$ by (x, y) with x in X and $y = (y_1, y_2, \ldots)$ in $\mathbb{R}^{\omega_0}$. The metric d on $X \times \mathbb{R}^{\omega_0}$ is given by the formula

$$\mathrm{d}\big((x,y),(z,w)\big) = \rho(x,z) + \sum_{i=1}^{\infty} \frac{1}{2^i} \frac{|y_i - w_i|}{1 + |y_i - w_i|} .$$

First we define a continuous mapping f from A to $\mathbb{R}^{\omega_0}$ by defining its component mappings $f_i \colon A \to \mathbb{R}$ by $f_i(x) = \frac{1}{\rho(x, X \setminus U_i)}$. As $A \subset U_i$ for each i, the denominators never vanish. Now we consider the graph $G = \{(x, f(x)) : x \in A\}$ of f. It is well-known that A is homeomorphic to the graph G and that G is a closed subset of $A \times \mathbb{R}^{\omega_0}$. In view of our tactic, it is sufficient to show that G is a closed subset of $X \times \mathbb{R}^{\omega_0}$. To this end we let (p_n, q_n), $n = 1, 2, \ldots$, be a sequence in G that converges to a point (p, q) in $X \times \mathbb{R}^{\omega_0}$. If $p \notin A$, then $p \notin U_i$ for some i. For this i we have $f_i(p_n)$, $n = 1, 2, \ldots$, is a divergent sequence because $\rho(p_n, X \setminus U_i) \to 0$ as $n \to \infty$. In particular, the sequence $f_i(p_n)$, $n = 1, 2, \ldots$, does not converge to the component q_i of q; this denies the continuity of f_i. So we must have $p \in A$ and, because f is continuous, the sequence $q_n = f(p_n)$, $n = 1, 2, \ldots$, must converge to $f(p)$. It follows that $q = f(p)$ and hence $(p, q) \in G$. This shows that G is closed and the proof is completed.

As a bonus we get the formula

$$\rho_1(u,v) = \rho(u,v) + \sum_{i=1}^{\infty} \frac{1}{2^i} \frac{|f_i(u) - f_i(v)|}{1 + |f_i(u) - f_i(v)|}, \qquad u,\ v \in A,$$

for a complete metric ρ_1 on A.

The following lemma is a preparation for Lavrentieff's theorem which will play a crucial role in our development of the theory.

7.2. Lemma. *Suppose that Y is a metric space with a complete metric ρ and that A is a subset of a metrizable space X. If $f\colon A \to Y$ is a continuous map, then there is a subset C of X and there is a continuous map $F\colon C \to Y$ such that C is a G_δ-set of X containing A and F is an extension of f, that is, $F|A = f$.*

Proof. Recall that the *diameter* $\delta(B)$ of a subset B of Y is defined by

$$\delta(B) = \sup\{\, \rho(y_1, y_2) : (y_1, y_2) \in B \times B \,\}.$$

Choose a metric for X and also recall that for each x in $\mathrm{cl}_X(A)$ the *oscillation* $\mathrm{osc}\,(x)$ of f at x is defined by

$$\mathrm{osc}\,(x) = \inf\{\, \delta\big(\mathrm{cl}_Y(f[\mathrm{S}_\varepsilon(x)])\big) : \varepsilon > 0 \,\}.$$

As f is continuous on A, we have $\mathrm{osc}\,(x) = 0$ for each x in A. The set $\mathrm{cl}_X(A)$ being a G_δ-set of X we have that

$$C = \{\, x \in \mathrm{cl}_X(A) : \mathrm{osc}\,(x) = 0 \,\}$$

is also a G_δ-set of X. (It is to be observed that for each n in $\mathbb{N}$ the set $\{\, x \in \mathrm{cl}_X(A) : \mathrm{osc}\,(x) < \frac{1}{n+1} \,\}$ is open in the subspace $\mathrm{cl}_X(A)$.) The map F will be defined as follows. Let $x \in C$. Consider the set $V(x) = \bigcap\{\, \mathrm{cl}_Y(f[\mathrm{S}_\varepsilon(x)]) : \varepsilon > 0 \,\}$. Let us show that this set is a singleton. From $\mathrm{osc}\,(x) = 0$ it follows that $\delta\big(\mathrm{cl}_Y(f[\mathrm{S}_\varepsilon(x)])\big) \to 0$ as $\varepsilon \to 0$. Thus $V(x)$ contains at most one point. Because Y is complete, the set $V(x)$ is nonempty, whence a singleton. Now define F by assigning its value $F(x)$ to be the unique member of $V(x)$ for each x in $V(x)$. The continuity of F is readily established. Finally it is obvious that $F|A = f$.

Now we turn to Lavrentieff's theorem.

7.3. Theorem. *Suppose that X and Y are complete metric spaces. Then every homeomorphism between subspaces A and B of X and Y respectively can be extended to a homeomorphism between G_δ-sets of X and Y.*

Proof. Let $f\colon A \to B$ be a homeomorphism and let $g\colon B \to A$ be its inverse. As $\mathrm{cl}_X(A)$ and $\mathrm{cl}_Y(B)$ are complete, we may assume without loss of generality that A and B are dense in X and Y respectively. By Lemma 7.2 there is a continuous extension $F\colon A_0 \to Y$ of f such that A_0 is a G_δ-set of X containing A. Similarly there is a continuous extension $G\colon B_0 \to X$ of g such that B_0 is a G_δ-set of Y containing B. Let $A_1 = A_0 \cap F^{-1}[B_0]$ and $B_1 = B_0 \cap G^{-1}[A_0]$. Clearly A_1 is a G_δ-set of A_0. As A_0 is a G_δ-set of X, so is A_1. Similarly B_1 is a G_δ-set of Y. And obviously $F|A_1$ is an extension of f and $G|B_1$ is an extension of g. We shall prove that $F|A_1$ and $G|B_1$ are inverses of each other. The composition GF is defined on A_1. As the restriction $(GF)|A$ equals the identity on the set A and A is dense in A_1 the function GF is the identity on A_1. It follows that $F[A_1] \subset G^{-1}[A_1] \subset G^{-1}[A_0]$. As $F[A_1] \subset B_0$ obviously holds, we have $F[A_1] \subset B_1$. Similarly FG is the identity on B_1 and $G[B_1] \subset A_1$. The theorem follows.

We now present a few applications of Lavrentieff's theorem. The first one is the classical characterization of completeness.

7.4. Theorem. *A metrizable space X is complete if and only if X is a G_δ-set of every metrizable space Y that contains X as a subspace.*

Because of this theorem, complete spaces are sometimes called *absolute G_δ-spaces*, that is, a space that is a G_δ-set of every metrizable space that contains it.

Proof. The sufficiency part follows from Theorem 7.1 because any space can be embedded in a complete space Y. To prove the necessity part, we may assume without loss of generality that X has a complete metric. Let $\widetilde{Y}$ be any complete extension of Y. The embedding of X into Y will be denoted by e. By Lavrientieff's theorem, the map e can be extended to a homeomorphism $\tilde{e}$ between G_δ-sets. Obviously $e = \tilde{e}$ and $e[X]$ is a G_δ-set of $\widetilde{Y}$, whence of Y.

The second application is a lemma that will prove to be very important. A topological property $\mathcal{P}$ will be called *hereditary with* *respect to G_δ-sets* if for each space X having property $\mathcal{P}$ every G_δ-set of X has property $\mathcal{P}$. According to Theorem 7.1, completeness is an example of a property that is hereditary with respect to G_δ-sets.

7.5. Lemma. *Let $\mathcal{P}$ be a topological property that is hereditary with respect to G_δ-sets. If some complete metric extension of a space X has a property $\mathcal{P}$, then every complete extension of X contains a complete extension of X with property $\mathcal{P}$. Similarly, if a space X has a complete metric extension Y with* $\operatorname{ind}(Y \setminus X) \leq n$, *then every complete metric extension of X contains a complete extension Y' of X with* $\operatorname{ind}(Y' \setminus X) \leq n$.

Proof. Let Z be a complete extension of X with property $\mathcal{P}$ and let Y be any complete metric extension of X. By Lavrentieff's theorem, the identity map of X can be extended to a homeomorphism between G_δ-sets A of Z and B of Y. Because $\mathcal{P}$ is hereditary with respect to G_δ-sets, the subspace A has property $\mathcal{P}$. The space B has property $\mathcal{P}$ because $\mathcal{P}$ is a topological property. And the space B is complete by Theorem 7.1. Thereby the space B is the required extension of X contained in Y.

Due to Lemma 7.5, complete extensions will be easier to handle than compactifications. This is also illustrated by the following proof of Tumarkin's [1926] extension theorem.

7.6. Theorem. *Suppose that X is a separable subspace of a metric space Y and let* $\operatorname{ind} X \leq n$. *Then there exists a G_δ-set X_1 of Y with $X \subset X_1$ and* $\operatorname{ind} X_1 \leq n$. *In particular, every separable metrizable space X has a complete extension X_1 with the same small inductive dimension as X.*

Proof. We shall assume the additional condition that Y is complete because the general case can be reduced to the complete case. One makes the reduction by first applying the special case to some complete extension $\widetilde{Y}$ of Y so as to obtain a set $\widetilde{X}_1$ and then applying the subspace theorem to the intersection of $\widetilde{X}_1$ and Y.

As $\mathrm{cl}_Y(X)$ is a G_δ-set of Y, we may also assume that X is dense in Y. Observe that Y is a separable space under this assumption. The proof will be by induction on n. The statement being obvious for $n = -1$ we shall assume that the statement has been established for all spaces that are less than n-dimensional. Let X be such that $\operatorname{ind} X \leq n$. By Proposition 3.6 there exists a base $\mathcal{B}$ for the open sets of X such that $\operatorname{ind} \mathrm{B}_X(U) \leq n-1$ for every U in $\mathcal{B}$. For each U in $\mathcal{B}$ we select a fixed open set V of Y with $V \cap X = U$. For any V obtained in this way the diameter $\delta(U)$ of $U = V \cap X$ is equal to the diameter $\delta(V)$. Let $\mathbf{V}$ denote the collection of all V obtained in this way. For each k in $\mathbb{N}$ we let $\mathbf{V}_k = \{ V \in \mathbf{V} : \delta(V) \leq \frac{1}{k+1} \}$ and $E_k = \bigcup \{ V : V \in \mathbf{V}_k \}$. Each space E_k is separable, whence Lindelöf. It follows that there is a countable subcollection $\{ V_{kj} : j \in \mathbb{N} \}$ of $\mathbf{V}_k$ such that $E_k = \bigcup \{ V_{kj} : j \in \mathbb{N} \}$. Let $Z = \bigcap \{ E_k : k \in \mathbb{N} \}$ and $W_{kj} = Z \cap V_{kj}$ for each k and j in $\mathbb{N}$. Obviously $X \subset Z \subset Y$ holds and Z is a G_δ-set of Y. Since each point of Z is a member of elements of $\mathcal{B}_1$ with arbitrarily small diameters, the collection $\mathcal{B}_1 = \{ W_{kj} : k \in \mathbb{N},\ j \in \mathbb{N} \}$ is a base for the open sets of Z. Consider the natural numbers k and j. The open set $U_{kj} = W_{kj} \cap X$ is in $\mathcal{B}$ and $\mathrm{B}_X(U_{kj}) \subset \mathrm{B}_Z(W_{kj})$ holds. By the induction hypothesis there is a G_δ-set C_{kj} of $\mathrm{B}_Z(W_{kj})$ with $\mathrm{B}_X(U_{kj}) \subset C_{kj}$ and $\operatorname{ind} C_{kj} \leq n-1$. The set $\mathrm{B}_Z(W_{kj}) \setminus C_{kj}$ is an F_σ-set of Z because $\mathrm{B}_Z(W_{kj})$ is closed. Using these F_σ-sets, we define the space X_1 by

$$X_1 = Z \setminus \left(\bigcup \{ \mathrm{B}_Z(W_{kj}) \setminus C_{kj} : k \in \mathbb{N},\ j \in \mathbb{N} \} \right).$$

It is readily established that X_1 is a G_δ-set of Z, whence of Y, and that the collection $\{ W_{kj} \cap X_1 : k \in \mathbb{N},\ j \in \mathbb{N} \}$ is a base for the open sets of X_1 such that $\mathrm{B}_{X_1}(W_{kj} \cap X_1) \subset C_{kj}$ for all k and j in $\mathbb{N}$. By the subspace theorem and Proposition 3.6 it follows that $\operatorname{ind} X_1 \leq n$.

With this recapitulation of the basics of complete extensions we can now give an outline of the theory of the completeness degree.

7.7. Definition. To every metrizable space X one assigns the *small inductive completeness degree* $\operatorname{icd} X$ as follows.

(i) $\operatorname{icd} X = -1$ if and only if X is complete.
(ij) For each n in $\mathbb{N}$, $\operatorname{icd} X \leq n$ if for each point p in X and for each closed set G with $p \notin G$ there exists a partition S between p and G such that $\operatorname{icd} S \leq n-1$.

7.8. Definition. To every metrizable space X one assigns the *large inductive completeness degree* $\operatorname{Icd} X$ as follows.

(i) $\operatorname{Icd} X = -1$ if and only if X is complete.
(ij) For each n in $\mathbb{N}$, $\operatorname{Icd} X \leq n$ if for each pair of disjoint closed sets F and G there exists a partition S between F and G such that $\operatorname{Icd} S \leq n-1$.

Although it would be nice to see some examples at this point we shall postpone our presentation of spaces with prescribed values of icd and Icd because the required computations will become much easier after the main theorem (Theorem 7.11) has been established.

7.9. Propositions. Just as before, we shall first collect a few propositions whose proofs are quite easy. Obviously the functions icd and Icd are topological invariants. By means of a straightforward induction proof one can get the following comparisons of some of the inductive invariants that have been defined up to now.

A. *For every metrizable space X,*

$$\operatorname{icd} X \leq \operatorname{cmp} X \leq \operatorname{ind} X,$$
$$\operatorname{icd} X \leq \operatorname{Icd} X \leq \mathcal{K}\text{-}\operatorname{Ind} X \leq \operatorname{Ind} X.$$

By copying the proof of the subspace theorem, we obtain the next result.

B. *For every G_δ-set Y of a metrizable space X,*

$$\operatorname{icd} Y \leq \operatorname{icd} X.$$

A similar result for Icd (Theorem 7.14) will be proved later with the help of the main theorem. The following is a generalization of Proposition 3.6.

C. *Suppose that X is a metrizable space and n is a natural number. Then* $\operatorname{icd} X \leq n$ *if and only if there exists a base $\mathcal{B}$ for the open sets of X such that* $\operatorname{icd} \mathrm{B}(U) \leq n-1$ *for every U in $\mathcal{B}$.*

Our first goal is to resolve the analogue of de Groot's compactification problem for the completeness degrees. To this end, we define the corresponding deficiency.

7.10. Definition. With $\mathcal{C}$ denoting the class of complete metrizable spaces, the *completeness deficiency* of a metrizable space X is defined by

$$\mathcal{C}\text{-}\operatorname{def} X = \min \{ \operatorname{Ind}(Y \setminus X) : X \subset Y,\ Y \in \mathcal{C} \}.$$

In the introduction to the coincidence theorem it was mentioned that the functions ind and Ind do not agree on the class of metrizable spaces. This fact should give rise to two completeness deficiency numbers, one for ind and the other for Ind. At this moment we will not worry about this because we shall restrict our attention to separable metrizable spaces for the remainder of the section; the discussion of dimension theory outside of the realm of separable metrizable spaces is yet to come.

7.11. Theorem (Main theorem for completeness degree). *For every separable metrizable space X,*

$$\operatorname{icd} X = \operatorname{Icd} X = \mathcal{C}\text{-def}\, X.$$

Proof. Proposition 7.9.A gives us $\operatorname{icd} X \leq \operatorname{Icd} X$. It remains to be shown that $\operatorname{Icd} X \leq \mathcal{C}\text{-def}\, X$ and $\mathcal{C}\text{-def}\, X \leq \operatorname{icd} X$ hold.

Proof of $\operatorname{Icd} X \leq \mathcal{C}\text{-def}\, X$. The proof is similar to that of Theorem 6.6 but somewhat easier. It is by induction on $\mathcal{C}\text{-def}\, X$. For the inductive step, let $\mathcal{C}\text{-def}\, X \leq n$ and let Y be any complete extension of X with $\operatorname{Ind}(Y \setminus X) = \mathcal{C}\text{-def}\, X \leq n$. Suppose that F and G are disjoint closed sets of X. With $Z = Y \setminus D$ where $D = \operatorname{cl}_Y(F) \cap \operatorname{cl}_Y(G)$, the space Z is complete as it is an open subspace of Y. By the subspace theorem, $\operatorname{Ind}(Z \setminus X) \leq n$. Note that $\operatorname{cl}_Z(F) \cap \operatorname{cl}_Z(G) = \emptyset$. By Proposition 4.6 there is a partition S between $\operatorname{cl}_Z(F)$ and $\operatorname{cl}_Z(G)$ in Z with $\operatorname{Ind}\big(S \cap (Z \setminus X)\big) \leq n - 1$. It follows that $S \cap X$ is a partition between F and G in X with $\mathcal{C}\text{-def}\, S \leq n - 1$. From the induction hypothesis we have $\operatorname{Icd}(S \cap X) \leq n - 1$. Hence $\operatorname{Icd} X \leq n$ holds.

Proof of $\mathcal{C}\text{-def}\, X \leq \operatorname{icd} X$. The proof is by induction on $\operatorname{icd} X$. It is very similar to the proof of Theorem 7.6. For the inductive step we assume that $\operatorname{icd} X \leq n$ and that Y is a complete extension of X. By Proposition 7.9.C there exists a base $\mathcal{B}$ for the open sets of X such that $\operatorname{icd} \mathrm{B}_X(U) \leq n - 1$ for every U in $\mathcal{B}$. For each U in $\mathcal{B}$ we select a fixed open set V of Y such that $V \cap X = U$. (Clearly the diameter $\delta(V)$ of V is the same as that of $U = V \cap X$.) Let $\mathbf{V}$ denote the collection of all V obtained in this way. Define the collections $\mathbf{V}_k = \{\, V \in \mathbf{V} : \delta(V) \leq \frac{1}{k+1} \,\}$ and $E_k = \bigcup\{\, V : V \in \mathbf{V}_k \,\}$ for each k in $\mathbb{N}$. Since each space E_k is separable, there is a countable subcollection $\{\, V_{kj} : j \in \mathbb{N} \,\}$ of $\mathbf{V}_k$ with $E_k = \bigcup\{\, V_{kj} : j \in \mathbb{N} \,\}$. Now define $Z = \bigcap\{\, E_k : k \in \mathbb{N} \,\}$ and $W_{kj} = Z \cap V_{kj}$ for k and j in $\mathbb{N}$. Obviously $X \subset Z \subset Y$ holds and Z is a G_δ-set of Y. Observe that $\mathcal{B}_1 = \{\, W_{kj} : k \in \mathbb{N},\ j \in \mathbb{N} \,\}$ is a base for the open sets of Z. Consider the natural numbers k and j. The open set $U_{kj} = W_{kj} \cap X$

is in $\mathcal{B}$ and $\mathrm{B}_X(U_{kj}) \subset \mathrm{B}_Z(W_{kj})$ holds. (It is at the next step that the proof differs from that of Theorem 7.6.) By the induction hypothesis and Lemma 7.5, there is a G_δ-set C_{kj} of $\mathrm{B}_Z(W_{kj})$ with $\operatorname{ind}\left(C_{kj} \setminus \mathrm{B}_X(U_{kj})\right) \leq n-1$. Clearly the set $\mathrm{B}_Z(W_{kj}) \setminus C_{kj}$ is an F_σ-set of Z. Using these F_σ-sets, we define the space X_1 by

$$X_1 = Z \setminus \left(\bigcup\{\, \mathrm{B}_Z(W_{kj}) \setminus C_{kj} : k \in \mathbb{N},\ j \in \mathbb{N}\,\}\right).$$

It is readily established that X_1 is a complete extension of X and that the collection $\{\, W_{kj} \cap (X_1 \setminus X) : k \in \mathbb{N},\ j \in \mathbb{N}\,\}$ is a base for the open sets of $X_1 \setminus X$. Also one readily verifies

$$\mathrm{B}_{X_1 \setminus X}(W_{kj} \cap (X_1 \setminus X)) \subset \mathrm{B}_Z(W_{kj}) \cap (X_1 \setminus X) \subset C_{kj} \setminus \mathrm{B}_X(U_{kj})$$

for k and j in $\mathbb{N}$. So $\operatorname{ind}\left(\mathrm{B}_{X_1 \setminus X}(W_{kj} \cap (X_1 \setminus X))\right) \leq n-1$ holds. By Proposition 3.6 we have $\operatorname{ind}(X_1 \setminus X) \leq n$, whence $\mathcal{C}\text{-def}\, X \leq n$. This completes the proof of the inductive step.

We can now present the promised examples.

7.12. Example. Recall that $\mathbb{Q}$ is the space of rational numbers and $\mathbb{I}$ is the interval $[-1, 1]$. We shall prove that

$$\operatorname{icd}(\mathbb{Q} \times \mathbb{I}^n) = \operatorname{icd}(\mathbb{Q} \times \mathbb{R}^n) = n, \qquad n \in \mathbb{N}.$$

As an aside, we note that the computations of the compactness degree $\operatorname{cmp}(\mathbb{Q} \times \mathbb{I}^n) = n$ in Example 5.10.d can be concluded with the aid of Proposition 7.9.A. The computation of the above values of the inductive completeness degree follows. The completion $\mathbb{R}^{n+1}$ of $\mathbb{Q} \times \mathbb{R}^n$ is obtained by adding $\mathbb{P} \times \mathbb{R}^n$ to $\mathbb{Q} \times \mathbb{R}^n$, where $\mathbb{P}$ is the space of irrational numbers. It follows that $\mathcal{C}\text{-def}(\mathbb{Q} \times \mathbb{R}^n) \leq n$. By the main theorem, $\operatorname{icd}(\mathbb{Q} \times \mathbb{R}^n) \leq n$. And from Proposition 7.9.B we have $\operatorname{icd}(\mathbb{Q} \times \mathbb{I}^n) \leq \operatorname{icd}(\mathbb{Q} \times \mathbb{R}^n)$. We shall complete the computations by showing $\mathcal{C}\text{-def}(\mathbb{Q} \times \mathbb{I}^n) \geq n$. The space $\mathbb{R} \times \mathbb{I}^n$ is a completion of $\mathbb{Q} \times \mathbb{I}^n$ and, in view of Lemma 7.5, there is a completion Z of $\mathbb{Q} \times \mathbb{I}^n$ such that

$$\mathbb{Q} \times \mathbb{I}^n \subset Z \subset \mathbb{R} \times \mathbb{I}^n \quad \text{and} \quad \operatorname{ind}\left(Z \setminus (\mathbb{Q} \times \mathbb{I}^n)\right) = \mathcal{C}\text{-def}(\mathbb{Q} \times \mathbb{I}^n).$$

Let p denote the natural projection of $\mathbb{R} \times \mathbb{I}^n$ onto $\mathbb{R}$. Since $\mathbb{I}^n$ is compact, the mapping p is closed. Moreover, the set $(\mathbb{R} \times \mathbb{I}^n) \setminus Z$

is an F_σ-set of $\mathbb{R} \times \mathbb{I}^n$ because Z is an absolute G_δ. So we write $\mathbb{R} \times \mathbb{I}^n \setminus Z$ as the union of a collection $\{ F_i : i \in \mathbb{N} \}$ where each F_i is closed in $\mathbb{R} \times \mathbb{I}^n$. Then we have

$$(*) \qquad \mathbb{P} \cap (\mathbb{R} \setminus p[Z]) \subset p[(\mathbb{R} \times \mathbb{I}^n) \setminus Z] = \bigcup \{ p[F_i] : i \in \mathbb{N} \} \subset \mathbb{P}.$$

It is easily seen that each $p[F_i]$ is a nowhere dense subset of the space $\mathbb{P}$. Hence, by Baire's category theorem applied to $\mathbb{P}$, there is a point y in $\mathbb{P} \setminus \bigcup \{ p[F_i] : i \in \mathbb{N} \}$. It follows from $(*)$ that $p^{-1}[y] \subset Z$. So $\{ y \} \times \mathbb{I}^n \subset Z \setminus (\mathbb{Q} \times \mathbb{I}^n)$ holds, whence $\mathcal{C}\text{-def}\,(\mathbb{Q} \times \mathbb{I}^n) \geq n$.

7.13. Example. Let X be a separable, complete and dense in itself space. It is to be observed that X contains a topological copy of the Cantor set. Suppose that $\operatorname{ind} X = n$ with $n \geq 2$. We shall exhibit a subset Y of X with $\operatorname{icd} Y \geq n - 2$. The construction of Y is related to a construction found in the proof of the Bernstein theorem concerning the existence of totally imperfect sets. The construction employs the following two easily verified facts. First, the cardinality of the space X as well as the cardinality of the collection of uncountable F_σ-sets of X is $\mathfrak{c}$ (the cardinality of the set $\mathbb{R}$). Second, by the Cantor-Bendixson theorem each uncountable F_σ-set of X also has cardinality $\mathfrak{c}$. Denoting the initial ordinal number of $\mathfrak{c}$ by Ω, we arrange the points of X into a transfinite sequence $\{ x_\alpha : \alpha < \Omega \}$ and the family of uncountable F_σ-sets into a transfinite sequence $\{ F_\alpha : \alpha < \Omega \}$. Then we define two transfinite sequences $\{ y_\alpha : \alpha < \Omega \}$ and $\{ z_\alpha : \alpha < \Omega \}$ as follows. Assuming that $\{ y_\alpha : \alpha < \beta \}$ and $\{ z_\alpha : \alpha < \beta \}$ have been defined for $\beta < \Omega$, we let y_β and z_β denote the first and second element respectively of the sequence $\{ x_\alpha : \alpha < \Omega \}$ that are in the set $F_\beta \setminus \bigcup \{ \{ y_\alpha, z_\alpha \} : \alpha < \beta \}$. The existence of y_β and z_β follows from the fact F_β has cardinality $\mathfrak{c}$ while the set $\bigcup \{ \{ y_\alpha, z_\alpha \} : \alpha < \beta \}$ has cardinality less than $\mathfrak{c}$. Now we define Y to be $\{ y_\alpha : \alpha < \Omega \}$. It follows from the way the set Y has been constructed that neither Y nor $X \setminus Y$ contains an uncountable F_σ-set. Let us show that $\operatorname{ind}(X \setminus Y) \geq n - 1$. From Theorem 7.6 due to Tumarkin, there exists a complete extension Z of $X \setminus Y$ with $X \setminus Y \subset Z \subset X$ and $\operatorname{ind} Z = \operatorname{ind}(X \setminus Y)$. The set $X \setminus Z$ is an F_σ-set of X contained in Y. It follows that $X \setminus Z$ is countable, whence zero-dimensional. From the identity $X = (X \setminus Z) \cup Z$ and the addition theorem, the inequality $\operatorname{ind}(X \setminus Y) = \operatorname{ind} Z \geq n - 1$ follows. Now let us show that

$\mathcal{C}$-def $Y \geq n-2$. By Lemma 7.5 there is a complete extension Y_1 of Y with $Y \subset Y_1 \subset X$ and $\operatorname{ind}(Y_1 \setminus Y) = \mathcal{C}$-def Y. The set $X \setminus Y_1$ is an F_σ-set of X that is contained in $X \setminus Y$ and hence is zero-dimensional. Because $X \setminus Y = (X \setminus Y_1) \cup (Y_1 \setminus Y)$, from the addition theorem we get $\operatorname{ind}(Y_1 \setminus Y) \geq n-2$. By the main theorem, $\operatorname{icd} Y \geq n-2$.

The strength of the main theorem will be illustrated with a few applications.

7.14. Theorem. *For every G_δ-set Y of a separable metrizable space X,*

$$\operatorname{Icd} Y \leq \operatorname{Icd} X.$$

Proof. This is an easy consequence of Proposition 7.9.B and the coincidence of icd and Icd from the main theorem.

7.15. Theorem (Addition theorem). *Let X be a separable metrizable space. If $X = Y \cup Z$, then*

$$\operatorname{icd} X \leq \operatorname{icd} Y + \operatorname{icd} Z + 1.$$

In particular, the small inductive completeness degree icd *of a separable metrizable space cannot be increased by the adjunction of a complete space or a point.*

Proof. Let X^* be a complete extension of X. By Lemma 7.5 and the main theorem, there are G_δ-sets Y^* and Z^* of X^* containing Y and Z respectively that satisfy $\operatorname{icd} Y = \operatorname{Ind}(Y^* \setminus Y)$ and $\operatorname{icd} Z = \operatorname{Ind}(Z^* \setminus Z)$. The set $Y^* \cup Z^*$ is a G_δ-set of X^*, whence a complete extension of $Y \cup Z$. We have

$$\begin{aligned} \operatorname{icd} X &\leq \operatorname{Ind}\big((Y^* \cup Z^*) \setminus (Y \cup Z)\big) \leq \operatorname{Ind}\big((Y^* \setminus Y) \cup (Z^* \setminus Z)\big) \\ &\leq \operatorname{Ind}(Y^* \setminus Y) + \operatorname{Ind}(Z^* \setminus Z) + 1 = \operatorname{icd} Y + \operatorname{icd} Z + 1. \end{aligned}$$

In a similar way we obtain the following result.

7.16. Theorem (Intersection theorem). *If Y and Z are subsets of a separable metrizable space X, then*

$$\operatorname{icd}(Y \cap Z) \leq \operatorname{icd} Y + \operatorname{icd} Z + 1.$$

Proof. With X^*, Y^* and Z^* as defined in the preceding proof we have

$$\begin{aligned}\operatorname{icd}(Y \cap Z) &\leq \operatorname{Ind}\big((Y^* \cap Z^*) \setminus (Y \cap Z)\big) \\ &\leq \operatorname{Ind}\big((Y^* \setminus Y) \cup (Z^* \setminus Z)\big) \leq \operatorname{icd} Y + \operatorname{icd} Z + 1.\end{aligned}$$

There is no countable sum theorem for icd. The space $\mathbb{Q} \times \mathbb{I}^n$ of Example 7.12 will serve as a counterexample since it is the countable union of compact components. However, there is a locally finite sum theorem for icd.

7.17. Theorem (Locally finite sum theorem). *Let* $\boldsymbol{F}$ *be a locally finite closed cover of a separable metrizable space* X. *Then*

$$\operatorname{icd} X = \sup \{ \operatorname{icd} F : F \in \boldsymbol{F} \}.$$

Proof. In view of Proposition 7.9.B, only $\sup \{ \operatorname{icd} F : F \in \boldsymbol{F} \} \geq \operatorname{icd} X$ needs to be shown. Let Y be any complete extension of X. For each point x in X we select an open neighborhood U_x of x in Y such that U_x meets only finitely many elements of $\boldsymbol{F}$. Then $W = \bigcup \{ U_x : x \in X \}$ is an open subset of Y with $X \subset W$ and the collection $\boldsymbol{F}$ is locally finite in the subspace W. Since W is a complete extension of X, we may assume $Y = W$. As a locally finite collection is closure preserving, we have $Y = \bigcup \{ \operatorname{cl}_Y(F) : F \in \boldsymbol{F} \}$. By the main theorem and Lemma 7.5, there is for each F in $\boldsymbol{F}$ a completion G_F of F with $F \subset G_F \subset \operatorname{cl}_Y(F)$ and $\operatorname{icd} F = \operatorname{Ind}(G_F \setminus F)$. Obviously $\operatorname{cl}_Y(F) \setminus G_F$ is an F_σ-set of Y. Let us show that the subset $H = \bigcup \{ \operatorname{cl}_Y(F) \setminus G_F : F \in \boldsymbol{F} \}$ of Y is also an F_σ-set. The collection $\{ \operatorname{cl}_Y(F) : F \in \boldsymbol{F} \}$ is locally finite in Y. As the space Y is separable, this collection must be countable. Thereby H is an F_σ-set of Y. It follows that $Y \setminus H$ is a complete extension of X. Now to complete the proof, observe that

$$(Y \setminus H) \setminus X \subset \bigcup \{ G_F : F \in \boldsymbol{F} \} \setminus X \subset \bigcup \{ G_F \setminus F : F \in \boldsymbol{F} \}.$$

We have

$$\operatorname{ind}\big((Y \setminus H) \setminus X\big) \leq \sup \{ \operatorname{ind}(G_F \setminus F) : F \in \boldsymbol{F} \}$$

by the coincidence theorem, the countable sum theorem and the subspace theorem of ind. As $\operatorname{icd} X = \mathcal{C}\text{-def}\, X \leq \operatorname{ind}\big((Y \setminus H) \setminus X\big)$, the theorem will follow.

It is still an open question whether or not the following decomposition theorem for the completeness degree holds true:

> $\operatorname{icd} X \leq n$ *if and only if* $X = X_0 \cup X_1 \cdots \cup X_n$ *for* $n+1$ *sets* X_i *with* $\operatorname{icd} X_i \leq 0$, $i = 0, 1, \ldots, n$.

Observe that the sufficiency will follow from the addition theorem for icd. (The decomposition can be made when $\operatorname{icd} X = \operatorname{ind} X$.) However, the "dual" of the decomposition statement, called the structure theorem, is readily established.

7.18. Theorem (Structure theorem). *Suppose that X is a separable metrizable space. Then $\operatorname{icd} X \leq n$ if and only if for some complete extension Y of X the set X can be represented as the intersection of $n+1$ subsets Y_i of Y, $i = 0, 1, \ldots, n$, with $\operatorname{icd} Y_i \leq 0$ for each i.*

The easy proof will be omitted.

We shall conclude this section with a characterization of icd that is suggested by the definition of Skl and the proof of Corollary 6.12. The characterization will require the following definition which is analogous to that of Skl.

7.19. Definition. Let X be a separable metrizable space. The *completeness order dimension* of the space X, denoted $\mathcal{C}\text{-Odim}\, X$, is defined as follows. $\mathcal{C}\text{-Odim}\, X \leq n$ if X has a base $\mathcal{B} = \{\, U_i : i \in \mathbb{N}\,\}$ for the open sets such that for any $n+1$ different indices $i_0, \ldots, i_n$ the intersection $\mathrm{B}\,(U_{i_0}) \cap \cdots \cap \mathrm{B}\,(U_{i_n})$ is complete.

7.20. Theorem. *For every separable metrizable space X,*

$$\operatorname{icd} X = \mathcal{C}\text{-Odim}\, X.$$

Proof. The proof will smoothly follow the steps that were taken in Section 6. The proof of the inequality $\operatorname{icd} X \leq \mathcal{C}\text{-Odim}\, X$ is achieved by a simple replacement of "compact" with "complete" in the proof of Lemma 6.10. The same replacement in the proof of Lemma 6.11 will yield the inequality $\mathcal{C}\text{-Odim}\, X \leq \mathcal{C}\text{-def}\, X$. The proof of the theorem is completed by an application of the main theorem.

The results of this section have been generalized in two directions. First, they hold for more general spaces X; second, they hold for classes of spaces $\mathcal{P}$ more general than the class $\mathcal{C}$ of complete spaces. The investigation of extensions of spaces along the same lines as in this section will be continued in Section V.2. Finally the discussion of order-type characterizations that are analogous to Theorem 7.20 will be continued in Section V.3.

8. The covering dimension dim

In addition to the small inductive dimension and the large inductive dimension there is yet another dimension function called the covering dimension. Unlike the compactness degree, the completeness degrees discussed in the previous section have many similarities with the two inductive dimension functions. In [1968] Aarts indicated that there is a natural generalization of covering dimension that also fits perfectly into the theory of completeness just as the covering dimension fits into the theory of dimension. So this seems to be the proper moment to present the covering dimension.

The covering dimension was first introduced by Čech in [1933]. It has its roots in a covering property studied by Lebesgue in [1911]. Lebesgue was searching for a topological property that would distinguish the n-dimensional cube $\mathbb{I}^n$ and the m-dimensional cube $\mathbb{I}^m$. In fact, Lebesgue was working on the same problem as Brouwer (see Section 2). The covering dimension uses the primitive idea of the order of a collection.

8.1. Definition. Let $\boldsymbol{D} = \{ D_\alpha : \alpha \in A \}$ be an indexed collection in a topological space X and let p be a point in X. The *order of* $\boldsymbol{D}$ *at* p is the number of indices α in A (possibly ∞) for which $p \in D_\alpha$ holds; this number is denoted by $\operatorname{ord}_p \boldsymbol{D}$. The *order of* $\boldsymbol{D}$ is defined by $\operatorname{ord} \boldsymbol{D} = \sup \{ \operatorname{ord}_p \boldsymbol{D} : p \in X \}$.

When the collection $\boldsymbol{D}$ has no explicit indexing we shall formally index the collection as $\{ D : D \in \boldsymbol{D} \}$. Then the order of $\boldsymbol{D}$ becomes the largest number n such that the collection $\boldsymbol{D}$ contains n distinct sets with a nonempty intersection.

The order of indexed collections has already been implicitly used in earlier sections, notably Section 6. The importance of the indexing set of the collection was shown to be critical when one considered the collection of boundaries that coresponded to the given collection of sets, that is, the two collections

$$\{ U_\alpha : \alpha \in A \} \quad \text{and} \quad \{ \mathrm{B}(U_\alpha) : \alpha \in A \}.$$

The discussion in Example 6.13 addressed the pitfalls that can arise when the indexing set A of the two collections was permitted to be changed to another indexing set for the collection of boundaries. The next example is a continuation of that example.

8.2. Example. For the Example 6.13 one can readily compute the two values for the indexed collections.

$$\operatorname{ord}\{ \mathrm{B}(U_0), \mathrm{B}(U_1), \mathrm{B}(U_2), \mathrm{B}(U_3) \} = 3,$$
$$\operatorname{ord}\{ \mathrm{B}(U_0), \mathrm{B}(U_1), \mathrm{B}(U_2) \} = 2.$$

When the indexing is neglected, the collections of boundaries are the same because $\mathrm{B}(U_1) = \mathrm{B}(U_3)$.

The following is a rewording of Corollary 6.12.

8.3. Proposition. *Let X be a separable metrizable space. Then* $\operatorname{ind} X \leq n$ *if and only if there is a base $\mathcal{B} = \{ U_i : i \in \mathrm{N} \}$ for the open sets of X with* $\operatorname{ord}\{ \mathrm{B}(U_i) : i \in \mathrm{N} \} \leq n$.

Recall the definitions: A collection $\boldsymbol{D}$ in a topological space X is a *cover* if $X = \bigcup \boldsymbol{D}$. A collection $\boldsymbol{D}$ is called *open* if every member of $\boldsymbol{D}$ is open. A collection $\boldsymbol{D}$ is said to be a *refinement* of a collection $\boldsymbol{E}$ if for each D in $\boldsymbol{D}$ there exists an E in $\boldsymbol{E}$ with $D \subset E$. (It is to be noted that if $\boldsymbol{D}$ is a refinement of $\boldsymbol{E}$, then the inclusion $\bigcup \boldsymbol{D} \subset \bigcup \boldsymbol{E}$ holds with the possibility of a proper inclusion being permitted.) We shall now define a special type of refinement.

8.4. Definition. Let $\boldsymbol{D} = \{ D_\alpha : \alpha \in A \}$ be an indexed cover of a space X. A collection $\boldsymbol{E} = \{ E_\alpha : \alpha \in A \}$ is called a *shrinking* of $\boldsymbol{D}$ if $\boldsymbol{E}$ is a cover of X and $E_\alpha \subset D_\alpha$ holds for every α in A.

We now define the covering dimension.

8.5. Definition. Let X be a space. The *covering dimension* of X, denoted $\dim X$, is defined as follows. $\dim X = -1$ if and only if $X = \emptyset$. For n in N, $\dim X \leq n$ if for each finite open cover $\boldsymbol{U}$ of X there exists an open cover $\boldsymbol{V}$ of X such that $\boldsymbol{V}$ is a refinement of $\boldsymbol{U}$

and ord $\mathbf{V} \leq n+1$ holds. As usual, $\dim X = n$ means $\dim X \leq n$ holds but $\dim X \leq n-1$ fails.

The following proposition exhibits a condition that is equivalent to the definition of dim.

8.6. Proposition. *Let X be a space and let n be in $\mathbb{N}$. Then $\dim X \leq n$ if and only if every finite open cover $\mathbf{U}$ of X has an open shrinking $\mathbf{V}$ of order not exceeding $n+1$.*

Proof. The sufficiency is obvious. To prove the necessity let us consider an open cover $\mathbf{U} = \{\, U_i : i = 0, 1, \ldots, k \,\}$. As $\dim X \leq n$, there exists an open cover $\mathbf{W}$ such that $\mathbf{W}$ is a refinement of $\mathbf{U}$ and ord $\mathbf{W} \leq n+1$. For each W in $\mathbf{W}$ we choose an index i such that $W \subset U_i$ holds and denote this choice by $\alpha(W) = i$. With V_i defined as the open set $\bigcup\{\, W \in \mathbf{W} : \alpha(W) = i \,\}$ one can easily verify that the collection $\mathbf{V} = \{\, V_i : i = 0, 1, \ldots, k \,\}$ is a shrinking of $\mathbf{U}$ such that ord $\mathbf{V} \leq n+1$.

The three dimension functions ind, Ind and dim reflect different geometrical structures of normal spaces. We have already seen that the functions ind and Ind coincide for separable metrizable spaces (Theorem 4.4). We shall prove that ind and dim also coincide for separable metrizable spaces and thereby show an intimate connection between these geometric structures. We need the following three lemmas.

8.7. Lemma. *Suppose that X is a subspace of a metrizable space Y and let $\mathbf{U} = \{\, U_\alpha : \alpha \in A \,\}$ be an open collection in X. Then there exists an open collection $\mathbf{V} = \{\, V_\alpha : \alpha \in A \,\}$ in Y such that $U_\alpha \subset V_\alpha \cap X$ for every α in A and* ord $\mathbf{U} =$ ord $\mathbf{V}$.

Proof. To each open set U of the subspace X we associate the open set U^* of Y defined by $U^* = \{\, y \in Y : d(y, U) < d(y, X \setminus U) \,\}$. (Recall that $d(y, \emptyset) = \infty$ and hence $\emptyset^* = \emptyset$.) Let us prove the identity $(U \cap V)^* = U^* \cap V^*$. This will be done in two steps.

First we assume $y \in (U \cap V)^*$. It readily follows that $d(y, U) \leq d(y, U \cap V) < d\big(y, (X \setminus U) \cup (X \setminus V)\big) \leq d(y, X \setminus U)$. So, $y \in U^*$. And $y \in V^*$ by symmetry. Therefore, $(U \cap V)^* \subset U^* \cap V^*$.

Next assume $y \in U^* \cap V^*$. Obviously,

$$\begin{aligned} d\big(y, X \setminus (U \cap V)\big) &= d\big(y, (X \setminus U) \cup (X \setminus V)\big) \\ &= \min\{\, d(y, X \setminus U),\ d(y, X \setminus V) \,\}. \end{aligned}$$

Without loss of generality we assume

$$d(y, X \setminus U) \leq d(y, X \setminus V).$$

Because $y \in U^*$, we have $d(y, U) < d(y, X \setminus U)$. Now

$$d(y, U) = \min\{\, d(y, U \cap V),\ d(y, U \setminus V)\,\}.$$

As $d(y, U \setminus V) \geq d(y, X \setminus V) \geq d(y, X \setminus U) > d(y, U)$, we have

$$d(y, U) = d(y, U \cap V).$$

Consequently $d(y, U \cap V) < d(y, X \setminus U) = d\big(y, X \setminus (U \cap V)\big)$ holds. So, $y \in (U \cap V)^*$. Thereby $(U \cap V)^* = U^* \cap V^*$ is proved. Clearly the finite intersection formula $(U_0 \cap \cdots \cap U_k)^* = U_0^* \cap \cdots \cap U_k^*$ now holds.

We complete the proof of the lemma by defining $V_\alpha = U_\alpha^*$ for each α in A.

8.8. Lemma. *Every finite open cover of a normal space X has a closed shrinking.*

Proof. It is somewhat easier to prove the following equivalent statement.

> *Every finite closed collection $\{\, F_i : i = 0, 1, \ldots, n\,\}$ with empty intersection has an open collection $\{\, V_i : i = 0, 1, \ldots, n\,\}$ such that $F_i \subset V_i$ holds for every i and $\bigcap\{\, V_i : i = 0, 1, \ldots, n\,\} = \emptyset$.*

This statement is proved as follows. The sets F_0 and $F_1 \cap \cdots \cap F_n$ are disjoint and closed. By the normality of X there exists an open set V_0 with $F_0 \subset V_0$ and $\mathrm{cl}\,(V_0) \cap (F_1 \cap \cdots \cap F_n) = \emptyset$. So we can replace the closed set F_0 with the bigger set $\mathrm{cl}\,(V_0)$ and at the same time preserve the empty intersection of the collection. Repeating the argument a finite number of times, we complete the proof.

This shrinking lemma is well known; its proof has been included because similar arguments will be found later. Its generalization to locally finite collections is also known, but a proof will not be given here. The reader is asked to provide a proof or consult standard texts in topology.

8.9. Lemma. *Suppose that X is a normal space with* $\dim X \le n$. *Let* $\{ F_i : i = 0, 1, \ldots, k \}$ *and* $\{ U_i : i = 0, 1, \ldots, k \}$ *be a closed collection and an open collection respectively such that* $F_i \subset U_i$ *for every* i. *Then there are two open collections*

$$\{ V_i : i = 0, 1, \ldots, k \} \quad \text{and} \quad \{ W_i : i = 0, 1, \ldots, k \}$$

such that the following hold.

$$F_i \subset V_i \subset \mathrm{cl}\,(V_i) \subset W_i \subset U_i, \qquad i = 0, 1, \ldots, k,$$
$$\mathrm{ord}\,\{ W_i \setminus \mathrm{cl}\,(V_i) : i = 0, 1, \ldots, k \} \le n.$$

The property which appears in the conclusion of the lemma will actually characterize the inequality $\dim X \le n$. Such a characterization will be discussed in a general setting in Section II.5. We shall give a direct proof of this characterization for separable metrizable spaces immediately after the proof of Theorem 8.10.

Proof. Consider $\mathbf{D} = \bigwedge\{ \{U_i,\ X \setminus F_i \} : i = 0, 1, \ldots, k \}$. (Recall that the members of $\mathbf{D}$ are of the form $\bigcap\{ A_i : i = 0, 1, \ldots, k \}$ where either $A_i = U_i$ or $A_i = X \setminus F_i$ for each i.) Being an open cover of X, it has an open shrinking $\mathbf{V} = \{ L_j : j = 0, 1, \ldots, l \}$ with $\mathrm{ord}\,\mathbf{V} \le n + 1$ (Proposition 8.6). And from Lemma 8.8, $\mathbf{V}$ has a closed shrinking $\{ K_j : j = 0, 1, \ldots, l \}$. Using the normality of the space, we can find open collections $\{ K_j^i : i = 0, 1, \ldots, k + 1 \}$ for $j = 0, 1, \ldots, l$ such that

$$\text{(1)} \qquad K_j \subset K_j^i \subset \mathrm{cl}\,(K_j^i) \subset L_j, \qquad i = 0, 1, \ldots, k + 1,$$

and

$$\text{(2)} \qquad \mathrm{cl}\,(K_j^i) \subset K_j^{i+1}, \qquad i = 0, 1, \ldots, k.$$

It should be noted that from (2) we have

$$\text{(3)} \quad \big(K_j^{i+1} \setminus \mathrm{cl}\,(K_j^i)\big) \cap \big(K_j^{i'+1} \setminus \mathrm{cl}\,(K_j^{i'})\big) = \emptyset, \qquad 0 \le i < i' \le k.$$

Now for i in $\{ 0, 1, \ldots, k \}$ we define

$$V_i = \bigcup\{ K_j^i : L_j \cap F_i \ne \emptyset \}$$

and

$$W_i = \bigcup \{ K_j^{i+1} : L_j \cap F_i \neq \emptyset \}.$$

Because $\{ K_j : j = 0, 1, \ldots, l \}$ is a cover of X and $\mathbf{V}$ is a refinement of $\mathbf{D}$, one readily proves

$$F_i \subset V_i \subset \mathrm{cl}\,(V_i) \subset W_i \subset U_i, \qquad i = 0, 1, \ldots, k.$$

This is the first formula which was to be proved.

By way of a contradiction we shall prove the second formula. Assume for some p in X and some distinct indices i_m, $m = 0, 1, \ldots, n$, that

$$p \in \bigcap \{ W_{i_m} \setminus \mathrm{cl}\,(V_{i_m}) : m = 0, 1, \ldots, n \}.$$

Note that we have

$$W_i \setminus \mathrm{cl}\,(V_i) \subset \bigcup \{ K_j^{i+1} \setminus \mathrm{cl}\,(K_j^i) : L_j \cap F_i \neq \emptyset \}, \qquad i = 0, 1, \ldots, k.$$

Thus, for suitable pairs (i_m, j_m), we have

$$p \in \bigcap \{ K_{j_m}^{i_m+1} \setminus \mathrm{cl}\,(K_{j_m}^{i_m}) : m = 0, 1, \ldots, n \}.$$

It follows from (3) that the indices j_m must be distinct. As X is covered by $\{ K_j : j = 0, 1, \ldots, l \}$, we have $p \in K_s$ for some s. It follows from (1) that s is distinct from each of the j_m's. Hence

$$p \in L_s \cap L_{j_0} \cap \cdots \cap L_{j_n},$$

that is,

$$\mathrm{ord}_p \, \mathbf{V} \geq n + 2.$$

This is a contradiction.

8.10. Theorem (Coincidence theorem). *For every separable metrizable space X,*

$$\dim X = \mathrm{ind}\, X = \mathrm{Ind}\, X.$$

Proof. The proof will be in two parts.

Proof of $\dim X \leq \mathrm{ind}\, X$. To begin, we shall prove the inequality under the additional assumption of $\mathrm{ind}\, X = 0$. Let $\mathbf{U}$ be a finite open cover of X. In view of Proposition 3.6 there is a cover $\boldsymbol{V}$ of X that consists of open-and-closed sets and refines $\boldsymbol{U}$. As X is separable, we may assume that $\boldsymbol{V}$ is countable; let $\boldsymbol{V} = \{ V_i : i \in \mathrm{N} \}$. For each i in N we let $W_i = V_i \setminus (V_0 \cup \cdots \cup V_{i-1})$. Then $\boldsymbol{W} = \{ W_i : i \in \mathrm{N} \}$ is a cover of X consisting of open-and-closed sets such that $\boldsymbol{W}$ refines $\boldsymbol{U}$ and $\operatorname{ord} \boldsymbol{W} = 1$. It follows that $\dim X = 0$.

For the general situation we argue as follows. Assume $\operatorname{ind} X = n$ is positive. By the decomposition theorem (Theorem 3.8) the space X can be partitioned into $n+1$ disjoint subsets X_i, $i = 0, 1, \ldots, n$, such that $\operatorname{ind} X_i = 0$ for each i. Let $\boldsymbol{U}$ be a finite open cover of X and let i be such that $0 \le i \le n$. Then the open cover $\{ U \cap X_i : U \in \boldsymbol{U} \}$ of X_i can be refined by an open cover $\boldsymbol{V}_i = \{ V_{i\alpha} : \alpha \in A_i \}$ of order 1. By Lemma 8.7 there is an open collection $\{ W_{i\alpha} : \alpha \in A_i \}$ of order 1 in X such that $W_{i\alpha} \cap X_i = V_{i\alpha}$. For each α in A_i we select a $U_{i\alpha}$ in $\boldsymbol{U}$ such that $V_{i\alpha} \subset U_{i\alpha} \cap X_i$. It is readily verified that the collection $\boldsymbol{V}_i^* = \{ W_{i\alpha} \cap U_{i\alpha} : \alpha \in A_i \}$ has order 1 and refines $\boldsymbol{U}$. Thus the open cover $\boldsymbol{V}^* = \boldsymbol{V}_0^* \cup \cdots \cup \boldsymbol{V}_n^*$ has order $n+1$ and refines $\boldsymbol{U}$.

Proof of $\operatorname{ind} X \le \dim X$. As in the proof of Corollary 6.12 we use the *order dimension* of the space X, that is, $\operatorname{Odim} X \le n$ if X has a base $\mathcal{B} = \{ U_i : i \in \mathrm{N} \}$ for the open sets such that for any $n+1$ different indices $i_0, \ldots, i_n$ we have $\mathrm{B}(U_{i_0}) \cap \cdots \cap \mathrm{B}(U_{i_n}) = \emptyset$. It has already been shown that $\operatorname{ind} X \le \operatorname{Odim} X$. So only $\operatorname{Odim} X \le \dim X$ remains to be proved. To this end let $\mathcal{B}'$ be any countable base for X. Consider the family $\boldsymbol{D} = \{ (C_i, D_i) : i \in \mathrm{N} \}$ of all pairs of elements of $\mathcal{B}'$ such that $\operatorname{cl}(C_i) \subset D_i$. For each pair of natural numbers i and j with $i \le j$ we shall inductively define open sets V_i^j and W_i^j such that, when $i \le j$,

$$C_i \subset V_i^j \subset \operatorname{cl}(V_i^j) \subset W_i^j \subset D_i\,,$$
$$\operatorname{cl}(V_i^j) \subset V_i^{j+1}, \quad \operatorname{cl}(W_i^{j+1}) \subset W_i^j$$

and, when $j \in \mathrm{N}$,

$$\operatorname{ord} \{ W_i^j \setminus \operatorname{cl}(V_i^j) : i = 0, 1, \ldots, j \} \le n.$$

Let V_0^0 and W_0^0 be open sets with $C_0 \subset V_0^0 \subset \operatorname{cl}(V_0^0) \subset W_0^0 \subset D_0$. Assuming that V_i^j and W_i^j have been defined for $i \le j \le s$ so that the above conditions are satisfied, we apply Lemma 8.9 to the collections

$$\{ \operatorname{cl}(V_i^s) : i = 0, 1, \ldots, s \} \cup \{ \operatorname{cl}(C_{s+1}) \}$$

and

$$\{ W_i^s : i = 0, 1, \ldots, s \} \cup \{ D_{s+1} \}$$

so as to obtain the open collections

$$\{ V_i^{s+1} : i = 0, 1, \ldots, s+1 \} \quad \text{and} \quad \{ W_i^{s+1} : i = 0, 1, \ldots, s+1 \}$$

satisfying the above conditions. Finally define $V_i = \bigcup \{ V_i^j : j \geq i \}$ and $\mathcal{B} = \{ V_i : i \in \mathbb{N} \}$. Note that $\mathrm{B}(V_i) \subset W_i^j \setminus \mathrm{cl}(V_i^j)$ when $j \geq i$. It readily follows that $\mathcal{B}$ is a base such that $\mathrm{B}(U_{i_0}) \cap \cdots \cap \mathrm{B}(U_{i_n})$ is empty for any $n+1$ different indices $i_0, \ldots, i_n$. So, $\mathrm{Odim}\, X \leq n$.

Let us return to our observation made after the statement of Lemma 8.9. As promised, we shall prove that the condition of the conclusion of that lemma characterizes the inequality $\dim X \leq n$ for separable metrizable spaces X. The characterization follows from the sequence of implications displayed below, where the numbers above the arrows refer to earlier proved statements.

$$\dim X \leq n \xrightarrow{(8.9)} \text{conclusion of Lemma 8.9}$$
$$\xrightarrow{(8.10)} \mathrm{Odim}\, X \leq n \xrightarrow{(6.12)} \mathrm{ind}\, X \leq n \xrightarrow{(8.10)} \dim X \leq n.$$

We have already commented earlier that ind and Ind do not coincide for metrizable spaces. But Ind and dim do coincide for general metrizable spaces. Further discussions on coincidence will appear in Sections II.7 and V.3.

9. The covering completeness degree

The similarity between dimension theory and completeness degree theory for separable metrizable space is very appealing. We have already mentioned in the last section that the covering dimension has its partner in completeness. We shall now introduce the covering completeness degree.

The key to our generalization is the proper adaptation of the notion of an open cover of X. Recall that $\mathcal{C}$ is the class of complete metrizable spaces.

9.1. Definition. Let X be a metrizable space. A *$\mathcal{C}$-kernel* of X is a complete space Y with $Y \subset X$. A *$\mathcal{C}$-border cover* of the space X is an open collection $\mathbf{U}$ of X such that $X \setminus \bigcup \mathbf{U}$ is complete; the set $X \setminus \bigcup \mathbf{U}$, which is a $\mathcal{C}$-kernel, is called the *enclosure* of $\mathbf{U}$.

The idea of considering objects like $\mathcal{C}$-border covers in the theory of extensions of spaces goes back to the [1965] paper of Smirnov. In that paper Smirnov gave a characterization of the compactness deficiency of a normal space in terms of the existence of a special sequence of seams (which are border covers with compact enclosures). This characterization of Smirnov will be discussed in detail in Section VI.6.

In light of the definition of $\mathcal{C}$-border cover the following definition of covering completeness degree is the natural extension of the definition of covering dimension.

9.2. Definition. Let X be a metrizable space. The *covering completeness degree* of X, denoted $\operatorname{ccd} X$, is defined as follows. $\operatorname{ccd} X = -1$ if and only if $X \in \mathcal{C}$. For n in $\mathbb{N}$, $\operatorname{ccd} X \leq n$ if for any finite $\mathcal{C}$-border cover $\mathbf{U}$ of X there exists a $\mathcal{C}$-border cover $\mathbf{V}$ of X such that $\mathbf{V}$ is a refinement of $\mathbf{U}$ and $\operatorname{ord} \mathbf{V} \leq n+1$. As usual, $\operatorname{ccd} X = n$ means $\operatorname{ccd} X \leq n$ holds but $\operatorname{ccd} X \leq n-1$ fails.

An obvious modification of the proof of Proposition 8.6 will yield the following lemma.

9.3. Lemma. *Suppose that X is a metrizable space and n is in $\mathbb{N}$. Then $\operatorname{ccd} X \leq n$ if and only if for every finite $\mathcal{C}$-border cover $\mathbf{U} = \{ U_i : i = 1, \ldots, k \}$ of X there exists a $\mathcal{C}$-border cover $\mathbf{V} = \{ V_i : i = 1, \ldots, k \}$ with enclosure G and $\operatorname{ord} \mathbf{V} \leq n+1$ such that G contains the enclosure of $\mathbf{U}$ and $\mathbf{V}$ shrinks the collection $\{ U_i \setminus G : i = 1, \ldots, k \}$ in the subspace $X \setminus G$.*

Though ccd is quite different from Icd in its definition we shall show in Section II.10 that ccd and Icd coincide for general metrizable spaces. Here we shall give a direct proof of this coincidence for the separable case. We do this because the proof of the coincidence theorem for dimension (Theorem 8.10) can be adapted to obtain the analagous one for the three completeness dimension functions.

Before proceding to the proof, we shall adopt the following modulo notation.

Modulo notation. Suppose that F is a $\mathcal{C}$-kernel of X. For subsets A and B of X and a collection $\mathbf{V}$ of subsets of X,

(1) $A \subset B \mod F$ means $A \setminus F \subset B \setminus F$,
(2) $A \cap B = \emptyset \mod F$ means $A \cap B \subset F$,
(3) $\operatorname{ord} \mathbf{V} \leq n \mod F$ means $\operatorname{ord} \{ V \setminus F : V \in \mathbf{V} \} \leq n$.

For the proof we shall need to establish the following variation of the preparatory Lemma 8.9.

9.4. Lemma. *Let n be a natural number and let X be a metrizable space with $\operatorname{ccd} X \leq n$. Suppose that G is a closed $\mathcal{C}$-kernel of X and that $\{ F_i : i = 0, 1, \ldots, k \}$ and $\{ U_i : i = 0, 1, \ldots, k \}$ are respectively closed and open collections of X such that $F_i \subset U_i \mod G$ for each i. Then there is a closed $\mathcal{C}$-kernel H of X and there are two open collections $\{ V_i : i = 0, 1, \ldots, k \}$ and $\{ W_i : i = 0, 1, \ldots, k \}$ such that the following conditions hold.*

$$F_i \subset V_i \subset \operatorname{cl}_X(V_i) \subset W_i \subset U_i \mod H, \qquad i = 0, 1, \ldots, k,$$
$$\operatorname{ord} \{ W_i \setminus \operatorname{cl}_X(V_i) : i = 0, 1, \ldots, k \} \leq n \mod H,$$
$$G \subset H.$$

The conditions

$$V_i \subset X \setminus H \text{ and } W_i \subset X \setminus H, \qquad i = 0, 1, \ldots, k,$$

may also be imposed.

Proof. Consider the $\mathcal{C}$-border cover

$$\mathbf{D} = \bigwedge \Big\{ \{ (U_i \setminus G), \ ((X \setminus F_i) \setminus G) \} : i = 0, 1, \ldots, k \Big\}.$$

There is a finite $\mathcal{C}$-border cover $\mathbf{V} = \{ L_j : j = 0, 1, \ldots, l \}$ with enclosure H such that $\mathbf{V}$ refines $\mathbf{D}$ and $\operatorname{ord} \mathbf{V} \leq n + 1$ (Lemma 9.3). Clearly the $\mathcal{C}$-kernel H contains G and is closed. In the normal subspace $X \setminus H$, the finite open collection $\mathbf{V}$ has a closed (in the subspace $X \setminus H$) shrinking $\{ K_j : j = 0, 1, \ldots, l \}$. The calculations of Lemma 8.9 applied to the subspace $X \setminus H$ will yield open collections $\{ V_i : i = 0, 1, \ldots, k \}$ and $\{ W_i : i = 0, 1, \ldots, k \}$ in $X \setminus H$ such that

$$F_i \setminus H \subset V_i \subset \operatorname{cl}_{X \setminus H}(V_i) \subset W_i \subset U_i \setminus H, \qquad i = 0, 1, \ldots, k,$$

and

$$\operatorname{ord}\{\, W_i \setminus \operatorname{cl}_{X\setminus H}(V_i) : i = 0, 1, \ldots, k \,\} \leq n.$$

As $\operatorname{cl}_{X\setminus H}(V_i) = \operatorname{cl}_X(V_i) \setminus H$ holds for each i, we have

$$F_i \subset V_i \subset \operatorname{cl}_X(V_i) \subset W_i \subset U_i \mod H, \qquad i = 0, 1, \ldots, k,$$
$$\operatorname{ord}\{\, W_i \setminus \operatorname{cl}_X(V_i) : i = 0, 1, \ldots, k \,\} \leq n \mod H$$

and

$$V_i \subset X \setminus H \text{ and } W_i \subset X \setminus H, \qquad i = 0, 1, \ldots, k.$$

Thus the proof is completed.

With this preparatory lemma we can prove the coincidence theorem.

9.5. Theorem (Coincidence theorem). *For every separable metrizable space X,*

$$\operatorname{ccd} X = \operatorname{icd} X = \operatorname{Icd} X.$$

Proof. As in the proof of the coincidence theorem for dimension we shall give the proof in two parts.

Proof of $\operatorname{ccd} X \leq \operatorname{icd} X$. This part of the proof will differ from that of Theorem 8.10 because it is not known that there is a decomposition theorem for icd. Suppose $\operatorname{icd} X \leq n$. By Theorem 7.11 and the coincidence theorem for dimension, there is a complete extension Y of X with $\dim(Y \setminus X) = \operatorname{icd} X \leq n$. Let $\boldsymbol{U} = \{\, U_i : i = 0, 1, \ldots, k \,\}$ be a finite $\mathcal{C}$-border cover X with enclosure F. From $\boldsymbol{U}$ we form an open collection $\boldsymbol{U}^* = \{\, U_i^* : i = 0, 1, \ldots, k \,\}$ in Y by selecting an open set U_i^* of Y such that $U_i^* \cap X = U_i$. Denote the subspace $X \cup (\bigcup \boldsymbol{U}^*)$ of Y by Y_1. Clearly the equality $Y_1 = F \cup (\bigcup \boldsymbol{U}^*)$ also holds. It follows that Y_1 is complete by Theorem 7.15 and that $\dim(Y_1 \setminus X) \leq n$ by the subspace theorem for dimension. By Proposition 8.6 the open cover $\{\, U_i^* \cap (Y_1 \setminus X) : i = 0, 1, \ldots, k \,\}$ of $Y_1 \setminus X$ has an open shrinking $\boldsymbol{V} = \{\, V_i : i = 0, 1, \ldots, k \,\}$ with $\operatorname{ord} \boldsymbol{V} \leq n + 1$. By Lemma 8.7 we have an open collection $\boldsymbol{W} = \{\, W_i : i = 0, 1, \ldots, k \,\}$ in the subspace Y_1 with $\operatorname{ord} \boldsymbol{W} \leq n + 1$ such that $W_i \cap (Y_1 \setminus X) = V_i$ for each i. Furthermore, $W_i \subset U_i$ may be assumed. Since $Y_1 \setminus \bigcup \boldsymbol{W} = X \setminus \bigcup \boldsymbol{W}$ holds, we have that $\{\, W_i \cap X : i = 0, 1, \ldots, k \,\}$ is a $\mathcal{C}$-border cover of X that refines $\boldsymbol{U}$ and has order not exceeding $n+1$. Thereby we have shown $\operatorname{ccd} X \leq n$.

Proof of $\operatorname{icd} X \leq \operatorname{ccd} X$. In view of Theorem 7.20, it will suffice to prove $\mathcal{C}$-Odim $X \leq \operatorname{ccd} X$. (Although the proof appears to be similar to the second part of the proof of Theorem 8.10, the nature of the construction forces complications that result from the need to consider $\mathcal{C}$-kernels.) Assume $\operatorname{ccd} X \leq n$ and let $\mathcal{B}_1$ be any countable base for X. Then let $\{(C_i, D_i) : i \in \mathbb{N}\}$ be the family of all pairs of elements of $\mathcal{B}_1$ such that $\operatorname{cl}(C_i) \subset D_i$. For each pair of natural numbers i and j with $i \leq j$ and for each natural number k we shall inductively define open sets V_i^j and W_i^j and closed $\mathcal{C}$-kernels G_k such that, when $i \leq j$,

$$V_i^j \subset X \setminus G_j \quad \text{and} \quad W_i^j \subset X \setminus G_j,$$

$$C_i \subset V_i^j \subset \operatorname{cl}(V_i^j) \subset W_i^j \subset D_i \mod G_j,$$

$$\operatorname{cl}(V_i^j) \subset V_i^{j+1} \mod G_{j+1} \quad \text{and} \quad \operatorname{cl}(W_i^{j+1}) \subset W_i^j \mod G_{j+1},$$

and

$$G_k \subset G_{k+1}, \qquad k = 0, 1, \dots,$$

$$\operatorname{ord}\{W_i^k \setminus \operatorname{cl}(V_i^k) : i = 0, 1, \dots, k\} \leq n \mod G_k, \quad k = 0, 1, \dots.$$

Let V_0^0 and W_0^0 be open sets with $C_0 \subset V_0^0 \subset \operatorname{cl}(V_0^0) \subset W_0^0 \subset D_0$ and $G_0 = \emptyset$. Suppose for some s that the open sets V_i^j and W_i^j have been defined for $i \leq j \leq s$ and the closed $\mathcal{C}$-kernels G_k have been defined for $k \leq s$ so that the above conditions are satisfied. Then we apply Lemma 9.4 to the $\mathcal{C}$-kernel G_s and the collections

$$\{\operatorname{cl}(V_i^s) \setminus G_s : i = 0, 1, \dots, s\} \cup \{\operatorname{cl}(C_{s+1}) \setminus G_s\}$$

and

$$\{W_i^s \setminus G_s : i = 0, 1, \dots, s\} \cup \{D_{s+1} \setminus G_s\}$$

so as to obtain the closed $\mathcal{C}$-kernel G_{s+1} and the open collections $\{V_i^{s+1} : i = 0, 1, \dots, s+1\}$ and $\{W_i^{s+1} : i = 0, 1, \dots, s+1\}$ satisfying the above conditions. From the inclusion

$$C_i \subset V_i^i \cup (G_i \cap D_i)$$

we find that the interior of $V_i^i \cup (G_i \cap D_i)$, denoted by V_i^*, satisfies

$$C_i \subset V_i^* \subset D_i \quad \text{and} \quad V_i^i = V_i^* \setminus G_i, \qquad i \in \mathbb{N}.$$

Moreover, when $i \leq k \leq j$ we have $V_i^j \subset W_i^k \mod G_j$ and consequently

$$V_i^j \setminus G_j \subset W_i^k \subset D_i.$$

Finally we define $\mathcal{B} = \{ V_i : i \in \mathbb{N} \}$ by

$$V_i = V_i^* \cup \left(\bigcup \{ V_i^j \setminus G_j : j > i \}\right), \qquad i \in \mathbb{N}.$$

It easily follows that $\mathcal{B}$ is a base for the open sets. Notice that

$$\mathrm{B}\,(V_i) \subset \left(W_i^k \setminus \mathrm{cl}\,(V_i^k)\right) \cup G_{k+1}, \qquad k \geq i.$$

Thus it follows that

$$\operatorname{ord}\{ \mathrm{B}\,(V_i) : i = 0, 1, \ldots, k \} \leq n \mod G_{k+1}, \qquad k \in \mathbb{N}.$$

We have shown for any $n+1$ different indices $i_0, \ldots, i_n$ that the intersection $\mathrm{B}\,(V_{i_0}) \cap \cdots \cap \mathrm{B}\,(V_{i_n})$ is contained in some G_k. It follows that this intersection is a complete space. We have established $\mathcal{C}\text{-}\mathrm{Odim}\, X \leq n$.

Investigations along the lines of coverings will be continued in Section II.5 and in the Chapters IV and V.

10. The σ-compactness degree

Even with the splitting of de Groot's compactification problem by way of the three functions Cmp, $\mathcal{K}$-Ind and Skl, very little progress had occurred in the first 25 years of its history. Negative facts such as the failure of a finite sum theorem for cmp had been established. In [1967] Nagata suggested a different approach to the compactification problem. He proposed that a σ-compactness degree be defined analogously to cmp (and icd) and that the intermediate problem of characterizing the σ-compactness deficiency by means of the σ-compactness degree be studied. The failure of the finite sum theorem should disappear in this context, a very pleasing prospect.

Continuing our historical perspective, we shall present the outcome of the program proposed by Nagata in [1967] as carried out by Aarts and Nishiura in [1973]. It was found that the notion of extension was not as natural as the notion of a kernel for classes like that of σ-compact spaces . This discovery led to a negative solution of Nagata's problem (see Example 10.13).

The class of all σ-compact spaces will be denoted by $\mathcal{S}$.

10.1. Definition. To every regular space X one assigns the *small inductive σ-compactness degree* $\mathcal{S}\text{-ind}\,X$ as follows.

(i) $\mathcal{S}\text{-ind}\,X = -1$ if and only if $X \in \mathcal{S}$.

(ij) For each natural number n, $\mathcal{S}\text{-ind}\,X \leq n$ if for each point p in X and for each closed set G of X with $p \notin G$ there is a partition S between p and G such that $\mathcal{S}\text{-ind}\,S \leq n-1$.

10.2. Definition. To every normal space X one assigns the *large inductive σ-compactness degree* $\mathcal{S}\text{-Ind}\,X$ as follows.

(i) $\mathcal{S}\text{-Ind}\,X = -1$ if and only if $X \in \mathcal{S}$.

(ij) For each natural number n, $\mathcal{S}\text{-Ind}\,X \leq n$ if for each pair of disjoint closed sets F and G of X there is a partition S between F and G such that $\mathcal{S}\text{-Ind}\,S \leq n-1$.

10.3. Propositions. Again we shall postpone the presentation of examples until after the main result (Theorem 10.5). First we shall collect a few propositions, the proofs of which are easy. Obviously the functions $\mathcal{S}$-ind and $\mathcal{S}$-Ind are topological invariants.

A. (a) *For every regular space X, $\mathcal{S}\text{-ind}\,X \leq \operatorname{cmp} X \leq \operatorname{ind} X$.*

(b) *For every normal space X, $\mathcal{S}\text{-Ind}\,X \leq \mathcal{K}\text{-Ind}\,X \leq \operatorname{Ind} X$.*

(c) *For every normal space X, $\mathcal{S}\text{-ind}\,X \leq \mathcal{S}\text{-Ind}\,X$.*

There is the following variant of the subspace theorem. The first step of the inductive proof is the observation that an F_σ-set of a σ-compact space is σ-compact.

B. *For every F_σ-set Y of a regular space X,*

$$\mathcal{S}\text{-ind}\,Y \leq \mathcal{S}\text{-ind}\,X.$$

There is a similar result for $\mathcal{S}$-Ind. A simple version of it will be proved later with the help of the main result of this section.

The following is a generalization of Proposition 3.6.

C. *Suppose that X is a regular space. For each natural number n, $\mathcal{S}$-ind $X \le n$ if and only if there exists a base $\mathcal{B}$ for the open sets such that $\mathcal{S}$-ind $\mathrm{B}(U) \le n-1$ for every U in $\mathcal{B}$.*

In the context of the class $\mathcal{S}$ of σ-compact spaces it has been stated that the notion of kernel was more natural than that of extension. Let us now make this more explicit.

10.4. Definition. A subset Y of a space X is called an *$\mathcal{S}$-kernel* of X if Y is σ-compact. The *$\mathcal{S}$-Surplus* of a metrizable space X is defined by

$$\mathcal{S}\text{-Sur}\, X = \min \{ \operatorname{Ind}(X \setminus Y) : Y \text{ is a } \mathcal{S}\text{-kernel of } X \}.$$

10.5. Theorem (Main theorem). *For every separable metrizable space X,*

$$\mathcal{S}\text{-ind}\, X = \mathcal{S}\text{-Ind}\, X = \mathcal{S}\text{-Sur}\, X.$$

Proof. It has been noticed already that $\mathcal{S}$-ind $X \le \mathcal{S}$-Ind X. We shall establish $\mathcal{S}$-Ind $X \le \mathcal{S}$-Sur X and $\mathcal{S}$-Sur $X \le \mathcal{S}$-ind X.

Proof of $\mathcal{S}$-Ind $X \le \mathcal{S}$-Sur X. The proof is by induction on the value of $\mathcal{S}$-Sur X. The inductive step follows. Suppose that X is a space with $\mathcal{S}$-Sur $X \le n$. Let F and G be disjoint closed sets and let Y be an $\mathcal{S}$-kernel of X with $\operatorname{Ind}(X \setminus Y) \le n$. By Proposition 4.6 there is a partition S between F and G with $\operatorname{Ind}\big(S \cap (X \setminus Y)\big) \le n-1$. We have $S \cap Y \in \mathcal{S}$ because S is a closed subset of X. Consequently, $\mathcal{S}$-Sur $S \le n-1$, whence $\mathcal{S}$-Ind $S \le n-1$ by the induction hypothesis. Thus, $\mathcal{S}$-Ind $X \le n$.

Proof of $\mathcal{S}$-Sur $X \le \mathcal{S}$-ind X. The initial step of the induction proof is obvious. For the inductive step, let $\mathcal{S}$-ind $X \le n$. In view of Proposition 10.3.C, there exists a base $\mathcal{B}$ for the open sets of X such that $\mathcal{S}$-ind $\mathrm{B}(U) \le n-1$ for every U in $\mathcal{B}$. Indeed, we may assume that $\mathcal{B}$ is countable. Let $\{ U_i : i = 0, 1, \dots \}$ be an enumeration of $\mathcal{B}$. By the induction hypothesis we have $\mathcal{S}$-Sur $\mathrm{B}(U_i) \le n-1$ for every i in $\mathbb{N}$. So there are disjoint sets Y_i and Z_i with $\mathrm{B}(U_i) = Y_i \cup Z_i$ such that $\operatorname{Ind} Y_i \le n-1$ and $Z_i \in \mathcal{S}$. Write $X = Z \cup B \cup A$ where $Z = \bigcup\{ Z_i : i \in \mathbb{N} \}$, $B = \bigcup\{ \mathrm{B}(U_i) \setminus Z : i \in \mathbb{N} \}$ and $A = X \setminus (B \cup Z)$.

Obviously Z is in $\mathcal{S}$. Also $\operatorname{ind} A \leq 0$ holds as $\{ U_i \cap A : i \in \mathbb{N} \}$ is a base for the open sets of A with $\mathrm{B}_A(U_i \cap A) = \emptyset$ for every i. Since $\mathrm{B}(U_i \setminus Z) \subset Y_i$, we have $\operatorname{ind}(\mathrm{B}(U_i) \setminus Z) \leq n - 1$. By the sum theorem for dimension, $\operatorname{ind} B \leq n - 1$. Consequently $\operatorname{ind}(A \cup B) \leq n$ holds by the addition theorem. Finally, by the coincidence theorem for dimension, it follows that $\operatorname{Ind}(A \cup B) \leq n$. Thus $\mathcal{S}\text{-}\operatorname{Sur} X \leq n$ has been shown.

Now let us show that for each n in $\mathbb{N}$ there exists a space with σ-compactness degree n.

10.6. Example. Recall that $\mathbb{P}$ is the set of irrational numbers and $\mathbb{I}$ is the interval $[-1, 1]$. For each natural number n,

$$\mathcal{S}\text{-}\operatorname{ind}(\mathbb{P} \times \mathbb{I}^n) = \mathcal{S}\text{-}\operatorname{ind}(\mathbb{P} \times \mathbb{R}^n) = n.$$

To prove this we first observe $\mathcal{S}\text{-}\operatorname{ind}(\mathbb{P} \times \mathbb{R}^n) \leq \operatorname{ind}(\mathbb{P} \times \mathbb{R}^n) = n$. So the equality will be established on showing $\mathcal{S}\text{-}\operatorname{Sur}(\mathbb{P} \times \mathbb{I}^n) \geq n$. Denoting the projection of $\mathbb{P} \times \mathbb{I}^n$ onto $\mathbb{P}$ by p, we have that the image $p[C]$ of any $\mathcal{S}$-kernel C of $\mathbb{P} \times \mathbb{I}^n$ is also an $\mathcal{S}$-kernel of $\mathbb{P}$. By the Baire category theorem we find that $\mathbb{P} \setminus p[C]$ is not empty. It follows that $p^{-1}[(\mathbb{P} \setminus p[C])] \subset (\mathbb{P} \times \mathbb{I}^n) \setminus C$. Hence, $\mathcal{S}\text{-}\operatorname{Sur}(\mathbb{P} \times \mathbb{I}^n) \geq n$.

In passing, we observe that the equality $\operatorname{cmp}(\mathbb{P} \times \mathbb{I}^n) = n$ follows with the aid of Proposition 10.3.A.

10.7. Remark. A consequence of Lemma 7.5 and Theorem 10.5 is the following observation.

Suppose that X is a compact metrizable space and Y and Z are disjoint subsets of X with $X = Y \cup Z$. Then $\mathcal{C}\text{-}\operatorname{def} Y = \mathcal{S}\text{-}\operatorname{Sur} Z$.

The observation together with Example 7.12 and Theorem 10.5 will result in another proof of the assertion in the previous example. Also in Example 7.13 we have exhibited a subset Y of a compact metrizable space X of dimension n with $\operatorname{icd} Y \geq n - 2$. In view of Theorems 7.11 and 10.5 we have $\mathcal{S}\text{-}\operatorname{ind}(X \setminus Y) \geq n - 2$.

The close analogies between the σ-compactness degree and the dimension function will be exhibited next. The main theorem (Theorem 10.5) is a key element in the exposition.

10.8. Theorem. *For every F_σ-set Y of a separable metrizable space X,*

$$\mathcal{S}\text{-}\operatorname{Ind} Y \leq \mathcal{S}\text{-}\operatorname{Ind} X.$$

Proof. See Proposition 10.3.B and Theorem 10.5.

10.9. Theorem (Decomposition theorem). *Let X be a separable metrizable space. Then $\mathcal{S}$-ind $X \leq n$ if and only if X is the union of $n+1$ sets X_i, $i = 0, 1, \ldots, n$, with $\mathcal{S}$-ind $X_i \leq 0$ for every i.*

Proof. Let $\mathcal{S}$-ind $X \leq n$. By the main theorem there is a subset C of X such that C is σ-compact and $\operatorname{Ind}(X \setminus C) \leq n$. From the coincidence and decomposition theorems for dimension we have a partitioning of $X \setminus C$ into $n+1$ at most 0-dimensional sets Y_i, $i = 0, 1, \ldots, n$. The main theorem gives $\mathcal{S}$-ind $(C \cup Y_i) \leq 0$ for every i. This proves the necessity. For the sufficiency we suppose that $X = X_0 \cup X_1 \cup \cdots \cup X_n$, where $\mathcal{S}$-ind $X_i \leq 0$ for $i = 0, 1, \ldots, n$. Let $X_i = Y_i \cup Z_i$ with $Y_i \in \mathcal{S}$ and $\operatorname{Ind} Z_i \leq 0$. Using the σ-compact subset $Y = Y_0 \cup Y_1 \cup \cdots \cup Y_n$ of X, we have $\mathcal{S}$-Sur $X \leq n$ because $\operatorname{Ind}(X \setminus Y) \leq \operatorname{Ind}(Z_0 \cup Z_1 \cup \cdots \cup Z_n) \leq n$. It follows from the main theorem that $\mathcal{S}$-ind $X \leq n$.

The addition theorem for $\mathcal{S}$-ind is deduced in the same way as the one for dimension.

10.10. Theorem (Addition theorem). *Let X be a separable metrizable space. If $X = Y \cup Z$, then*

$$\mathcal{S}\text{-ind}\, X \leq \mathcal{S}\text{-ind}\, Y + \mathcal{S}\text{-ind}\, Z + 1.$$

10.11. Theorem (Countable sum theorem). *Suppose that $\{ F_i : i \in \mathbb{N} \}$ is a countable closed cover of a separable metrizable space X. Then*

$$\mathcal{S}\text{-ind}\, X \leq \sup \{ \mathcal{S}\text{-ind}\, F_i : i \in \mathbb{N} \}.$$

Proof. For each i we let F_i be the union $C_i \cup Y_i$ with $C_i \in \mathcal{S}$ and $\operatorname{ind} Y_i \leq \mathcal{S}$-ind $F_i = \mathcal{S}$-Sur F_i and then let $C = \bigcup \{ C_i : i \in \mathbb{N} \}$. The collection $\{ F_i \setminus C : i \in \mathbb{N} \}$ is a countable closed cover of $X \setminus C$ with $F_i \setminus C \subset Y_i$ for each i. By the countable sum theorem of dimension we have $\operatorname{ind}(X \setminus C) \leq \sup \{ \operatorname{ind}(F_i \setminus C) : i \in \mathbb{N} \}$. The theorem follows.

It has been clearly shown that both the σ-compactness degree and the completeness degree admit analogies to dimension. But a dichotomy is also becoming apparent. The function $\mathcal{S}$-ind is a prototype of the functions of the inductive dimensional type which are studied in Chapter III and the function ccd (see Section 9) is a prototype of the functions of covering dimensional type which are studied in Chapter IV and V. Also, $\mathcal{S}$-kernels and $\mathcal{S}$-Surplus are keys to the theory of $\mathcal{S}$-ind, and complete extensions and $\mathcal{C}$-def are keys to the theory of ccd. This dichotomy will be underscored by the next two examples.

10.12. Example. Let X be a metrizable space. Recall that Y is a *$\mathcal{C}$-kernel* of X if Y is complete space that is contained in X. We define the *$\mathcal{C}$-surplus* of X, denoted $\mathcal{C}\text{-sur}\, X$, by

$$\mathcal{C}\text{-sur}\, X = \inf\{\dim(X \setminus Y) : Y \text{ is a } \mathcal{C}\text{-kernel of } X\}.$$

In contrast to the main theorem for $\mathcal{S}$ we shall exhibit a space X for which

$$\operatorname{icd} X < \mathcal{C}\text{-sur}\, X.$$

The space X is the subspace $(\mathbb{R} \times \mathbb{Q}) \cup (\mathbb{Q} \times \mathbb{R}) = \mathbb{R}^2 \setminus \mathbb{P}^2$ of $\mathbb{R}^2$. It is clear that $\mathcal{C}\text{-def}\, X \leq 0$. Thus $\operatorname{icd} X \leq 0$ holds by Theorem 7.11. As X is of the first Baire category, we have $X \notin \mathcal{C}$. Observe that each dense subspace of a first Baire category space is also a first Baire category space. Consequently no $\mathcal{C}$-kernel of X is dense in X. Because $\dim U = 1$ for every nonempty open subset U of X, it follows that $\mathcal{C}\text{-sur}\, X \geq 1$. Obviously, $\mathcal{C}\text{-sur}\, X \leq \operatorname{ind} X = 1$. So $\mathcal{C}\text{-sur}\, X = 1$ holds.

10.13. Example. Let X be a separable metrizable space. We say that a space Y is a an *$\mathcal{S}$-hull* of X if Y is a metrizable σ-compact space that contains the space X. And the *$\mathcal{S}$-deficiency* of X is defined as

$$\mathcal{S}\text{-def}\, X = \inf\{\dim(Y \setminus X) : Y \text{ is an } \mathcal{S}\text{-hull of } X\}.$$

In [1965] Nagata conjectured that $\mathcal{S}\text{-ind}\, X = \mathcal{S}\text{-def}\, X$. We shall present an example of a separable metrizable space X with

$$\mathcal{S}\text{-ind}\, X = 0 \quad \text{and} \quad \mathcal{S}\text{-def}\, X = 1.$$

Observe also that this is in contrast to the main theorem for $\mathcal{C}$. Let X be the subspace of $\mathbb{R}^3$ given by $X = (\mathbb{R} \times \mathbb{R} \times \mathbb{Q}) \cup (\mathbb{P} \times \mathbb{P} \times \mathbb{P})$.

The space X is of the second Baire category since its dense subset $\mathbb{P} \times \mathbb{P} \times \mathbb{P}$ is so. It follows that X cannot be σ-compact, whence $\mathcal{S}$-ind $X \geq 0$. On the other hand, $\mathbb{R} \times \mathbb{R} \times \mathbb{Q}$ is σ-compact. By the main theorem we find $\mathcal{S}\text{-ind}\, X = \mathcal{S}\text{-Sur}\, X \leq 0$. So, $\mathcal{S}\text{-ind}\, X = 0$.

To help us compute the value of $\mathcal{S}\text{-def}\, X$ we note that $\{0\} \times \mathbb{R} \times \mathbb{Q}$ is a closed subspace of X. Thus by Example 7.12 and Proposition 5.9.C we have $1 = \operatorname{cmp}(\mathbb{R} \times \mathbb{Q}) \leq \operatorname{cmp} X$. (It is not difficult to prove further that $\operatorname{cmp} X = 1$.) Let us show that $\operatorname{ind}(Y \setminus X) \geq 1$ for any $\mathcal{S}$-hull Y of X. We may assume that X is dense in Y. As Y is the countable union of compact sets and X is of the second category, there is a compact subset Z of Y with nonvoid interior. It follows that $Z \cap X$ contains a copy of X, say W, that is open in X. As $\operatorname{cmp} W = 1$, by Proposition 5.9.E we have $\operatorname{cmp}(Z \cap X) \geq 1$. It follows from Theorem 5.8 that $\operatorname{ind}\big(\operatorname{cl}_Z((Z \cap X) \setminus X)\big) \geq 1$. We have shown that $\operatorname{ind}(Y \setminus X) \geq 1$ holds. Consequently, $\mathcal{S}\text{-def}\, X \geq 1$ by the coincidence theorem for dimension. The equality now follows easily.

Results similar to those of this section can be shown to hold for more general spaces X and for classes of spaces $\mathcal{P}$ that are more general than $\mathcal{S}$. The investigation along the these lines will be continued in Chapter III.

11. Pol's example

In [1982] Pol gave an example of a space X with $\operatorname{cmp} X = 1$ and $\operatorname{def} X = 2$ and thereby resolved de Groot's compactification problem in the negative. We shall describe Pol's example and Hart's generalization which was constructed in [1985].

Agreement. *Throughout the section, n is a natural number such that $n \geq 2$.*

Pol's example is a modification of an example constructed by Luxemburg in [1973] of a compact metric space with noncoinciding transfinite dimensions. The descriptions will require the following notation.

11.1. Notation. As usual, $\mathbb{I}$ will denote the interval $[-1, 1]$ and x_i will denote the i-th coordinate of a point x in $\mathbb{I}^n$. Define for $1 \leq i \leq n$ the sets

$$F_i = \{\, x \in \mathbb{I}^n : x_i = -1 \,\},$$

$$
\begin{aligned}
G_i &= \{\, x \in \mathbb{I}^n : x_i = +1 \,\}, \\
K_i &= \{\, x \in \mathbb{I}^n : -1 \le x_i \le -\tfrac{1}{3} \,\}, \\
L_i &= \{\, x \in \mathbb{I}^n : +\tfrac{1}{3} \le x_i \le +1 \,\}.
\end{aligned}
$$

The combinatorial boundary of $\mathbb{I}^n$ will be denoted by $\partial\,\mathbb{I}^n$. That is,

$$\partial\,\mathbb{I}^n = \{\, x \in \mathbb{I}^n : x_j = -1 \text{ or } x_j = 1 \text{ for some } j \,\}.$$

Finally let C be the subset of $\mathbb{R}$ defined by

$$C = \{\, \tfrac{1}{k+1} : k \in \mathbb{N} \,\} \cup \{\, 0 \,\}.$$

To construct examples such as Pol's, we must build a feature into the construction that will push up the value of the compactness deficiency as well as another feature that will keep down the value of the compactness degree. Let us first concentrate on the feature that forces the compactness deficiency to exceed n.

11.2. Proposition. *For m in $\mathbb{N}$ and j in $\{\, n+1, \dots, 2n-1 \,\}$ let E_j^m be a partition between K_j and L_j in $\mathbb{I}^{2n-1}$ and let*

$$E_m = \bigcap\{\, E_j^m : j = n+1, \dots, 2n-1 \,\}.$$

Define the subspace X of $\mathbb{I}^{2n-1} \times C$ to be

$$X = \big((\partial\,\mathbb{I}^{2n-1}) \times \{0\}\big) \cup \big(\bigcup\{\, E_m \times \{\, \tfrac{1}{m+1} \,\} : m \in \mathbb{N} \,\}\big).$$

Then $\operatorname{def} X \ge \operatorname{Cmp} X \ge n$.

Proof. We shall make use of the invariant Comp that was introduced in Example 6.7. It will suffice to prove that $\operatorname{Comp} X \le n-1$ fails. To this end, suppose that it does not fail and consider the n pairs of opposite faces $(F_i \times \{0\}, G_i \times \{0\})$, $i = 1, \dots, n$, in the space $(\partial\,\mathbb{I}^{2n-1}) \times \{0\}$. There are partitions S_i in X between $F_i \times \{0\}$ and $G_i \times \{0\}$, $i = 1, \dots, n$, such that $S_1 \cap \dots \cap S_n$ is compact. Let $Y = \mathbb{I}^{2n-1} \times C$. Since X is a subspace of Y, by Lemma 4.5 there are partitions T_i between $F_i \times \{0\}$ and $G_i \times \{0\}$ in Y such that $T_i \cap X = S_i$, $i = 1, \dots, n$. So $T_1 \cap \dots \cap T_n \cap X$ will be compact also. As the sets $F_i \times \{0\}$ and $G_i \times \{0\}$ are compact, there is an m' such that each T_i is a partition between $F_i \times \{\, \frac{1}{m+1} \,\}$ and $G_i \times \{\, \frac{1}{m+1} \,\}$

in Y for $m \geq m'$ and $i = 1, \ldots, n$. As E_j^m are partitions between F_j and G_j in $\mathbb{I}^{2n-1}$ for $j = n+1, \ldots, 2n-1$ and for all m, we also have that the sets $E_j^m \times \{\frac{1}{m+1}\}$ are partitions between $F_j \times \{\frac{1}{m+1}\}$ and $G_j \times \{\frac{1}{m+1}\}$ in Y. So for $m \geq m'$ we have that $\mathbb{I}^{2n-1} \times \{\frac{1}{m+1}\}$ is contained in Y and that

$$T_i \cap (\mathbb{I}^{2n-1} \times \{\tfrac{1}{m+1}\}), \qquad i = 1, \ldots, n,$$

and

$$E_j^m \times \{\tfrac{1}{m+1}\}, \qquad j = n+1, \ldots, 2n-1,$$

are, for $k = 1, \ldots, 2n-1$, partitions in $\mathbb{I}^{2n-1} \times \{\frac{1}{m+1}\}$ between

$$F_k \times \{\tfrac{1}{m+1}\} \quad \text{and} \quad G_k \times \{\tfrac{1}{m+1}\}.$$

With the aid of Theorem 4.9 we choose points p_m such that

$$p_m \in X \cap T_1 \cap \cdots \cap T_n \cap (E_m \times \{\tfrac{1}{m+1}\}), \qquad m \geq m'.$$

It will follow from the definition of the subsets K_j and L_j of $\mathbb{I}^{2n-1}$ for $j \geq n-1$ that the sequence $\{p_m : m \in \mathbb{N}\}$ cannot have any convergent subsequence in X. Thus we have $T_1 \cap \cdots \cap T_n \cap X$ is not compact. This is a contradiction.

The following is a key lemma in Pol's construction. The lemma will build into the construction the feature that keeps the value of the compactness degree down. The lemma itself is a special case of Lemma 3.1 of Luxemburg [1981].

11.3. Lemma. *In the cube $\mathbb{I}^3$ there is a base $\mathcal{B} = \{U_i : i \in \mathbb{N}\}$ for the open sets and for each m in $\mathbb{N}$ there is a partition E_m between K_3 and L_3 such that $\mathrm{B}(U_i) \cap E_m$ is a finite disjoint union of compact sets of diameter not exceeding $\frac{1}{m+1}$ when $i \leq m$. Moreover,* ind $E_m \leq 2$ *can be imposed for each m.*

Proof. Let $\mathcal{B} = \{U_i : i \in \mathbb{N}\}$ be the base for the open sets of $\mathbb{I}^3$ given by Corollary 6.12. Then the intersection $\mathrm{B}(U_{i_0}) \cap \cdots \cap \mathrm{B}(U_{i_3})$ is empty for any four distinct indices $i_0, \ldots, i_3$. As $\mathbb{I}^3$ is compact, we may assume that $\delta(U_i)$ converges to 0 as i tends to ∞ and that $\delta(U_i)$ are bounded by $\frac{1}{3}$.

Let us construct E_m for each m in $\mathbb{N}$. For $i \leq m$ we select finite covers $\mathcal{B}_i$ of $\mathbb{I}^3$ consisting of elements of $\mathcal{B} \setminus \{ U_0, \ldots, U_m \}$ such that

$$\delta(U) \leq \tfrac{1}{m+1}, \qquad U \in \mathcal{B}_i$$

and

$$\mathcal{B}_i \cap \mathcal{B}_j = \emptyset, \qquad j < i.$$

Then define

$$C_i = \mathrm{B}(U_i) \cap \big(\bigcup\{ \mathrm{B}(U) : U \in \mathcal{B}_i \}\big), \qquad i \leq m.$$

Because $\operatorname{ord}\{ \mathrm{B}(U_i) : i \in \mathbb{N} \} \leq 3$, the sets C_i are pairwise disjoint. Also no set C_i meets both K_3 and L_3. Let

$$K = K_3 \cup \big(\bigcup\{ C_i : i \leq m \text{ and } C_i \cap K_3 \neq \emptyset \}\big),$$
$$L = L_3 \cup \big(\bigcup\{ C_i : i \leq m \text{ and } C_i \cap K_3 = \emptyset \}\big).$$

Because the sets C_i are pairwise disjoint, the sets K and L are disjoint. Let E_m be any partition between K and L with $\operatorname{ind} E_m \leq 2$. That E_m has the required property easily follows from the observation that the subset $\mathrm{B}(U_i) \cap E_m$ of $X \setminus \big(\bigcup\{ \mathrm{B}(U) : U \in \mathcal{B}_i \}\big)$ is the finite disjoint union of open sets of diameter not exceeding $\frac{1}{m+1}$ whenever $i \leq m$.

Remark. We have invoked Corollary 6.12 to guarantee the existence of a base $\mathcal{B} = \{ U_i : i \in \mathbb{N} \}$ such that any four boundaries of elements of $\mathcal{B}$ have empty intersection. Since we are dealing with $\mathbb{I}^3$, there is an elementary construction. It is a nice exercise to show that there is a base with this property consisting of "little" cubes.

11.4. Example (Pol [1982]). The example is a space X such that

$$\operatorname{cmp} X = 1 \quad \text{and} \quad \operatorname{Cmp} X = \operatorname{def} X = 2.$$

It will be a subspace of $\mathbb{I}^4$. By Lemma 11.3 there exists a base $\mathcal{B} = \{ U_i : i \in \mathbb{N} \}$ for the open subsets of $\mathbb{I}^3$ and for each m in $\mathbb{N}$ there exists a partition E_m between K_3 and L_3 in $\mathbb{I}^3$ with $\operatorname{ind} E_m \leq 2$ such that the set $\mathrm{B}(U_i) \cap E_m$ is the disjoint union of compact sets of diameter not exceeding $\frac{1}{m+1}$ when $i \leq m$. Define the space to be

$$X = \big((\partial\, \mathbb{I}^3) \times \{ 0 \}\big) \cup \big(\bigcup\{ E_m \times \{ \tfrac{1}{m+1} \} : m \in \mathbb{N} \}\big).$$

From Proposition 11.2 we have that $\operatorname{def} X \geq \operatorname{Cmp} X \geq 2$. As the space X is two-dimensional, it follows from Theorem 5.11 that $\operatorname{def} X = 2$.

To show that $\operatorname{cmp} X = 1$ it will suffice to show that $\operatorname{cmp} X \leq 1$ because de Groot's theorem yields $\operatorname{cmp} X \not\leq 0$. That is, we must show that each point x of X has arbitrarily small open neighborhoods whose boundaries are rim-compact. There are compact neighborhoods for each point x not in $(\partial\,\mathbb{I}^3) \times \{0\}$. So let us consider a point $x = (y, 0)$ in $(\partial\,\mathbb{I}^3) \times \{0\}$. The open ball in $\mathbb{I}^3$ with radius ε centered at a point w will be denoted by $\mathrm{S}_\varepsilon(w)$. Recall that X is a subset of $\mathbb{I}^4$. A neighborhood base of x is formed by the open sets $V_i = \big(U_i \times [0, \frac{1}{m+1}]\big) \cap X$ with $y \in U_i \in \mathcal{B}$ and $m \in \mathbb{N}$. In view of Proposition 5.9.D, it suffices to verify $H_i = \big(\mathrm{B}_{\mathbb{I}^3}(U_i) \times [0, \frac{1}{m+1}]\big) \cap X$ is rim-compact. Observe that H_i is a subset of

$$\big((\mathrm{B}_{\mathbb{I}^3}(U_i) \cap \partial\,\mathbb{I}^3) \times \{0\}\big) \cup \big(\bigcup\{\, (\mathrm{B}_{\mathbb{I}^3}(U_i) \cap E_m) \times \{\tfrac{1}{m+1}\} : m \in \mathbb{N}\,\}\big).$$

To prove that H_i is rim-compact it is sufficient to prove that a point $z = (w, 0)$ in $\big((\partial\,\mathbb{I}^3) \times \{0\}\big) \cap H_i$ has arbitrarily small neighborhoods in H_i with compact boundary. Let $0 < \varepsilon < \frac{1}{i}$. Because the set $(\mathrm{B}_{\mathbb{I}^3}(U_i) \cap E_m) \times \{\frac{1}{m+1}\}$ is a finite disjoint union of compact sets of diameter not exceeding $\frac{1}{m+1}$ for each m with $i \leq m$, we can find an open set W of X such that

$$\big((\mathrm{S}_\varepsilon(w) \cap \partial\,\mathbb{I}^3) \times \{0\}\big) \cup \big(\bigcup\{\, \mathrm{S}_{\varepsilon-\frac{1}{m}}(w) \times \{\tfrac{1}{m}\} : m > \tfrac{1}{\varepsilon}\}\big) \subset W,$$
$$W \subset \big(\mathrm{S}_\varepsilon(w) \times \{0\}\big) \cup \big(\bigcup\{\, \mathrm{S}_{\varepsilon+\frac{1}{m}}(w) \times \{\tfrac{1}{m}\} : m > \tfrac{1}{\varepsilon}\}\big)$$

and

$$\mathrm{B}_X(W) \cap H_i \subset (\partial\,\mathbb{I}^3) \times \{0\}.$$

We have shown that H_i is rim-compact.

In Pol's example the gap between $\operatorname{cmp} X$ and $\operatorname{def} X$ is one. In [1988a] Kimura published examples of spaces for which the gap between cmp and def is arbitrarily large. Hart indicated in [1985] how such spaces can be constructed by a modification of Pol's construction which will be given next. The following is a preparatory lemma.

11.5. Lemma. *In a normal space X, let $\mathbf{F}$ be a finite closed collection with order not exceeding k. Then for $i = 1, \ldots, k$ there* *are collections* $\boldsymbol{H}_i$ *of disjoint closed sets such that* $\boldsymbol{H}_i$ *refines* $\boldsymbol{F}$ *for every* i *and* $\bigcup \boldsymbol{F} = \bigcup\{ (\bigcup \boldsymbol{H}_i) : i = 1, \ldots, k \}$.

Proof. We may assume $X = \bigcup \boldsymbol{F}$. The proof is by induction on k. As the case $k = 1$ is obvious, we shall discuss the inductive step. For each subset $\boldsymbol{G}$ of $\boldsymbol{F}$ we define the set $K_{\boldsymbol{G}} = \bigcap\{ F : F \in \boldsymbol{G} \}$ and we denote the number of elements of $\boldsymbol{G}$ by $|\boldsymbol{G}|$. The collection $\{ K_{\boldsymbol{G}} : \boldsymbol{G} \subset \boldsymbol{F},\ |\boldsymbol{G}| = k \}$ is disjoint as $k \geq 2$. For each $\boldsymbol{G}$ with $\boldsymbol{G} \subset \boldsymbol{F}$ and $|\boldsymbol{G}| = k$ there exists an open set $U_{\boldsymbol{G}}$ such that $K_{\boldsymbol{G}} \subset U_{\boldsymbol{G}}$ holds and the collection $\{ \mathrm{cl}(U_{\boldsymbol{G}}) : \boldsymbol{G} \subset \boldsymbol{F},\ |\boldsymbol{G}| = k \}$ is disjoint (see the proof of Lemma 8.8). The required collection $\boldsymbol{H}_k$ is defined to be $\{ \mathrm{cl}(U_{\boldsymbol{G}}) : \boldsymbol{G} \subset \boldsymbol{F},\ |\boldsymbol{G}| = k \}$. To get the remaining $k - 1$ collections let $U = \bigcup\{ U_{\boldsymbol{G}} : \boldsymbol{G} \subset \boldsymbol{F},\ |\boldsymbol{G}| = k \}$ and define the closed collection $\boldsymbol{F}_{k-1} = \{ F \setminus U : F \in \boldsymbol{F},\ F \setminus U \neq \emptyset \}$. As $\operatorname{ord} \boldsymbol{F}_{k-1} \leq k - 1$ holds, the induction hypothesis applies and the collections $\boldsymbol{H}_i$ exist for the remaining $i = 1, \ldots, k - 1$.

The first step towards generalizing Pol's example is the following adaptation of Lemma 11.3.

11.6. Lemma. *In the* $(2n-1)$*-dimensional cube* $\mathbb{I}^{2n-1}$ *there is a base* $\mathcal{B} = \{ U_i : i \in \mathbb{N} \}$ *for the open sets and for each* m *in* $\mathbb{N}$ *there are partitions* E_j^m *between* K_j *and* L_j *for* j *in* $\{ n+1, \ldots, 2n-1 \}$ *such that the set*

$$\mathrm{B}(U_i) \cap \left(\bigcap\{ E_j^m : j = n+1, \ldots, 2n-1 \}\right)$$

is a finite disjoint union of compact sets of diameter not exceeding $\frac{1}{m+1}$ *when* $i \leq m$.

Proof. The proof of Lemma 11.3 will be modified. By Corollary 6.12 there is a base $\mathcal{B} = \{ U_i : i \in \mathbb{N} \}$ for the open sets of $\mathbb{I}^{2n-1}$ such that the intersection $\mathrm{B}(U_{i_0}) \cap \cdots \cap \mathrm{B}(U_{i_{2n-1}})$ is empty for any set of 2n indices satisfying $i_0 < \cdots < i_{2n-1}$. We may assume further that $\delta(U_i) < \frac{1}{3}$ for all i and that $\delta(U_i) \to 0$ as $i \to \infty$. Let us construct the partitions E_j^m, $j = n+1, \ldots, 2n-1$. For $i \leq m$ we choose a finite cover $\mathcal{B}_i$ of $\mathbb{I}^{2n-1}$ with elements of $\mathcal{B} \setminus \{ U_0, \ldots, U_m \}$ such that

$$\delta(U) \leq \tfrac{1}{m+1}, \quad U \in \mathcal{B}_i,$$

and

$$\mathcal{B}_i \cap \mathcal{B}_j = \emptyset, \quad i < j \leq m.$$

For $i \leq m$ we let

$$C_i = \mathrm{B}(U_i) \cap \left(\bigcup\{\mathrm{B}(U) : U \in \mathcal{B}_i\}\right).$$

Then $\operatorname{ord}\{\mathrm{B}(U_i) : i \in \mathbb{N}\} \leq 2n - 1$ gives $\operatorname{ord}\{C_i : i \leq m\} \leq n - 1$. By Lemma 11.5, $\{C_i : i \in \mathbb{N}\}$ has disjoint closed refinements $\boldsymbol{H}_i$, $i = 1, \dots, n - 1$, such that

$$\bigcup\{C_i : i \leq m\} = \bigcup\{\bigcup \boldsymbol{H}_i : i = 1, \dots, n - 1\}.$$

Now for each j in $\{n + 1, \dots, 2n - 1\}$ we define

$$K_j^* = K_j \cup \left(\bigcup\{H : H \in \mathbf{H}_{j-n},\ H \cap K_j \neq \emptyset\}\right),$$
$$L_j^* = L_j \cup \left(\bigcup\{H : H \in \mathbf{H}_{j-n},\ H \cap K_j = \emptyset\}\right).$$

As in the proof of Lemma 11.3 the sets K_j^* and L_j^* are disjoint and the partitions E_j^m between K_j^* and L_j^* will yield the desired partitions.

11.7. Example. The example is a space X for which

$$\operatorname{cmp} X = 1 \quad \text{and} \quad \operatorname{def} X \geq \operatorname{Cmp} X \geq n.$$

The example will be a subset of $\mathbb{I}^{2n}$. Choose the base $\mathcal{B}$ and the sets E_j^m, $j = n + 1, \dots, 2n - 1$, as provided by Lemma 11.6 and let

$$E_m = \bigcap\{E_j^m : j = n + 1, \dots, 2n - 1\}.$$

The space is defined to be

$$X = \left((\partial\, \mathbb{I}^{2n-1}) \times \{0\}\right) \cup \left(\bigcup\{E_m \times \{\tfrac{1}{m+1}\} : m \in \mathbb{N}\}\right).$$

By Proposition 11.2, $\operatorname{Cmp} X \geq n$. The equality $\operatorname{cmp} X = 1$ is proved in the exact same way as in Example 11.4.

Using this example, we can construct spaces for which the gap between cmp and def has any prescribed value, including ∞. The following space will achieve the gap n. Let $X \oplus Z$ denote the topological sum of a space X with $\operatorname{def} X - n = k \geq 1$ and $\operatorname{cmp} X = 1$ and the space $Z = \mathbb{Q} \times \mathbb{I}^k$. Then $\operatorname{cmp}(X \oplus Z) = k$ and $\operatorname{def}(X \oplus Z) = n + k$. Consequently, $\operatorname{def}(X \oplus Z) - \operatorname{cmp}(X \oplus Z) = n$.

12. Kimura's theorem

Although de Groot's compactification problem was resolved by Pol's example, the example did not shed any light on the validity of $\operatorname{Cmp} X = \operatorname{def} X$ or $\operatorname{Skl} X = \operatorname{def} X$. The possibility of these identities is the result of two splittings of the compactification problem. Our historical perspective will now be concluded with a discussion of these splittings.

As we have seen, the compactification problem had the splittings

$$\operatorname{cmp} X \leq \operatorname{Cmp} X \leq \operatorname{def} X \quad \text{and} \quad \operatorname{cmp} X \leq \operatorname{Skl} X \leq \operatorname{def} X.$$

In [1985], Aarts, Bruijning and van Mill established the following connection between these two splittings.

12.1. Theorem. *For every separable metrizable space X,*

$$\operatorname{Cmp} X \leq \operatorname{Skl} X.$$

Proof. The proof of the theorem follows a pattern similar to that of the coincidence theorem for dimension. The proof is by induction on $\operatorname{Skl} X$. As a consequence of Theorems 6.9 and 5.7 we have for $n \leq 0$ that $\operatorname{Skl} X \leq n$ if and only if $\operatorname{def} X \leq n$. So by Theorem 6.6, $\operatorname{Skl} X \leq n$ if and only if $\operatorname{Cmp} X \leq n$ when $n \leq 0$. The inductive step for $n \geq 1$ is proved as follows. Let X be a separable metrizable space with $\operatorname{Skl} X \leq n$. Suppose that F and G are disjoint closed sets of X. By the definition of $\operatorname{Skl} X \leq n$ there is a base $\mathcal{B} = \{ U_i : i \in \mathbb{N} \}$ for the open sets of X such that the intersection $\mathrm{B}(U_{i_0}) \cap \cdots \cap \mathrm{B}(U_{i_n})$ is compact for any $n+1$ different indices $i_0, \ldots, i_n$ from $\mathbb{N}$. Consider the family $\{ (C_k, D_k) : k \in \mathbb{N} \}$ of all pairs of elements of $\mathcal{B}$ with $\operatorname{cl}(C_k) \subset D_k$ such that $\operatorname{cl}(D_k) \cap G = \emptyset$ or $D_k \cap F = \emptyset$. Let $V_k = D_k \setminus \bigcup \{ \operatorname{cl}(C_m) : m = 0, 1, \ldots, k-1 \}$ for each k in $\mathbb{N}$. The collection $\boldsymbol{V} = \{ V_k : k \in \mathbb{N} \}$ is a locally finite cover of X. For each k in $\mathbb{N}$ we have $\mathrm{B}(V_k) \subset \mathrm{B}(C_0) \cup \cdots \cup \mathrm{B}(C_{k-1}) \cup \mathrm{B}(D_k)$. Let W be the open set $\bigcup \{ V_k : \operatorname{cl}(D_k) \cap G = \emptyset \}$. Clearly $F \subset W$ holds. Because the collection $\boldsymbol{V}$ is locally finite, we have $\operatorname{cl}(W) \cap G = \emptyset$ and $\mathrm{B}(W) \subset \bigcup \{ \mathrm{B}(V_k) : \operatorname{cl}(D_k) \cap G = \emptyset \}$. It follows that $\mathrm{B}(W)$ is a partition between F and G. We shall show that $\operatorname{Cmp} \mathrm{B}(W) \leq n-1$. Observe that each $\mathrm{B}(V_k)$ is the union of a finite collection of closed subsets of the boundaries of elements of $\mathcal{B}$. It follows that $\mathrm{B}(W)$

is the union of a locally finite collection $\{E_j : j \in N_0\}$ where N_0 is a subset of $\mathbb{N}$ and each E_j is a nonempty closed subset of $\mathrm{B}(U_j)$. (Here and also in the rest of the proof, the original indexing of $\mathcal{B}$ is being retained.) We shall select a subset N_1 of $\mathbb{N} \setminus N_0$ such that $\{U_i \cap \mathrm{B}(W) : i \in N_1\}$ is a base for $\mathrm{B}(W)$ that witnesses the fact that $\mathrm{Skl}\,\mathrm{B}(W) \leq n-1$. This is done in two steps. The first step follows. For each j in N_0 we choose a point p_j from E_j to form the set $P = \{p_j : j \in N_0\}$. Observe that P is closed. Define N_2 to be $\{i \in \mathbb{N} : U_i \in \mathcal{B},\ \mathrm{B}(U_i) \cap P = \emptyset\}$. It is easily seen that $\mathcal{B}^* = \{U_i \in \mathcal{B} : i \in N_2\}$ is a base for the open sets of X and that the boundary $\mathrm{B}(U)$ for each U in $\mathcal{B}^*$ is distinct from the boundary $\mathrm{B}(U_j)$ for every j in N_0. So in particular $N_0 \cap N_2 = \emptyset$. For the second step let $\mathbf{G}$ be an open cover of X such that each element of $\mathbf{G}$ meets at most finitely many members of $\{E_j : j \in N_0\}$. Let N_1 be the set of all i in N_2 such that $\mathrm{cl}(U_i) \subset G$ for some G in $\mathbf{G}$. We shall show that $\{U_i \cap \mathrm{B}(W) : i \in N_1\}$ is a base for $\mathrm{B}(W)$ that witnesses the fact that $\mathrm{Skl}\,\mathrm{B}(W) \leq n-1$. To this end, let $i_0, \ldots, i_{n-1}$ be n distinct indices from N_1 and, with ∂ as the boundary operator in $\mathrm{B}(W)$, let

$$S = \partial\left(U_{i_0} \cap \mathrm{B}(W)\right) \cap \cdots \cap \partial\left(U_{i_{n-1}} \cap \mathrm{B}(W)\right).$$

Then S is a closed subset of $R = \mathrm{B}(U_{i_0}) \cap \cdots \cap \mathrm{B}(U_{i_{n-1}}) \cap \mathrm{B}(W)$. As each $\mathrm{B}(U_{i_j})$ is a subset of some G in $\mathbf{G}$ and each member of $\mathbf{G}$ meets at most finitely many members of $\{E_j : j \in N_0\}$, it follows that $R = \bigcup\{R \cap E_k : k \in N'\}$ for some finite subset N' of N_0. For each k in N' the set $R \cap E_k$ is a closed subset of $\mathrm{B}(U_k)$. Because the indices $k, i_0, \ldots, i_{n-1}$ are distinct, the set $R \cap \mathrm{B}(U_k)$ is compact. It follows that R and, consequently, S are compact. We have proved that $\mathrm{Skl}\,\mathrm{B}(W) \leq n-1$. Then $\mathrm{Cmp}\,\mathrm{B}(W) \leq n-1$ by the induction hypothesis. It follows that $\mathrm{Cmp}\,X \leq n$ holds.

Remark. The use of the set P in the above proof allows us to avoid the indexing pitfall that was discussed in Example 6.13. This method of avoiding the pitfall will be used again in Theorem V.4.3. A second method will be used in Section VI.5.

The following corollary is of interest. It will be generalized in Section V.4.

12.2. Corollary. *For $n \geq 1$ let X be a separable metrizable* *space with* $\operatorname{Skl} X \leq n$. *Then between any two disjoint closed sets of* X *there is a partition* S *such that* $\operatorname{Skl} S \leq n-1$.

Now that we know the inequality $\operatorname{Cmp} X \leq \operatorname{Skl} X$ holds for every separable metrizable space X it is clear that the invariant Skl is a better candidate than Cmp for internally characterizing the compactness deficiency.

In [1988] Kimura made a major breakthrough by proving the following theorem.

12.3. Theorem (Kimura [1988]). *For every separable metrizable space* X,

$$\operatorname{Skl} X = \operatorname{def} X.$$

We shall present a proof of Kimura's theorem in Section 5 of Chapter VI, a chapter which is largely devoted to compactifications. In [1990] the picture of the compactness invariants was completed by Kimura with the following example.

12.4. Example (Kimura). There is a separable metric space X with

$$\operatorname{cmp} X = \operatorname{Cmp} X = 1 \quad \text{and} \quad 2 \leq \operatorname{def} X \leq 3.$$

The reader is referred to Kimura's [1990] paper for a description of the space X.

13. Guide to dimension theory

In this chapter we have presented the most important dimension functions, namely the small inductive dimension ind, the large inductive dimension Ind and the covering dimension dim. The topics discussed in this book were not picked at random but selected with the theory of dimension-like functions in mind. This had the effect that some aspects of dimension theory, notably dimensional properties of topological completions and compactifications, were emphasized while other aspects, notably product theorems and the dimension of subsets of $\mathbb{R}^n$, did not get the attention they deserved. So for a balanced overview of dimension theory the reader should consult books on dimension theory proper. For a long time Hurewicz and Wallman [1941] had been the one and only main reference. More recent books on dimension theory that can be consulted are Alexandroff and Pasynkov [1973], Nagata [1965], Nagami [1970], Pears [1975] and Engelking [1978]. Also, several text books on general topology have introductory chapters on dimension theory, viz., Kuratowski [1958] and [1961], Engelking [1977] and van Mill [1989].

No prior knowledge of dimension theory is required for a successful reading of this book. In a sense the material of Chapter I can be considered to be a first introduction to dimension theory. Except for the sum and decompositions theorems, we have proved many results in great detail. Chapter II deals with mappings into spheres and dimension. It is an ab ovo introduction of the generalized theory of large inductive dimension (modulo $\mathcal{P}$) and covering dimension (modulo $\mathcal{P}$) and the characterization of these dimension functions by the existence of mappings into spheres. This theory is helpful in computing the values of the dimension-like functions in various situations. In Chapter III the fundamental sum and decomposition theorems are discussed for those dimension-like functions that are of inductive type. As the sum and decomposition theorems for dimension are included, this discussion provides a firm base for Chapter I. The covering type functions are discussed in Chapter IV. Excluding compactifications, the theory of extensions and dimension is given in Chapter V. The developed theory uses the basic inductive dimension as a main tool. (Here basic refers to bases.) The final chapter is devoted to the interplay of dimension theory and compactification theory.

To place the book in proper perspective, a simultaneous browsing on a book in dimension theory is recommended.

14. Historical comments and unsolved problems

The chapter was motivated by an historical theme. We touched upon the origins of dimension theory in Section 2. An interesting account on this subject is Duda [1979]. In the whole, our historical comments have been concentrated on the history of dimension-like functions rather than on a precise historical account of dimension theory. So we did not give, for example, a detailed history of the sum theorem. We refer the reader to books on dimension theory proper for this, notably Engelking [1978].

Much of the material of Section 5 is in de Groot's thesis [1942]. It can also be found in de Groot and Nishiura [1966].

The proofs in Section 6 are part of the folklore. The large inductive compactness degree Cmp was introduced in de Groot and Nishiura [1966]. In his [1960] paper Sklyarenko proposed the function Skl as a candidate for the internal characterization of the compactness defect. The splittings of the compactification problem were mentioned in Isbell [1964].

The presentation of properties of topological completeness found in Section 7 closely follows Engelking [1977]. Most of the ideas in this section beginning with Lemma 7.5 are from Aarts [1968].

The proof of Lemma 8.7 is due to Kuratowski [1958]. And the proofs of Lemma 8.9 and the coincidence theorem (Theorem 8.10) go back to Morita [1950].

As already mentioned in Section 9, the notion of a border cover first appeared in Smirnov [1965]. Although the first part of the proof of Theorem 9.5 is almost verbatim from Aarts [1968], the proof of the second part is new and follows the pattern of the proof of a corresponding result about ordinary dimension in Nagata [1965].

The results in Section 10 are due to Aarts and Nishiura [1973].

The main result in of Section 11, namely Example 11.4, is due to Pol [1982]. Example 11.7 is due to Hart [1985]; the first (published) space with these properties was constructed by Kimura [1988a].

Theorem 12.1 and Corollary 12.2 are due to Aarts, Bruijning and van Mill [1985]. Theorem 12.3 and Example 12.4 are due to Kimura in [1988] and [1990] respectively.

Unsolved problems

1. Is there a decomposition theorem for icd?

See the comments after Theorem 7.17.

2. In Example 5.10.f we have seen the failure of the point addition theorem for cmp. That example also showed the failure of the addition theorem. Although the examples of van Douwen [1973] and Przymusiński [1974] show that the point addition theorem for ind does not hold (see Example III.1.2), the following question remains open.

Does the addition theorem for ind hold for normal spaces?

An inductive proof will show that the addition theorem for ind holds for hereditarily normal spaces. See Theorem III.1.9.

3. The computation of $\operatorname{def}(\mathbb{I}^n \setminus E^{n-1}) = n - 1$ was given in Example 5.10.e., where

$$E^{n-1} = \{ (x_1, \ldots, x_{n-1}, 1) : -1 < x_i < 1,\ i = 1, \ldots, n-1 \}$$

is an open face of $\mathbb{I}^n$. The value of $\operatorname{cmp}(\mathbb{I}^n \setminus E^{n-1})$ is not known for $n \geq 4$ (see de Groot and Nishiura [1966]).

What are the values of $\operatorname{cmp}(\mathbb{I}^n \setminus E^{n-1})$ for $n \geq 4$?

This question has been proposed as Problem 417 in van Mill and Reed [1990]. The equality $\operatorname{Cmp}(\mathbb{I}^n \setminus E^{n-1}) = n - 1$ was established in Example 6.7.

4. The inequalities $\operatorname{cmp} X \leq \operatorname{Cmp} X \leq \operatorname{def} X$ are shown to be sharp by Pol's and Kimura's examples. But, Pol's example has $\operatorname{Cmp} X = \operatorname{def} X$ and Kimura's example has $\operatorname{cmp} X = \operatorname{Cmp} X$.

Exhibit a separable metrizable space X such that the inequalities $\operatorname{cmp} X < \operatorname{Cmp} X < \operatorname{def} X$ hold.

5. Related to problem 4 is the following gap question. (See Example 12.4.)

For each n with $n \geq 1$, does there exist a separable metrizable space X such that $\operatorname{def} X - \operatorname{Cmp} X = n$?

The question for the corresponding gap formula $\operatorname{def} X - \operatorname{cmp} X = n$ has been answered in the affirmative in Example 11.7.

6. There is an obvious sharpening of the gap statement in Example 11.7.

For k and m in $\mathbb{N}$ with $k < m$, does there exist a separable metrizable space X such that $\operatorname{cmp} X = k$ and $\operatorname{def} X = m$?

In [1986] Kimura has constructed for all m and k with $m \geq k$ a countably compact, completely regular space X_{mk} such that $\operatorname{cmp} X_{mk} = k$ and $\operatorname{def} X_{mk} = m$. But the example is not metrizable.

Chart 1. The Absolute Borel Classes

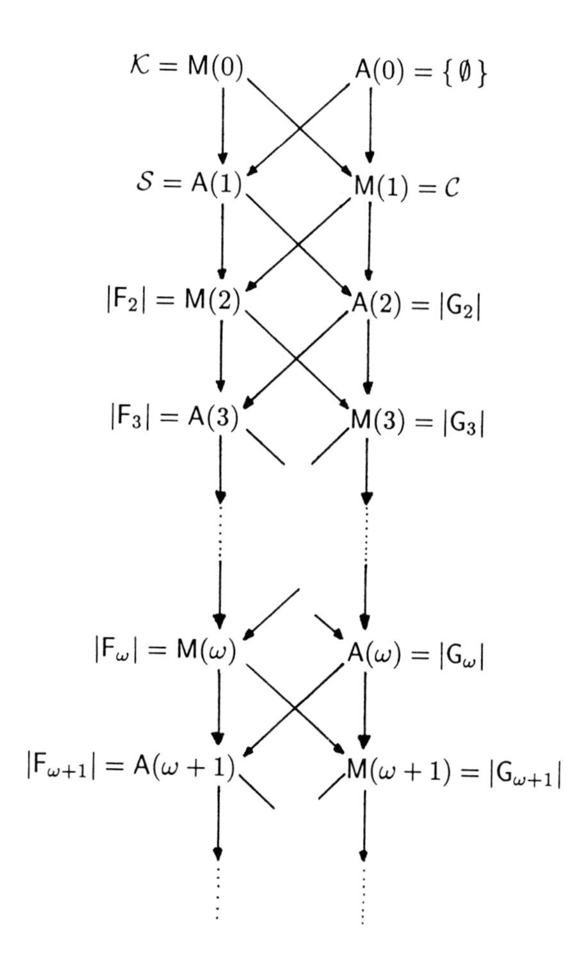

The chart summarizes the inclusion relations among the absolute Borel classes in the universe $\mathcal{M}_0$ of separable metrizable spaces. The arrows indicate inclusion.

CHAPTER II

MAPPINGS INTO SPHERES

It is time to change the perspective. The historical survey of the first chapter was essentially limited to separable metrizable spaces and to the dimension theory and its generalizations spawned by the conjecture of de Groot, namely the modulo theory for the specific classes $\mathcal{K}$, $\mathcal{C}$ and $\mathcal{S}$. The focus of the new perspective will be the discovery of relationships between the various dimensions modulo a general class $\mathcal{P}$ and mappings into spheres. It will afford a quick exposure to the general problems and techniques found in the theory. The chapter will end with applications to the absolute Borel classes, the natural generalizations of the classes $\mathcal{K}$, $\mathcal{C}$ and $\mathcal{S}$. A discussion of the absolute additive, multiplicative and ambiguous Borel classes of order α will be given in Section 9. The chart on the opposite page outlines the connections among the various absolute Borel classes in the setting of separable metrizable spaces.

At this juncture it will be useful to make an agreement on how general the spaces under consideration will be. The definitions in Chapter I were made for general topological spaces. But as the discussion progressed it became evident that the spaces should be T_1-spaces in order to connect ind and Ind. Indeed, the requirements increased from T_1-spaces to regular spaces and then to normal spaces. So the following minimal agreement is made.

Agreement. *Every topological space will be assumed to be a T_1-space.*

As the chapter develops the class of spaces under discussion will gradually be narrowed with further agreements. By the beginning of Section 4 all spaces will be assumed to be hereditarily normal spaces. The discussion up to that point will expose the rationale behind the general restrictions that will be placed on the spaces for the remainder of the book.

1. Classes and universe

The early development of $\operatorname{ind} X$, $\operatorname{Ind} X$ and $\dim X$ occurred in the setting of separable metrizable spaces X. Instead of the word "setting" the expression "universe of discourse" would be more appropriate. The first generalizations of dimension theory concerned the extension of the theory to larger universes. The natural definitions were introduced for arbitrary topological spaces. But the theory did not extend—many counterexamples to the basic theorems were discovered. Thus the spaces X had to be limited in order to derive a coherent theory which reflected the theory for the universe of separable metrizable spaces. The first successful extension was to the universe of metrizable spaces (but with the loss of the equality $\operatorname{ind} X = \operatorname{Ind} X$). Other universes such as the class of perfectly normal spaces and the class invented by Dowker called the totally normal spaces (see Chapter III where these universes will be discussed further) have also been shown to yield satisfactory theories. It is now time to define the notions of *classes of spaces* and *universe of discourse*.

1.1. Definition. A *topologically invariant class* (*class* for short) is a collection of spaces X with the property that if Y is homeomorphic to X then Y is also a member of the class. A *universe of discourse* (*universe* for short and always denoted by $\mathcal{U}$) is a class that defines the spaces under discussion.

Once the universe of discourse has been specified, usually at the beginning of each section, all spaces in the subsequent discussion are to belong to the universe. As classes enter into the discussion, only the members of the class that belong to the universe will be considered.

A topological property will define a class of spaces and conversely a class of spaces will define a topological property. The following are some obvious examples of classes of spaces that will appear in the book. By agreement they are contained in the class of all T_1-spaces.

1.2. Examples.

(1) $\mathcal{T} = \{ X : X \text{ is a topological space} \}$
(2) $\mathcal{R} = \{ X : X \text{ is a regular space} \}$
(3) $\mathcal{R}_c = \{ X : X \text{ is a completely regular space} \}$
(4) $\mathcal{N} = \{ X : X \text{ is a normal space} \}$

(5) $\mathcal{N}_H = \{ X : X \text{ is a hereditarily normal space} \}$
(6) $\mathcal{N}_T = \{ X : X \text{ is a totally normal space} \}$
(7) $\mathcal{N}_P = \{ X : X \text{ is a perfectly normal space} \}$
(8) $\mathcal{M} = \{ X : X \text{ is metrizable} \}$
(9) $\mathcal{M}_0 = \{ X : X \text{ is a separable metrizable space} \}$
(10) $\{ \emptyset \}$
(11) $\mathcal{K} = \{ X : X \text{ is compact} \}$
(12) $\mathcal{C} = \{ X : X \text{ is a complete metrizable space} \}$
(13) $\mathcal{S} = \{ X : X \text{ is } \sigma\text{-compact} \}$
(14) $\mathcal{L} = \{ X : X \text{ is locally compact} \}$

In Chapter I we have seen how the dimension functions ind, Ind and dim are generalized to the functions cmp, icd, Icd, ccd, $\mathcal{S}$-ind and $\mathcal{S}$-Ind. The inductive generalizations were made by replacing the empty space with spaces from the classes $\mathcal{K}$, $\mathcal{C}$ and $\mathcal{S}$ respectively to begin the induction. The natural way to generalize these inductive processes is to use an arbitrary class $\mathcal{P}$ to define the functions $\mathcal{P}$-ind, $\mathcal{P}$-Ind, etc.

When the universe $\mathcal{U}$ is used in conjunction with a class $\mathcal{P}$, one naturally changes the class $\mathcal{P}$ to the smaller class $\mathcal{P} \cap \mathcal{U}$. For the universe $\mathcal{M}_0$ and the class $\mathcal{K}$, the inductive dimensions cmp and $\mathcal{K}$-Ind fortunately agree with the respective inductive dimensions using the smaller class $\mathcal{K} \cap \mathcal{M}_0$ in place of $\mathcal{K}$. The situation will be the same for the universe $\mathcal{U}$ and a class $\mathcal{P}$ under rather mild conditions on $\mathcal{U}$ and $\mathcal{P}$. This will be made evident in the next section. Also in Sections 3, 5 and 8 we shall be dealing with other topologically invariant functions associated with classes $\mathcal{P}$ such as surplus, deficiency and covering dimensions that could be affected by the introduction of the universe $\mathcal{U}$. Again we shall see that the introduction of the universe $\mathcal{U}$ (and hence the introduction of the smaller class $\mathcal{P} \cap \mathcal{U}$) will not affect our results when the same very mild conditions on the universe $\mathcal{U}$ and the class $\mathcal{P}$ are assumed.

At the end of Section 2 the minimal conditions that will be needed for our development are formulated into an agreement that will be in effect for the remainder of the book. It will become clear from the discussion of that section as to why such conditions are needed. The role of the universe will show its greatest influence in Chapter III when we embark upon a search for an optimal universe in which a reasonable dimension theory can be developed.

2. Inductive dimension modulo a class $\mathcal{P}$

Agreement. *The empty space $\emptyset$ is a member of every class.*

This section introduces the inductive approach to the extension of dimension modulo $\mathcal{P}$ to an arbitrary class. Obviously the definitions of these extensions are motivated by the usual small and large inductive dimensions ind and Ind. The small and large inductive dimensions were defined in terms of partitions—see the Definitions I.3.2 and I.4.1. The extensions of these definitions and their relationships to the universe of discourse will be established here. Since the aim is to extend dimension theory, every class $\mathcal{P}$ is assumed to have the empty space as a member. Consequently, the class $\mathcal{P}$ will be nonempty.

The following is the natural extension of the definition of the small inductive dimension ind.

2.1. Definition. Let $\mathcal{P}$ be a class of spaces and X be a topological space. One assigns the *small inductive dimension modulo $\mathcal{P}$*, denoted $\mathcal{P}\text{-ind}\, X$, as follows.

(i) $\mathcal{P}\text{-ind}\, X = -1$ if and only if X is in $\mathcal{P}$.

(ij) For each natural number n, $\mathcal{P}\text{-ind}\, X \leq n$ if for each point p and for each closed set G of X with $p \notin G$ there is a partition S between p and G such that $\mathcal{P}\text{-ind}\, S \leq n-1$.

(iij) For each natural number n, $\mathcal{P}\text{-ind}\, X = n$ if $\mathcal{P}\text{-ind}\, X \leq n$ and $\mathcal{P}\text{-ind}\, X \not\leq n-1$.

(iv) $\mathcal{P}\text{-ind}\, X = \infty$ if the inequality $\mathcal{P}\text{-ind}\, X \leq n$ does not hold for any natural number n.

Since the class $\mathcal{P}$ is topologically invariant, the function $\mathcal{P}$-ind is also topologically invariant. Obviously, $\{\emptyset\}\text{-ind} = \text{ind}$; the notation ind will be the preferred one.

Our first proposition is an immediate consequence of the definition. Its proof is left to the reader.

2.2. Proposition. *If $\mathcal{P}\text{-ind}\, X < \infty$, then X is in $\mathcal{P}$ or X is a regular space.*

A consequence of the proposition is that there is no loss in assuming that all spaces are regular.

The easy inductive proof of the next proposition is left to the reader.

2.3. Proposition. *If $\mathcal{P}$ and $\mathcal{Q}$ are classes with $\mathcal{P} \supset \mathcal{Q}$, then* $\mathcal{P}$-ind $\leq \mathcal{Q}$-ind. *In particular,* $\mathcal{P}$-ind $\leq$ ind.

Let us now prepare for the generalization of Proposition I.3.6. That proposition showed a nice connection between $\operatorname{ind} X$ and the boundaries of the members of an appropriate base $\mathcal{B}$ for the open sets of X. The following example will illustrate the difficulties that can arise in generalizing this proposition if no conditions are placed on the class $\mathcal{P}$.

2.4. Example. For this example, let $\mathcal{P}$ be the class of spaces X with $\operatorname{ind} X = 1$ or $X = \emptyset$. (Note the exact equality!) It should be observed that a closed subspace Y of a space X in $\mathcal{P}$ need not be in $\mathcal{P}$. This class $\mathcal{P}$ will contrast the use of general partitions between points and closed sets and the use of partitions between points and closed sets by boundaries of open sets. Consider the space X which is the disjoint topological sum of $\mathbb{R}$ and $\mathbb{R}^2$. It is obvious that for each point p in X and each closed set G with $p \notin G$ there is a partition S between p and G with $\operatorname{ind} S = 1$. That is, S is in $\mathcal{P}$. Or, loosely speaking, points and closed sets can be separated by members of $\mathcal{P}$. Also X is not a member of $\mathcal{P}$. So, $\mathcal{P}$-ind $X = 0$. Since $\mathbb{R}$ has the property that the boundary of every bounded open set has ind equal to 0, some point p of X has the property that each neighborhood base $\mathcal{B}(p)$ has a U in $\mathcal{B}(p)$ such that $\mathrm{B}(U)$ is not in $\mathcal{P}$. So there is no base $\mathcal{B}$ for the open sets of X with $\mathrm{B}(U) \in \mathcal{P}$ for each U in $\mathcal{B}$. This example shows that Definition 2.1 will not lead to an analogue of Proposition I.3.6.

In order to capture the analogue of Proposition I.3.6 one will need the property that a closed subspace of a space in the class $\mathcal{P}$ is also in $\mathcal{P}$. We make the following definitions.

2.5. Definition. The universe is said to be *closed-monotone* if every closed subspace of each space in the universe is also a member of the universe.

2.6. Definition. For the universe of discourse, a class $\mathcal{P}$ is said to be *closed-monotone in $\mathcal{U}$* if $\mathcal{P}$ satisfies the condition:

If X is a space in $\mathcal{P} \cap \mathcal{U}$, then every closed subspace of X is in $\mathcal{P}$.

When a class $\mathcal{P}$ is closed-monotone in the universe $\mathcal{T}$ of all topological spaces, the class $\mathcal{P}$ will be called *absolutely closed-monotone.*

Observe that the classes $\mathcal{K}$, $\mathcal{C}$ and $\mathcal{S}$ introduced in Chapter I are absolutely closed-monotone. Further observe that if the universe is closed-monotone and the class $\mathcal{P}$ is closed-monotone in $\mathcal{U}$, then $\mathcal{P} \cap \mathcal{U}$ is absolutely closed-monotone. The following proposition is easily proved.

2.7. Proposition. *Let the universe be closed-monotone. The following statements are equivalent for a class* $\mathcal{P}$.

(a) $\mathcal{P}$ *is closed-monotone in* $\mathcal{U}$.
(b) $\mathcal{P}\text{-ind}\, Y \leq \mathcal{P}\text{-ind}\, X$ *holds for every closed subspace* Y *of each space* X *in the universe.*

Proof. Clearly statement (b) implies statement (a). For the converse we assume that statement (a) holds. We shall prove inductively the statement: If Y is a closed subspace of X and $\mathcal{P}\text{-ind}\, X \leq n$, then $\mathcal{P}\text{-ind}\, Y \leq n$. When $n = -1$, the statement is obvious. So assume that the statement is true for $n - 1$ and consider a closed subset Y of a space X with $\mathcal{P}\text{-ind}\, X \leq n$. Let p be a point of Y and G be closed in the subspace Y with $p \notin G$. Since Y is closed in X, the set G is also closed in X. Consequently there is a partition S between p and G in X with $\mathcal{P}\text{-ind}\, S \leq n - 1$. We have that S is a member of $\mathcal{U}$ because $\mathcal{U}$ is closed-monotone. By the induction hypothesis, $\mathcal{P}\text{-ind}\,(S \cap Y) \leq n - 1$. Since $S \cap Y$ is a partition between p and G in Y, it follows that $\mathcal{P}\text{-ind}\, Y \leq n$. Thus statement (b) holds.

The next example will aid the reader in seeing the connection between the closed monotonicity of the universe $\mathcal{U}$ and of the class $\mathcal{P}$ in $\mathcal{U}$.

2.8. Example. Though the definition of a class of spaces depends only on topological properties, the notion of closed-monotone class in $\mathcal{U}$ is dependent on the universe of discourse. To see this we consider the class $\mathcal{P}$ defined by separability and consider the universe $\mathcal{U} = \mathcal{R}_c$ of completely regular spaces. The Čech-Stone compactification $\beta\mathrm{N}$ of the space of natural numbers N is completely regular and is also in the class $\mathcal{P}$. Clearly $\mathcal{R}_c$ is closed-monotone. The closed set $\beta\mathrm{N} \setminus \mathrm{N}$ of $\beta\mathrm{N}$ is known to be nonseparable, that is, $\beta\mathrm{N} \setminus \mathrm{N} \notin \mathcal{P}$. So the class $\mathcal{P}$ is not closed-monotone in the universe of completely regular spaces. In contrast to this, the class $\mathcal{P}$ is closed-monotone in the closed-monotone universe $\mathcal{M}$ of metrizable spaces.

The following theorem is a useful complement to Proposition 2.7. It will illustrate why one passes from $\mathcal{P}$ to $\mathcal{P} \cap \mathcal{U}$ in the context of the universe of discourse. Observe that the theorem requires the class $\mathcal{P}$ to satisfy only the agreement made at the beginning of this section.

2.9. Theorem. *Let $\mathcal{Q}$ be an absolutely closed-monotone class. For every space X in $\mathcal{Q}$ and every class $\mathcal{P}$,*

$$\mathcal{P}\text{-ind}\, X = (\mathcal{P} \cap \mathcal{Q})\text{-ind}\, X.$$

Proof. We already know that $\mathcal{P}\text{-ind}\, X \leq (\mathcal{P} \cap \mathcal{Q})\text{-ind}\, X$ holds for any X. For the reverse inequality we shall prove inductively the statement: *If $X \in \mathcal{Q}$ and $\mathcal{P}\text{-ind}\, X \leq n$, then $(\mathcal{P} \cap \mathcal{Q})\text{-ind}\, X \leq n$.* The statement is obviously true for $n = -1$. So we assume that it is true for $n - 1$ and let X be a space in $\mathcal{Q}$ with $\mathcal{P}\text{-ind}\, X \leq n$. Suppose that p is a point and G is a closed set with $p \notin G$. Then there is a partition S between p and G such that $\mathcal{P}\text{-ind}\, S \leq n - 1$. Because X is in $\mathcal{Q}$ and $\mathcal{Q}$ is absolutely closed-monotone, S is in $\mathcal{Q}$. The induction hypothesis yields $(\mathcal{P} \cap \mathcal{Q})\text{-ind}\, S \leq n - 1$. Thereby we have $(\mathcal{P} \cap \mathcal{Q})\text{-ind}\, X \leq n$.

With the aid of Proposition 2.7 we can prove the following analogue of Proposition I.3.6.

2.10. Theorem. *Let the universe be closed-monotone and n be a natural number. Suppose that the class $\mathcal{P}$ is closed-monotone in $\mathcal{U}$ and X is a regular space in the universe. Then the following statements are equivalent.*

(a) $\mathcal{P}\text{-ind}\, X \leq n$.
(b) *There exists a base $\mathcal{B}$ for the open sets of X such that $\mathcal{P}\text{-ind}\, \mathrm{B}(U) \leq n - 1$ for every U in $\mathcal{B}$.*

Proof. The reader will recognize similarities between the proof of Proposition I.3.6 and the one given here.

Suppose $\mathcal{P}\text{-ind}\, X \leq n$. We must consider two cases. The first is that of $\mathcal{P}\text{-ind}\, X = -1$. In this case Proposition 2.7 will yield that any base $\mathcal{B}$ for the open sets has the required property because the regular space X is a member of the universe. Let $\mathcal{P}\text{-ind}\, X \geq 0$ for the second case. For a point p and an open neighborhood U of p there is a partition S between p and $X \setminus U$ with $\mathcal{P}$-ind $S \leq n-1$. That is, there are disjoint open sets V and W with $p \in V$ and $X \setminus U \subset W$ such that $X \setminus S = V \cup W$. Observed that S is in $\mathcal{U}$ because the universe is closed-monotone. Since $\mathrm{B}(V)$ is a closed subset of S, we have $\mathcal{P}$-ind $\mathrm{B}(V) \leq n-1$ by Proposition 2.7.

For the converse implication we let p be a point and G be a closed set with $p \notin G$. Since X is a regular space, there exist disjoint open sets V and W such that $p \in V$ and $G \subset W$. In the base $\mathcal{B}$ there is an open set U with $p \in U \subset V$. The set $\mathrm{B}(U)$ yields the required partition between p and G to establish $\mathcal{P}$-ind $X \leq n$.

Let us now give the natural extension of the definition of the large inductive dimension Ind.

2.11. Definition. Let $\mathcal{P}$ be a class of spaces and X be a topological space. One assigns the *large inductive dimension modulo* $\mathcal{P}$, denoted $\mathcal{P}$-Ind X, as follows.

(i) $\mathcal{P}$-Ind $X = -1$ if and only if X is in $\mathcal{P}$.

(ij) For each natural number n, $\mathcal{P}$-Ind $X \leq n$ if for each pair of disjoint closed sets F and G there exists a partition S between F and G in X such that $\mathcal{P}$-Ind $S \leq n-1$.

(iij) For each natural number n, $\mathcal{P}$-Ind $X = n$ if $\mathcal{P}$-Ind $X \leq n$ and $\mathcal{P}$-Ind $X \nleq n-1$.

(iv) $\mathcal{P}$-Ind $X = \infty$ if the inequality $\mathcal{P}$-Ind $X \leq n$ does not hold for any natural number n.

Remark. Just as we have found in Chapter I, the topological invariants that appear in dimension theory are often defined by means of inequalities. This leads to the need to define the values of these functions by requirements such as (iij) and (iv) in Definitions 2.1 and 2.11. Continuous repetition of these two conditions becomes somewhat monotonous. So from this point on we shall not give them explicitly and leave it to the reader to supply the missing parts.

Clearly $\mathcal{P}$-Ind is a topological invariant and $\{\emptyset\}$-Ind = Ind. As with ind, the preferred notation will be Ind.

Example. In [1966] Mardešić couched an interesting theorem in a form that used the function $\mathcal{P}$-Ind. The theorem concerned a property of ordered compacta. The class $\mathcal{P}$ in that paper was taken to be

$$\mathcal{P} = \{ X : \text{all components of } X \text{ are metrizable continua} \}$$

and the following theorem was proved.

Theorem. *Let X be a compact Hausdorff space. If X is the continuous image of an ordered compactum, then $\mathcal{P}$-Ind $X \leq 0$.*

Since the proof of this theorem is not directly related to our purposes, it will not be given here.

The easy inductive proof of the following theorem is left to the reader.

2.12. Theorem. *For every space X,*

$$\mathcal{P}\text{-ind}\, X \leq \mathcal{P}\text{-Ind}\, X.$$

The next two propositions also have easy proofs.

2.13. Proposition. *If $\mathcal{P}$ and $\mathcal{Q}$ are classes with $\mathcal{P} \supset \mathcal{Q}$, then $\mathcal{P}$-Ind $\leq \mathcal{Q}$-Ind. In particular, $\mathcal{P}$-Ind $\leq$ Ind.*

2.14. Proposition. *If $\mathcal{P}$-Ind $X < \infty$, then X is in $\mathcal{P}$ or X is a normal space.*

Thus there will be no loss in assuming that the universe of discourse is contained in the class $\mathcal{N}$ of normal spaces. This is similar to the situation for $\mathcal{P}$-ind in which the class of regular spaces was found to be important.

Exactly analogous to Theorem 2.9 is the following.

2.15. Theorem. *Let $\mathcal{Q}$ be an absolutely closed-monotone class. For every space X in $\mathcal{Q}$ and every class $\mathcal{P}$,*

$$\mathcal{P}\text{-Ind}\, X = (\mathcal{P} \cap \mathcal{Q})\text{-Ind}\, X.$$

Proof. The proof is a straightforward modification of the one for Theorem 2.9.

The subspace theorem for dimension (Theorem I.3.4) in the universe $\mathcal{M}_0$ of separable metrizable spaces has a straightforward proof. It should be no surprise that the subspace theorem does not generalize to $\mathcal{P}$-ind. For example, the function cmp fails to have a subspace theorem. But some form of the theorem can be recovered for $\mathcal{P}$-ind and $\mathcal{P}$-Ind. We shall give three of them.

2.16. Theorem. *Let the universe be closed-monotone. For classes $\mathcal{P}$ the following three conditions are equivalent.*

(a) *$\mathcal{P}$ is closed-monotone in $\mathcal{U}$.*

(b) *$\mathcal{P}$-ind $Y \leq \mathcal{P}$-ind X holds for every closed subspace Y of each space X in the universe.*

(c) *$\mathcal{P}$-Ind $Y \leq \mathcal{P}$-Ind X holds for every closed subspace Y of each space X in the universe.*

Proof. The equivalence of conditions (a) and (b) has been established in Proposition 2.7. That condition (c) implies condition (a) is easily proved. Finally, that condition (a) implies (c) is proved inductively in a manner similar to the induction found in the proof of Proposition 2.7. (See also Proposition I.4.3.)

The next theorem concerns the behavior of $\mathcal{P}$-ind for open subspaces. The example $\mathcal{K}$ of the class of compact spaces fails to have the property that open subspaces are members of the class. But a rather general open subspace theorem does hold for $\mathcal{P}$-ind.

2.17. Theorem. *Let $\mathcal{P}$ be any class of spaces. If $0 \leq \mathcal{P}$-ind X, then $\mathcal{P}$-ind $Y \leq \mathcal{P}$-ind X for each open subspace Y of X.*

Proof. Let $n = \mathcal{P}$-ind $X < \infty$. Suppose that p is a point of Y and G is a closed set in Y with $p \notin G$. Then p is not in the closed subset $G' = G \cup (X \setminus Y)$ of X. Since $\mathcal{P}$-ind $X \geq 0$, there is a partition S between p and G' in X with $\mathcal{P}$-ind $S \leq n-1$. Clearly S is a partition between p and G in Y. Thus, $\mathcal{P}$-ind $Y \leq n$.

2.18. Definition. The universe is said to be *monotone* if every subspace of each space in the universe is also a member of the universe. A class of spaces $\mathcal{P}$ is said to be *monotone in $\mathcal{U}$* if $\mathcal{P}$ satisfies the condition:

If X is a space in $\mathcal{P} \cap \mathcal{U}$, then every subspace of X is in $\mathcal{P}$.

When a class $\mathcal{P}$ is monotone in the universe $\mathcal{T}$ of all topological spaces, the class $\mathcal{P}$ will be called *absolutely monotone.*

Clearly the class $\{\emptyset\}$ is absolutely monotone. Indeed, the class of all spaces whose cardinality is bounded by a fixed cardinal number is absolutely monotone. In Chapter I we mentioned that rational curves are defined by using countable spaces. For the other extreme, the classes $\mathcal{K}$, $\mathcal{C}$ and $\mathcal{S}$ discussed in Chapter I are not monotone in any universe of interest.

2.19. Theorem. *Let the universe be monotone. For classes $\mathcal{P}$ the following conditions are equivalent.*

(a) *$\mathcal{P}$ is monotone in $\mathcal{U}$.*
(b) *$\mathcal{P}\text{-ind}\, Y \leq \mathcal{P}\text{-ind}\, X$ holds for every subspace Y of spaces X in the universe.*

Proof. The proof is similar to that of Theorem 2.7.

Suppose that condition (b) holds and let X be a space in $\mathcal{P} \cap \mathcal{U}$. If Y is a subspace of X, then Y is in $\mathcal{U}$ because the universe is monotone. From condition (b) we have $\mathcal{P}\text{-ind}\, Y \leq \mathcal{P}\text{-ind}\, X = -1$. That is, Y is in $\mathcal{P}$. Therefore the class $\mathcal{P}$ is monotone in $\mathcal{U}$.

Let us prove next that condition (a) implies (b). Let $\mathcal{P}$ be monotone in $\mathcal{U}$. We shall prove inductively the statement: If Y is a subspace of X and $\mathcal{P}\text{-ind}\, X \leq n$, then $\mathcal{P}\text{-ind}\, Y \leq n$. When $n = -1$, the statement is obvious from condition (a) because the universe is monotone. So we assume that the statement is true for $n - 1$ and consider a subspace Y of a space X with $\mathcal{P}\text{-ind}\, X \leq n$. That is, assume that the class $\mathcal{P}' = \{ X : \mathcal{P}\text{-ind}\, X \leq n - 1 \}$ is monotone in $\mathcal{U}$. We must now prove that the statement is true. But this is exactly the same situation that is found in the proof of the subspace theorem for ind given in Theorem I.3.4. Indeed, the exact same proof applied in the monotone universe will complete the induction.

Concerning the monotonicity of $\mathcal{P}$-Ind, it is known that Ind fails to have a general subspace theorem in the universe $\mathcal{N}_H$ of hereditarily normal spaces (see Pol and Pol [1979]). In Chapter III we shall investigate further the problem of monotonicity.

Dual to the subspace concept is that of the ambient space. Our spaces X are often contained in ambient spaces Y. So it will be convenient to have partition theorems for $\mathcal{P}$-ind and $\mathcal{P}$-Ind in the setting of ambient spaces.

2.20. Proposition. *Let the universe be closed-monotone and n be a natural number. Suppose that X is in $\mathcal{U}$ and $\mathcal{P}$ is a class of spaces. If X is a subspace of a hereditarily normal space Y, then $\mathcal{P}\text{-ind}\, X \leq n$ if and only if* (a) *$X \in \mathcal{P}$ or* (b) *for each point p of X and each closed set F of Y not containing p there is a partition S between p and F in Y with $\mathcal{P}\text{-ind}\,(S \cap X) \leq n - 1$. Moreover, if $\mathcal{P}$ is closed-monotone in $\mathcal{U}$, the condition* (a) *can be dropped.*

Proof. Suppose that $\mathcal{P}$-ind $X \leq n$ and $X \notin \mathcal{P}$ hold. Let p be a point in X and F be a closed set of Y with $p \notin F$. Let F' be a closed neighborhood of F such that $p \notin F'$. There is a partition S_0 between p and $F' \cap X$ in X with $\mathcal{P}$-ind $S_0 \leq n-1$. Note that S_0 is also a partition between p and F in the subspace $X \cup F$. By Lemma I.4.5 there is a partition S between p and F in Y with $S_0 = S \cap X$.

For the converse we assume that X is not in $\mathcal{P}$. Let p be a point of X and let G be a closed subset of X with $p \notin G$. Then p is not a member of $F = \mathrm{cl}_Y(G)$. Let S be a partition between p and F in the space Y with $\mathcal{P}$-ind $(S \cap X) \leq n-1$. The set $S \cap X$ is a partition between p and G in the space X. Hence, $\mathcal{P}$-ind $X \leq n$.

Let us turn to a $\mathcal{P}$-Ind analogue of Proposition I.4.6. Its statement will contain the notion of separated sets. Recall that two subsets A and B of a space X are *separated* if $\mathrm{cl}_X(A) \cap B = \emptyset = A \cap \mathrm{cl}_X(B)$.

2.21. Theorem. *Let the universe be closed-monotone and n be a natural number. Suppose that Y is in $\mathcal{U}$ and $\mathcal{P}$ is a class of spaces. If Y is a subspace of a hereditarily normal space X, then $\mathcal{P}$-Ind $Y \leq n$ implies* (a) $Y \in \mathcal{P}$ *or* (b) *for each pair of separated subsets F and G of X satisfying*

$$Y \cap \mathrm{cl}_X(F) \cap \mathrm{cl}_X(G) = \emptyset$$

there is a partition S between F and G in X with $\mathcal{P}$-Ind $(S \cap Y) \leq n-1$. Moreover, if $\mathcal{P}$ is closed-monotone in $\mathcal{U}$, the condition (a) *can be dropped.*

Proof. The proof uses the one for Proposition I.4.6 in the open subspace $X \setminus \big(\mathrm{cl}_X(F) \cap \mathrm{cl}_X(G)\big)$.

Let us turn now to the promised statement of the agreement that is to hold for the remainder of the book. It should be apparent that the development of any viable theory that generalizes dimension theory will require that the universe be closed-monotone and that the classes $\mathcal{P}$ be, at the least, closed-monotone in the universe of discourse. Thus the following agreement will be made for the remainder of the book. With this agreement, reference to the universe will be suppressed unless there are additional assumptions to be made about the universe.

Agreement. *For the remainder of the book, the following are assumed to hold. The universe is closed-monotone. Every class is topologically invariant, has the empty space as a member and is closed-monotone in $\mathcal{U}$. Unless indicated otherwise, X will always be a space in the universe.*

3. Kernels and surplus

The concept of a kernel of a set has appeared in many areas of mathematics. As in the definitions of $\mathcal{P}$-ind and $\mathcal{P}$-Ind, our notion of kernel will be phrased in such a way as to depend only on the class $\mathcal{P}$, namely in the context of the class $\mathcal{T}$ of all topological spaces.

3.1. Definition. For a class $\mathcal{P}$ and a space X in $\mathcal{T}$ a subset G of X is called a *$\mathcal{P}$-kernel of X* if G is in $\mathcal{P}$.

The definition of the next function will be in the context of the universe of discourse.

3.2. Definition. Let $\mathcal{P}$ be a class of spaces. For a space X in the universe the *$\mathcal{P}$-Surplus of X* is the extended integer

$$\mathcal{P}\text{-Sur}\, X = \min \{ \operatorname{Ind}(X \setminus G) : G \text{ is a } (\mathcal{P} \cap \mathcal{U})\text{-kernel of } X \}.$$

Note that the $\mathcal{P}$-Surplus is undefined for spaces not in the universe. Also observe that $X \setminus G$ is not necessarily in the universe when G is a $(\mathcal{P} \cap \mathcal{U})$-kernel of X.

The concept of surplus has already been encountered in Chapter I when the notion of σ-compactness degree was introduced. It should be observed that the definition of $\mathcal{P}$-Sur depends on the universe of discourse. This is in contrast to the definitions of $\mathcal{P}$-ind and $\mathcal{P}$-Ind as has been shown in Theorems 2.9 and 2.15. The definition is designed so as to yield the equality

$$\mathcal{P}\text{-Sur}\, X = (\mathcal{P} \cap \mathcal{U})\text{-Sur}\, X.$$

The preferred notation will be $\mathcal{P}$-Sur. Obviously if the universe $\mathcal{U}_0$ is contained in the universe $\mathcal{U}$, then

$$(\mathcal{P} \cap \mathcal{U})\text{-Sur}\, X \leq (\mathcal{P} \cap \mathcal{U}_0)\text{-Sur}\, X \quad \text{for} \quad X \in \mathcal{U}_0.$$

Also the $\mathcal{P}$-Surplus is undefined for spaces not in $\mathcal{U}_0$ when we are dealing with the smaller universe. Thus one must be careful with the function $\mathcal{P}$-Sur. To underscore these points, consider the following example.

3.3. Example. Consider the class $\mathcal{M}$ of metrizable spaces and the class $\mathcal{L}$ of all locally compact spaces. Both classes are absolutely closed-monotone. So they are candidates for being the universe according to our agreement at the end of Section 2. For the absolutely monotone class $\mathcal{P} = \{ X : X \text{ is countable} \}$ we have two different values of $\mathcal{P}\text{-Sur}\,\mathbb{R}$ according to whether the universe is $\mathcal{M}$ or is $\mathcal{L}$. Indeed, when $\mathcal{U} = \mathcal{M}$ we have $\mathcal{P}\text{-Sur}\,\mathbb{R} = 0$. But when $\mathcal{U} = \mathcal{L}$ we have $\mathcal{P}\text{-Sur}\,\mathbb{R} = 1$ because each G in $\mathcal{P} \cap \mathcal{L}$ has isolated points by Baire's theorem.

The next two propositions are obvious.

3.4. Proposition. *For a closed subset F of a space X, the set F is a $\mathcal{P}$-kernel of X if and only if it is a $(\mathcal{P} \cap \mathcal{U})$-kernel of X.*

3.5. Proposition. *If $\mathcal{P}$ and $\mathcal{Q}$ are classes with $\mathcal{P} \supset \mathcal{Q}$, then $\mathcal{P}\text{-Sur} \leq \mathcal{Q}\text{-Sur}$. In particular, $\mathcal{P}\text{-Sur}\,X \leq \operatorname{Ind} X$.*

The next theorem shows the connection between $\mathcal{P}$-Sur and $\mathcal{P}$-Ind.

3.6. Theorem. *For every hereditarily normal space X,*

$$\mathcal{P}\text{-Ind}\,X \leq \mathcal{P}\text{-Sur}\,X.$$

Proof. From Theorem 2.15 we have $\mathcal{P}\text{-Ind} = (\mathcal{P} \cap \mathcal{U})\text{-Ind}$. So the theorem will follow from the stronger inequality for the absolutely closed-monotone class $\mathcal{P} \cap \mathcal{U}$: $(\mathcal{P} \cap \mathcal{U})\text{-Ind} \leq (\mathcal{P} \cap \mathcal{U})\text{-Sur}$. The proof of this inequality is by induction. Obviously the case where $(\mathcal{P} \cap \mathcal{U})\text{-Sur}\,X = -1$ follows from the definitions. For the inductive step we assume that X is such that $(\mathcal{P} \cap \mathcal{U})\text{-Sur}\,X = n < \infty$. Let F and G be disjoint closed sets and let H be a $(\mathcal{P} \cap \mathcal{U})$-kernel of X with $\operatorname{Ind}(X \setminus H) \leq n$. From Theorem 2.21 applied to the class $\{\emptyset\}$ we have a partition S between F and G in X such that $\operatorname{Ind}\big(S \cap (X \setminus H)\big) \leq n - 1$. Since H is in $\mathcal{P} \cap \mathcal{U}$ and S is closed in X, we have $H \cap S \in \mathcal{P} \cap \mathcal{U}$ by the absolute closed monotonicity of $\mathcal{P} \cap \mathcal{U}$. So $(\mathcal{P} \cap \mathcal{U})\text{-Sur}\,S \leq n - 1$ holds. By the induction hypothesis we have $(\mathcal{P} \cap \mathcal{U})\text{-Ind}\,S \leq n - 1$. Thus $(\mathcal{P} \cap \mathcal{U})\text{-Ind}\,X \leq n$ has been shown.

The following closed subspace proposition for $\mathcal{P}$-Sur holds.

3.7. Proposition. *For every closed subspace Y of X,*

$$\mathcal{P}\text{-Sur}\, Y \leq \mathcal{P}\text{-Sur}\, X.$$

Proof. Let $\mathcal{P}\text{-Sur}\, X = n < \infty$. Then there is a $(\mathcal{P} \cap \mathcal{U})$-kernel G of X with $\operatorname{Ind}(X \setminus G) = n$. Since $Y \cap G$ is a closed subspace of G and G is in $\mathcal{P} \cap \mathcal{U}$, we have that $Y \cap G$ is in $\mathcal{P} \cap \mathcal{U}$. Because $\{\emptyset\}$ is absolutely closed-monotone and $Y \setminus G$ is closed in $X \setminus G$, we have $\operatorname{Ind}(Y \setminus G) \leq \operatorname{Ind}(X \setminus G) = n$. So, $\mathcal{P}\text{-Sur}\, Y \leq n$.

We shall close this section with an example that summarizes earlier calculations.

Example. The following have been computed in Chapter I.

$$0 = \operatorname{cmp} \mathbb{R}^n \leq \mathcal{K}\text{-Ind}\, \mathbb{R}^n \leq \mathcal{K}\text{-Sur}\, \mathbb{R}^n \leq \operatorname{Ind} \mathbb{R}^n = n.$$
$$-1 = \operatorname{icd} \mathbb{R}^n = \operatorname{Icd} \mathbb{R}^n = \mathcal{C}\text{-Sur}\, \mathbb{R}^n \leq \operatorname{Ind} \mathbb{R}^n = n.$$
$$-1 \leq \operatorname{cmp} X \leq \mathcal{K}\text{-Ind}\, X \leq \mathcal{K}\text{-Sur}\, X \leq \operatorname{Ind} X, \qquad X \in \mathcal{M}_0.$$
$$-1 \leq \operatorname{icd} X \leq \operatorname{Icd} X = \mathcal{C}\text{-Ind}\, X \leq \operatorname{Ind} X, \qquad X \in \mathcal{M}.$$

The inequalities involving the class $\mathcal{K}$ will be sharpened at the end of Section 6.

4. $\mathcal{P}$-Ind and mappings into spheres

Agreement. *In this section the universe will be contained in the class $\mathcal{N}_H$ of hereditarily normal spaces.*

This section is devoted to developing relationships between $\mathcal{P}$-Ind and mappings into spheres. For this development we need some terminology concerning topological operations in subspaces. (Some of the terms have been defined already in Section I.9.) The first half of the section will be devoted to a discussion of results not specifically related to dimension theory.

4.1. Terminology. In Section I.9 we adopted some modulo notation to take care of certain set calculations that occurred in subspaces of a space. For example, the notation $A \subset B \mod F$ was used to mean that $A \setminus F$ is a subset of $B \setminus F$. Loosely speaking, the notation says that A is a subset of B in the subspace $X \setminus F$. This modulo notation is a very convenient way of dealing with set properties in the subspace $X \setminus F$. Clearly one can devise many others besides those that were introduced in Section I.9, more than one can reasonably list. We will freely use this terminology. The reader should have no trouble in getting the meaning whenever we use it.

4.2. Propositions. The following two propositions concerning subspaces will be used repeatedly. Of course, these statements do not require that the spaces be hereditarily normal. The proofs will be left to the reader.

A. *If F is a closed subset of a space X and A is a subset of X, then*

$$\mathrm{cl}_{X\setminus F}(A \setminus F) = \mathrm{cl}_X(A) \setminus F.$$

B. *Let F be a subset of X and let V be an open subset of X. Then*

$$\mathrm{B}_{X\setminus F}(V \setminus F) \subset \mathrm{B}_X(V) \setminus F.$$

Moreover,

$$\mathrm{B}_{X\setminus F}(V \setminus F) = \mathrm{B}_X(V) \setminus F$$

when F is closed.

4.3. Proposition. *Let H be a closed subset of a space X and let $\mathbf{V}$ be a finite open cover of X modulo H. Then there is a closed collection $\mathbf{F}$ that shrinks $\mathbf{V}$ modulo H and that covers X modulo H. Moreover, it may be assumed that H is contained in each member of $\mathbf{F}$.*

Proof. Since H is closed, the proposition is an immediate consequence of Lemma I.8.8 applied to the space $X \setminus H$.

4.4. Proposition. *Let H be a closed subset of a space X and let $\mathbf{F} = \{ F_i : i = 0, 1, \dots, n \}$ be a collection of closed sets. Then there is a collection $\mathbf{W} = \{ W_i : i = 0, 1, \dots, n \}$ of open sets such that $\mathbf{W}$ is a swelling of $\mathbf{F}$ modulo H and such that the collection $\{ \mathrm{cl}\,(W_i) : i = 0, 1, \dots, n \}$ is combinatorially equivalent to $\mathbf{F}$ modulo H. That is,*

$$F_i \subset W_i \mod H, \qquad i = 0, 1, \dots, n,$$

and for any finite set of indices $i_1, \dots, i_m$

$$F_{i_1} \cap \cdots \cap F_{i_m} \neq \emptyset \mod H$$

if and only if

$$\mathrm{cl}\,(W_{i_1}) \cap \cdots \cap \mathrm{cl}\,(W_{i_m}) \neq \emptyset \mod H.$$

Moreover, if $\mathbf{U} = \{ U_i : i = 0, 1, \ldots, n \}$ *is an open swelling of* $\mathbf{F}$ *modulo* H, *then the collection* $\{ \operatorname{cl}(W_i) : i = 0, 1, \ldots, n \}$ *can be assumed to be a shrinking of* $\mathbf{U}$ *modulo* H *as well, i.e.,* $\operatorname{cl}(W_i) \subset U_i \mod H$ *for all* i.

Proof. The proof follows the lines of that of Lemma I.8.8. Beginning with the closed set F_0, one modifies the proof given there by considering all collections of indices $i_1, \ldots, i_m$ for which

$$F_0 \cap F_{i_1} \cap \cdots \cap F_{i_m} = \emptyset \mod H.$$

Then, working in the subspace $X \setminus H$, one finds an open set W_0 containing $F_0 \setminus H$ whose closure in the subspace is disjoint with each of the sets $(F_{i_1} \cap \cdots \cap F_{i_m}) \setminus H$. The remaining details are similar. The final statement is obvious.

Let us resume our discussion of $\mathcal{P}$-Ind. We shall develop relationships between $\mathcal{P}$-Ind and mappings into spheres that are similar to those found in dimension theory. The natural ones are those concerned with unstable values of mappings and with extensions of mappings into n-spheres. Let us begin with the following theorem. It is the $\mathcal{P}$-Ind analogue of Corollary I.4.7.

4.5. Theorem. *If* $\mathcal{P}$-Ind $X \leq n$, *then for any* $n+1$ *pairs* (F_i, G_i), $i = 0, 1, \ldots, n$, *of disjoint closed sets* F_i *and* G_i *there are partitions* S_i *between* F_i *and* G_i *in* X *such that* $\bigcap \{ S_i : i = 0, 1, \ldots, n \}$ *is a closed* $\mathcal{P}$*-kernel of* X.

Proof. The proof is exactly the same as the one for Corollary I.4.7 except that Theorem 2.21 replaces Proposition I.4.6.

The above theorem can be used to prove an "unstable value" theorem modulo a class $\mathcal{P}$. To this end we make the following definition.

4.6. Definition. Let $\mathcal{P}$ be a class of spaces and let X be in $\mathcal{T}$ and Y be a metric space. For a continuous map f of X into Y a point y of Y is called an *unstable value of* f *modulo* $\mathcal{P}$ if for each positive number ε there is a closed $\mathcal{P}$-kernel H of X and there is a continuous map g of $X \setminus H$ into Y such that

$$d(f(x), g(x)) < \varepsilon \quad \text{and} \quad g(x) \neq y, \qquad x \in X \setminus H.$$

A point y of Y is called a *stable value of* f *modulo* $\mathcal{P}$ if it is not an unstable value of f modulo $\mathcal{P}$.

4.7. Theorem. *Let f be a continuous map of X into $\mathbb{I}^{n+1}$. If $\mathcal{P}$-Ind $X \leq n$, then each point of $\mathbb{I}^{n+1}$ is an unstable value of f modulo $\mathcal{P}$.*

Proof. It is clear that every point of the boundary $\partial\, \mathbb{I}^{n+1}$ of $\mathbb{I}^{n+1}$ is an unstable value of f modulo $\mathcal{P}$. So only the interior points of $\mathbb{I}^{n+1}$ need to be considered. Also it is easily seen that the assertion for interior points will follow from the special case of the origin 0 of $\mathbb{I}^{n+1}$. Let us prove this special case. The map f has coordinate maps f_k of X into $\mathbb{I}$, $k = 0, 1, \ldots, n$. There is no loss in assuming $6\varepsilon\sqrt{n+1} < 1$. For each k define the disjoint closed sets $A_k = \{ x : f_k(x) \geq \varepsilon \}$ and $B_k = \{ x : f_k(x) \leq -\varepsilon \}$. By Theorem 4.5 there are partitions S_k between A_k and B_k such that $H = \bigcap \{ S_k : k = 0, 1, \ldots, n \}$ is a closed $\mathcal{P}$-kernel of X. By Proposition 4.4 there is a collection $\mathbf{W} = \{ W_k : k = 0, 1, \ldots, n \}$ of open sets such that $S_k \subset W_k \mod H$ and $\bigcap \{ \operatorname{cl}(W_k) : k = 0, 1, \ldots, n \} = \emptyset \mod H$. For each k let φ_k be a continuous map of $X \setminus H$ into $[-\varepsilon, \varepsilon]$ such that $W_k \supset \{ x : \varphi_k(x) = 0 \}$, $A_k = \{ x : \varphi_k(x) = \varepsilon \}$ and $B_k = \{ x : \varphi_k(x) = -\varepsilon \}$. Defining h_k by $h_k(x) = f_k(x)$ for x in $A_k \cup B_k$ and $h_k(x) = \varphi_k(x)$ for x not in $A_k \cup B_k$, we get continuous maps h_k of $X \setminus H$ into $\mathbb{I}$. Let h be the map of $X \setminus H$ into $\mathbb{I}^{n+1}$ whose coordinates are h_k. Then $\|f(x) - h(x)\| \leq 2\varepsilon\sqrt{n+1}$ for x in $X \setminus H$. One easily verifies $0 \notin h[X \setminus H]$.

In the next lemma it will be convenient to use the supremum norm $\|\cdot\|_\infty$ in $\mathbb{R}^{n+1}$.

4.8. Lemma. *If f is a continuous map of X into $\mathbb{I}^{n+1}$ and the origin 0 of $\mathbb{I}^{n+1}$ is an unstable value of f modulo $\mathcal{P}$, then there is a closed $\mathcal{P}$-kernel H of X and there is a continuous map g of $X \setminus H$ into $\mathbb{I}^{n+1}$ such that*

$$\|g(x)\|_\infty = 1, \qquad x \in X \setminus H,$$

and

$$f(x) = g(x), \qquad x \in f^{-1}[\partial\, \mathbb{I}^{n+1}] \setminus H.$$

Proof. For $\varepsilon = 1/4$ let H be the corresponding closed $\mathcal{P}$-kernel of X and let $h_0 : (X \setminus H) \to \mathbb{I}^{n+1}$ be the corresponding continuous map such that

(1) $\|f(x) - h_0(x)\|_\infty < 1/4$ and $\|h_0(x)\|_\infty \neq 0, \qquad x \in X \setminus H.$

Define M to be the open set

$$M = \{ x : x \notin H, \|f(x)\|_\infty > 3/4 \}. \tag{2}$$

Then $f^{-1}[\partial(\mathbb{I}^{n+1})] \setminus H \subset M \subset X \setminus H$. As $X \setminus H$ is a normal space, there is a continuous function Φ on $X \setminus H$ into $[0,1]$ with

$$\Phi(x) = 1, \qquad x \in f^{-1}[\partial(\mathbb{I}^{n+1})] \setminus H, \tag{3}$$

$$\Phi(x) = 0, \qquad x \in (X \setminus H) \setminus M. \tag{4}$$

Define h on $X \setminus H$ by the formula $h(x) = \Phi(x)f(x) + [1 - \Phi(x)]h_0(x)$. Then (1), (2), (3) and (4) will yield

$$h(x) = f(x), \qquad x \in f^{-1}[\partial(\mathbb{I}^{n+1})] \setminus H,$$

$$\|h(x)\|_\infty \neq 0, \qquad x \in X \setminus H.$$

We obtain the required map $g\colon (X \setminus H) \to \partial(\mathbb{I}^{n+1})$ by composing h with the the radial projection of $\mathbb{I}^{n+1} \setminus \{0\}$ onto $\partial(\mathbb{I}^{n+1})$.

The next theorem concerns a sufficient condition for the existence of extensions of continuous maps into spheres. In later sections (Sections 5 and 6) we shall define the covering dimension $\mathcal{P}$-dim and show that a corresponding extension theorem for $\mathcal{P}$-dim characterizes the function $\mathcal{P}$-dim. In this way we will be able to show that $\mathcal{P}$-dim is smaller than $\mathcal{P}$-Ind.

4.9. Theorem. *Let n be a natural number. If $\mathcal{P}$-Ind $X \leq n$, then for every closed set C of X and every continuous map f of C into $\mathbb{S}^n$ there exists a closed $\mathcal{P}$-kernel H of X with $H \subset X \setminus C$ such that f can be extended to $X \setminus H$.*

Proof. Let f be a continuous map of C into $\mathbb{S}^n$. We can regard $\mathbb{S}^n$ as the boundary of $\mathbb{I}^{n+1}$. Since $\mathbb{S}^n$ is an absolute neighborhood retract, the map f has a continuous extension to an open neighborhood U of C. We shall also call this extension f. Let V be a open set such that $C \subset V \subset \mathrm{cl}(V) \subset U$. Since $X \setminus V$ is closed, we have by Theorem 2.16 that $\mathcal{P}$-Ind $(X \setminus V) \leq n$ holds. Consider the continuous map $f|\mathrm{B}(V)$ on the closed subset $\mathrm{B}(V)$ of $X \setminus V$. As $\mathbb{I}^{n+1}$ is an absolute retract, we may assume that $f|\mathrm{B}(V)$ has been extended to a continuous map f of $X \setminus V$ into $\mathbb{I}^{n+1}$. We have from Theorem 4.7

that 0 is an unstable value of this extended map f modulo $\mathcal{P}$. By Lemma 4.8 there is a closed $\mathcal{P}$-kernel H of $X \setminus V$ and a continuous map g of $(X \setminus V) \setminus H$ into $\mathbb{I}^{n+1}$ such that

$$\begin{aligned} \|g(x)\|_\infty &= 1, && x \in (X \setminus V) \setminus H, \\ f(x) &= g(x), && x \in f^{-1}[\mathbb{S}^n] \cap (X \setminus V) \setminus H. \end{aligned}$$

Clearly, $H \subset X \setminus C$. The required extension is now easily constructed since f and g agree on the closed subset $\mathrm{B}(V) \setminus H$ of the space $X \setminus H$.

As an application of the last theorem, we shall prove an extension of a theorem that was first proved by Eilenberg in [1936].

4.10. Theorem. *Suppose C is a closed subset of a space X with* $\operatorname{Ind}(X \setminus C) \leq n$ *and k is a natural number with $k \leq n$. Then for each continuous map f of C into $\mathbb{S}^k$ there exists a closed set H contained in $X \setminus C$ with* $\operatorname{Ind} H < n - k$ *such that f can be extended over $X \setminus H$.*

Proof. According to our agreement, the universe of discourse will naturally be the class $\mathcal{N}_H$ of hereditarily normal spaces. Let $\mathcal{P}$ be the class of all spaces Y with $\operatorname{Ind} Y \leq n - k - 1$. Clearly the class $\mathcal{P}$ is absolutely closed-monotone. Moreover, we have the equivalence $\operatorname{Ind} Y \leq n + m - k$ if and only if $\mathcal{P}\text{-}\operatorname{Ind} Y \leq m$. Now Theorem 4.9 can be applied because $\mathcal{P}\text{-}\operatorname{Ind}(X \setminus C) \leq k$.

Theorem 4.9 can be used to compute lower bounds for $\mathcal{P}$-Ind as the next examples show.

4.11. Examples. The examples will use product spaces $X \times Y$ and the natural projection $\pi\colon X \times Y \to X$. The map π will be perfect when Y is compact, that is, a closed map with compact point inverses.

a. For every noncompact space X,

$$\mathcal{K}\text{-}\operatorname{Ind}(X \times \mathbb{I}^n) \geq n.$$

To see this we suppose that the inequality fails and consider the closed set $C = X \times \partial\, \mathbb{I}^n$ and the continuous map $f\colon C \to \partial\, \mathbb{I}^n$ determined by the natural projection π of $X \times \mathbb{I}^n$ onto $\mathbb{I}^n$. Since $\mathbb{S}^{n-1}$ is homeomorphic to $\partial\,\mathbb{I}^n$, by Theorem 4.9 we have a closed $\mathcal{K}$-kernel H with $H \subset (X \times \mathbb{I}^n) \setminus C$ and a continuous extension g of f from the set C to the set $(X \times \mathbb{I}^n) \setminus H$. As X is not compact, $X \setminus \pi[H] \neq \emptyset$. So for some point c in X we have $(\{c\} \times \mathbb{I}^n) \cap H = \emptyset$. The extension g is defined on the set $\{c\} \times \mathbb{I}^n$ and can be regarded as the identity map on the set $\partial\,\mathbb{I}^n$. The composition of g and the reflection of $\partial\,\mathbb{I}^n$ in the origin defines a map of $\mathbb{I}^n$ into itself without fixed ponts. A contradiction to the Brouwer fixed-point theorem has occurred.

b. Recall that $\mathcal{S}$ is the class of all σ-compact spaces. Since the continuous image of a σ-compact space is again σ-compact, we have by the same argument as in the above example that

$$\mathcal{S}\text{-Ind}\,(X \times \mathbb{I}^n) \geq n$$

for every non-σ-compact space X.

c. For the class $\mathcal{C}$ of complete metrizable spaces we claim that

$$\mathcal{C}\text{-Ind}\,(X \times \mathbb{I}^n) \geq n$$

for every noncompletely metrizable space X. The proof of this claim follows the same lines as that of the first example. This time the set H is a closed $\mathcal{C}$-kernel of $X \times \mathbb{I}^n$. As π is a perfect map, we also have that $\pi|H$ is a perfect map. From Henriksen and Isbell [1958] and Čech [1937] we infer that $\pi[H]$ is in $\mathcal{C}$. Consequently, $X \setminus \pi[H] \neq \emptyset$. The remainder of the proof is the same as in the other two examples.

5. Covering dimensions modulo a class $\mathcal{P}$

The covering completeness degree was introduced in Section I.9 for the class $\mathcal{C}$ of complete metrizable spaces. Its definition was a natural extension of the definition of the covering dimension dim of dimension theory. We shall now introduce the extension for an arbitrary class $\mathcal{P}$. Our discussion will be focused on generalizing the following characterization of dim due to Morita [1950a].

5.1. Theorem. *Let X be a normal space and let n be a natural number. Then $\dim X \leq n$ if and only if for every locally finite collection $\{U_\alpha : \alpha \in A\}$ of open sets and every collection $\{F_\alpha : \alpha \in A\}$ of closed sets such that $F_\alpha \subset U_\alpha$ for all α in A there exist collections $\{W_\alpha : \alpha \in A\}$ and $\{V_\alpha : \alpha \in A\}$ of open sets satisfying the* *inclusions* $F_\alpha \subset V_\alpha \subset \mathrm{cl}(V_\alpha) \subset W_\alpha \subset \mathrm{cl}(W_\alpha) \subset U_\alpha$ *for* α *in* A *and* $\mathrm{ord}\{\mathrm{cl}(W_\alpha) \setminus V_\alpha : \alpha \in A\} \leq n$.

Just as the functions $\mathcal{P}$-ind and $\mathcal{P}$-Ind of inductive dimensional types were defined for all topological spaces, the functions of covering dimensional types will also be defined for all topological spaces. We shall begin with the natural extension of the notion of border cover given in Chapter I for the class $\mathcal{C}$. (The reader is reminded that the empty space is a member of every class by the agreement made at the end of Section 2.)

5.2. Definition. Let $\mathcal{P}$ be a class of spaces and X be in $\mathcal{T}$. An open collection $\mathbf{V}$ in X is called a *$\mathcal{P}$-border cover* of X if $X \setminus \bigcup \mathbf{V}$ is in $\mathcal{P}$; the set $X \setminus \bigcup \mathbf{V}$ is called the *enclosure* of $\mathbf{V}$.

It is clear that the enclosure of a $\mathcal{P}$-border cover of X is a closed $\mathcal{P}$-kernel of X. Using $\mathcal{P}$-border covers, we now define two functions of covering dimensional types.

5.3. Definition. Let $\mathcal{P}$ be a class of spaces and X be a space in $\mathcal{T}$. Then the *small covering dimension of X modulo the class $\mathcal{P}$*, denoted $\mathcal{P}$-dim X, is defined as follows.

(i) $\mathcal{P}$-dim $X = -1$ if and only if X is in $\mathcal{P}$.

(ij) For each natural number n, $\mathcal{P}$-dim $X \leq n$ if every finite $\mathcal{P}$-border cover $\mathbf{V}$ of X has a $\mathcal{P}$-border cover refinement of order less than or equal to $n + 1$.

5.4. Definition. Let $\mathcal{P}$ be a class of spaces and X be a space in $\mathcal{T}$. Then the *large covering dimension of X modulo the class $\mathcal{P}$*, denoted $\mathcal{P}$-Dim X, is defined as follows.

(i) $\mathcal{P}$-Dim $X = -1$ if and only if X is in $\mathcal{P}$.

(ij) For each natural number n, $\mathcal{P}$-Dim $X \leq n$ if every $\mathcal{P}$-border cover $\mathbf{V}$ of X that is locally finite in the subspace $\bigcup \mathbf{V}$ has a $\mathcal{P}$-border cover refinement of order less than or equal to $n + 1$.

The discussion of this section will be concentrated on $\mathcal{P}$-dim. So we shall be dealing only with finite $\mathcal{P}$-border covers. The definition of $\mathcal{P}$-Dim uses a local finiteness condition rather than an arbitrary one. (Obviously these two conditions are equivalent for hereditarily paracompact spaces.) The reason for this bias is Theorem 5.1 which has locally finite collections in its statement. The function $\mathcal{P}$-Dim will be discussed in detail in Chapter IV.

5.5. Propositions. The following three propositions concern the class $\mathcal{T}$ of all topological spaces. The proofs of the first two are easy. For the third, see the proof of Proposition I.8.6.

A. $\mathcal{P}$-dim $\leq \mathcal{P}$-Dim.

B. *If $\mathcal{P}$ and $\mathcal{Q}$ are classes with $\mathcal{P} \supset \mathcal{Q}$, then $\mathcal{P}$-dim $\leq \mathcal{Q}$-dim and $\mathcal{P}$-Dim $\leq \mathcal{Q}$-Dim. In particular, $\mathcal{P}$-dim $\leq$ dim and $\mathcal{P}$-Dim $\leq$ Dim.*

C. *Let $n \in \mathbb{N}$. For classes $\mathcal{P}$ and spaces X in $\mathcal{T}$, $\mathcal{P}$-dim $X \leq n$ if and only if for each $\mathcal{P}$-border cover $\boldsymbol{U} = \{ U_i : i = 0, 1, \ldots, k \}$ of X there is a $\mathcal{P}$-border cover $\mathbf{V} = \{ V_i : i = 0, 1, \ldots, k \}$ with enclosure G such that $\mathbf{V}$ shrinks $\boldsymbol{U}$ modulo G and ord $\mathbf{V} \leq n + 1$. The corresponding equivalent formulation exists for $\mathcal{P}$-Dim.*

We shall prove next a closed monotonicity theorem for $\mathcal{P}$-dim. (Remember the agreement that was made at the end of Section 2).

5.6. Theorem. *For every closed subspace Y of a space X,*

$$\mathcal{P}\text{-dim}\, Y \leq \mathcal{P}\text{-dim}\, X.$$

Proof. Let $\mathcal{P}$-dim $X = n < \infty$ and let $\boldsymbol{U}$ be a finite $\mathcal{P}$-border cover of Y with enclosure F. Since Y is closed, the finite open collection $\boldsymbol{W} = \{ (X \setminus Y) \cup U : U \in \boldsymbol{U} \}$ is a $\mathcal{P}$-border cover of X with enclosure F. There is a $\mathcal{P}$-border cover $\boldsymbol{W}'$ of X with enclosure G' such that $\boldsymbol{W}'$ refines $\boldsymbol{W}$ and ord $\boldsymbol{W}' \leq n + 1$. Since $\mathcal{P}$ is closed-monotone in $\mathcal{U}$, it is easily shown that $\mathbf{V} = \{ Y \cap W' : W' \in \boldsymbol{W}' \}$ is a $\mathcal{P}$-border cover of Y with enclosure $Y \cap G'$. Consequently we have $\mathcal{P}$-dim $Y \leq n$.

The following theorem characterizes $\mathcal{P}$-dim.

5.7. Theorem. *Let n be a natural number. The following conditions are equivalent for a space X.*

(a) *$\mathcal{P}$-dim $X \leq n$.*

(b) *For each $\mathcal{P}$-border cover $\boldsymbol{U} = \{ U_i : i = 0, 1, \ldots, n + 1 \}$ of X with exactly $n + 2$ indices there exists a $\mathcal{P}$-border cover $\mathbf{V} = \{ V_i : i = 0, 1, \ldots, n + 1 \}$ such that ord $\mathbf{V} \leq n + 1$ and $\mathbf{V}$ shrinks $\boldsymbol{U}$ modulo G where G is the enclosure of $\mathbf{V}$.*

Proof. That (a) implies (b) is obvious from Proposition 5.5.C. So we shall prove (b) implies (a). It will be easier to prove the contrapositive. Suppose that the negation of condition (a) holds. We claim that there is a $\mathcal{P}$-border cover $\boldsymbol{Z} = \{ Z_i : i = 0, 1, \ldots, k \}$ of X such that $\operatorname{ord} \boldsymbol{Z} \geq n + 2$ is true and such that each $\mathcal{P}$-border cover $\boldsymbol{V} = \{ V_i : i = 0, 1, \ldots, k \}$ that shrinks $\boldsymbol{Z}$ modulo G is combinatorially equivalent to $\boldsymbol{Z}$ modulo G, where G is the enclosure of $\boldsymbol{V}$. By combinatorially equivalent modulo G we mean that for any increasing sequence $i_1, i_2, \ldots, i_m$ of natural numbers not exceeding k the following condition holds.

$$(1) \qquad \begin{gathered} V_{i_1} \cap V_{i_2} \cap \cdots \cap V_{i_m} \neq \emptyset \mod G \\ \text{if and only if} \\ Z_{i_1} \cap Z_{i_2} \cap \cdots \cap Z_{i_m} \neq \emptyset \mod G. \end{gathered}$$

Indeed, from the negation of (a) and Proposition 5.5.C there is a $\mathcal{P}$-border cover $\boldsymbol{Z} = \{ Z_i : i = 0, 1, \ldots, k \}$ such that $\operatorname{ord} \boldsymbol{V} \geq n + 2$ for every $\mathcal{P}$-border cover $\boldsymbol{V} = \{ V_i : i = 0, 1, \ldots, k \}$ that shrinks $\boldsymbol{Z}$ modulo G, where G is the enclosure of $\boldsymbol{V}$. If $\boldsymbol{Z}$ does not satisfy condition (1), one just replaces the $\mathcal{P}$-border cover $\boldsymbol{Z}$ with the $\mathcal{P}$-border cover $\{ V_i : i = 0, 1, \ldots, k \}$ and continues this process until a $\mathcal{P}$-border cover with the required property is obtained; as the number of subsets of $\{ 0, 1, \ldots, k \}$ is finite, the process will come to an end after finitely many such steps.

Since $n + 2 \leq \operatorname{ord} \boldsymbol{Z}$, rearranging if need be the members of $\boldsymbol{Z}$, we have

$$(2) \qquad \bigcap \{ Z_i : i = 0, 1, \ldots, n + 1 \} \neq \emptyset.$$

We shall show that the $\mathcal{P}$-border cover $\boldsymbol{U}$ whose elements are $U_i = Z_i$ for $i = 0, 1, \ldots, n$ and $U_{n+1} = \bigcup \{ Z_i : i = n + 1, \ldots, k \}$ has the property that every $\mathcal{P}$-border cover $\boldsymbol{V}$ with $n + 2$ elements that shrinks $\boldsymbol{U}$ modulo G, where G is the enclosure of $\boldsymbol{V}$, has $\operatorname{ord} \boldsymbol{V}$ exceeding $n + 2$; that is, condition (b) fails. To this end, consider a $\mathcal{P}$-border cover $\boldsymbol{V} = \{ V_i : i = 0, 1, \ldots, n + 1 \}$ such that $\boldsymbol{V}$ shrinks $\boldsymbol{U}$ modulo G, where G is the enclosure of $\boldsymbol{V}$. Then the open collection

$$\{ V_0, V_1, \ldots, V_n, V_{n+1} \cap Z_{n+1}, V_{n+1} \cap Z_{n+2}, \ldots, V_{n+1} \cap Z_k \}$$

is a $\mathcal{P}$-border cover with enclosure G that shrinks $\boldsymbol{Z}$ modulo G. By conditions (1) and (2) we have

$$\bigcap \boldsymbol{V} \supset \left(\bigcap \{ V_j : i = 0, 1, \ldots, n \} \right) \cap \left(V_{n+1} \cap Z_{n+1} \right) \neq \emptyset \mod G.$$

Thus we have completed the proof of (a) implies (b).

The following corollary is an immediate consequence of the last theorem.

5.8. Corollary. *For a normal space* X, $\dim X \leq 0$ *if and only if* $\operatorname{Ind} X \leq 0$.

We now prove a second characterization theorem. This theorem will yield a generalization of the finite collection version of Theorem 5.1.

5.9. Theorem. *Let n be a natural number. The following conditions are equivalent for hereditarily normal spaces X in the universe.*

(a) $\mathcal{P}$-$\dim X \leq n$.

(b) *For each closed $\mathcal{P}$-kernel G of X and for each pair of open and closed collections* $\mathbf{U} = \{ U_i : i = 0, 1, \ldots, k \}$ *and* $\mathbf{F} = \{ F_i : i = 0, 1, \ldots, k \}$ *of X such that* $F_i \subset U_i \mod G$ *holds for each i there exists a closed $\mathcal{P}$-kernel H containing G and there exist open collections* $\mathbf{V} = \{ V_i : i = 0, 1, \ldots, k \}$ *and* $\mathbf{W} = \{ W_i : i = 0, 1, \ldots, k \}$ *such that*

$$(b.1) \qquad \begin{aligned} &F_i \setminus H \subset V_i \subset W_i \subset U_i, && i = 0, 1, \ldots, k, \\ &\operatorname{cl}(V_i) \subset W_i \mod H, && i = 0, 1, \ldots, k, \\ &\operatorname{cl}(W_i) \subset U_i \mod H, && i = 0, 1, \ldots, k, \end{aligned}$$

and

$$(b.2) \qquad \operatorname{ord} \{ W_i \setminus \operatorname{cl}(V_i) : i = 0, 1, \ldots, k \} \leq n \mod H.$$

(c) *For each closed $\mathcal{P}$-kernel G of X and for each collection of $n+1$ pairs* (A_i, B_i), $i = 0, 1, \ldots, n$, *of closed sets A_i and B_i that are disjoint modulo G there is a closed $\mathcal{P}$-kernel H containing G such that* $\bigcap \{ S_i : i = 0, 1, \ldots, n \} = H$ *where for each i the set S_i is some partition between A_i and B_i modulo H.*

Proof. The proof of (a) implies (b) is verbatim the same as that for Lemmas I.8.9 and I.9.4.

Let us now prove that (b) implies (c). Let G be a closed $\mathcal{P}$-kernel and let $(A_0, B_0), \ldots, (A_n, B_n)$ be pairs of closed sets A_i and B_i such that $A_i \cap B_i = \emptyset \mod G$. For each i the sets $U_i = X \setminus (A_i \cup G)$ and $F_i = B_i$ are respectively open and closed and are such that $F_i \subset U_i \mod G$. By (b) there exists a closed $\mathcal{P}$-kernel H containing G and there exist two open collections $\mathbf{V} = \{ V_i : i = 0, 1, \ldots, n \}$ and $\mathbf{W} = \{ W_i : i = 0, 1, \ldots, n \}$ such that formulas (b.1) and (b.2) hold. By (b.1) the closed sets $\operatorname{cl}(V_i)$ and $X \setminus W_i$ are disjoint modulo H. Since X is hereditarily normal, there is a partition S_i' between $\operatorname{cl}(V_i)$ and $X \setminus W_i$ modulo H. Clearly $S_i = S_i' \cup H$ is a partition between A_i and B_i modulo H. From (b.2) we have the equality $H = \bigcap \{ S_i : i = 0, 1, \ldots, n \}$.

Finally let us prove (c) implies (a). This will be done by means of Theorem 5.7. Consider an $n+2$ element $\mathcal{P}$-border cover $\boldsymbol{U} = \{ U_i : i = 0, 1, \ldots, n+1 \}$ of X and denote its enclosure by G. In the subspace $X \setminus G$ let $\{ F_i : i = 0, 1, \ldots, n+1 \}$ be a closed shrinking of $\boldsymbol{U}$. The collection $\boldsymbol{F} = \{ F_i \cup G : i = 0, 1, \ldots, n+1 \}$ is a closed shrinking of $\boldsymbol{U}$ modulo G. Let $A_i = F_i \cup G$ and $B_i = X \setminus U_i$ for each i. Then $A_i \cap B_i = \emptyset \mod G$. By (c) there exists a closed collection $\{ S_i : i = 0, 1, \ldots, n \}$ and there exists a closed $\mathcal{P}$-kernel H such that S_i is a partition between A_i and B_i modulo H, $G \subset H$ and $\bigcap \{ S_i : i = 0, 1, \ldots, n \} = H$. So let V_i and W_i be disjoint open subsets of $X \setminus H$ such that $A_i \setminus H \subset V_i$, $B_i \setminus H \subset W_i$ and $S_i \setminus H = (X \setminus H) \setminus (V_i \cup W_i)$. Obviously V_i and W_i are open in X as well. Let $\mathbf{V} = \{ V_i : i = 0, 1, \ldots, n \}$ and $\mathbf{W} = \{ W_i : i = 0, 1, \ldots, n \}$. Observe that $(\bigcup \mathbf{V}) \cup (\bigcup \mathbf{W}) = \bigcup \{ V_i \cup W_i : i = 0, 1, \ldots, n \} = X \setminus H$ holds because $\bigcap \{ S_i : i = 0, 1, \ldots, n \} = H$. Then the inclusion

$$\begin{aligned}(\textstyle\bigcup \mathbf{V}) \cup (U_{n+1} \cap (\bigcup \mathbf{W})) &= ((\textstyle\bigcup \mathbf{V}) \cup U_{n+1}) \cap ((\bigcup \mathbf{V}) \cup (\bigcup \mathbf{W})) \\ &\supset \textstyle\bigcup \{ A_i \setminus H : i = 0, 1, \ldots, n+1 \} = X \setminus H\end{aligned}$$

results from $A_i \setminus H \subset V_i$ for $i = 0, 1, \ldots, n$ and $A_{n+1} \setminus H \subset U_{n+1}$. So $\boldsymbol{Z} = \{ Z_i : i = 0, 1, \ldots, n+1 \}$, where $Z_i = V_i$ for $i = 0, 1, \ldots, n$ and $Z_{n+1} = U_{n+1} \cap (\bigcup \mathbf{W})$, is a $\mathcal{P}$-border cover with enclosure H. Finally

$$\textstyle\bigcap \boldsymbol{Z} = (\bigcap \mathbf{V}) \cap (U_{n+1} \cap (\bigcup \mathbf{W})) \subset (\bigcap \mathbf{V}) \cap (\bigcup \mathbf{W}) = \emptyset,$$

and therefore $\operatorname{ord} \boldsymbol{Z} \leq n+1$. Thus (a) holds by Theorem 5.7.

A generalization of the finite collection version of Theorem 5.1 is next.

5.10. Theorem. *Let n be a natural number and let X be a hereditarily normal space in the universe. Then $\mathcal{P}$-$\dim X \leq n$ if and only if for each closed $\mathcal{P}$-kernel G of X and for each open collection $\{ U_i : i = 0, 1, \ldots, k \}$ and each closed collection $\{ F_i : i = 0, 1, \ldots, k \}$ such that $F_i \subset U_i \mod G$ holds for every i there exists a closed $\mathcal{P}$-kernel H and there exist open collections $\{ W_i : i = 0, 1, \ldots, k \}$ and $\{ V_i : i = 0, 1, \ldots, k \}$ such that $G \subset H$ and*

$$(*) \qquad \begin{aligned} &F_i \setminus H \subset V_i \subset W_i \subset U_i, && i = 0, 1, \ldots, k, \\ &\mathrm{cl}\,(V_i) \subset W_i \mod H, && i = 0, 1, \ldots, k, \\ &\mathrm{cl}\,(W_i) \subset U_i \mod H, && i = 0, 1, \ldots, k, \end{aligned}$$

and

$$(**) \qquad \mathrm{ord}\,\{ \mathrm{cl}\,(W_i) \setminus V_i : i = 0, 1, \ldots, k \} \leq n \mod H.$$

Proof. Only the condition (**) of this theorem differs from condition (b.2) of Theorem 5.9. Suppose $\mathcal{P}$-$\dim X \leq n$ holds. Then the condition (b) of Theorem 5.9 and the hereditary normality of X will yield the required closed $\mathcal{P}$-kernel H and the open collections of the theorem. Conversely, the required closed $\mathcal{P}$-kernel H and partitions S_i of condition (c) of Theorem 5.9 are easily constructed from the open collections of the theorem because of the condition (**).

Remark. The above proofs illustrate the naturalness of the definition of covering dimension modulo a class $\mathcal{P}$. When $\mathcal{P}$ is $\{ \emptyset \}$, the theorems reduce to those of the covering dimension dim. The only unfortunate aspects of the proofs are that the theorems for dim hold in the universe $\mathcal{N}$ of normal spaces while some of the proofs here have the additional requirement of hereditary normality. But the full force of hereditary normality is not needed and the following definition may provide a way out of this unfortunate situation. For a class $\mathcal{P}$ the universe $\mathcal{N}[\mathcal{P}]$, called the *normal universe modulo $\mathcal{P}$*, is defined as

$$\mathcal{N}[\mathcal{P}] = \{ X : X \setminus G \text{ is normal for each closed } \mathcal{P}\text{-kernel } G \text{ of } X \}.$$

Observe that the space $X \setminus G$ in the definition is not assumed to be in the universe $\mathcal{N}[\mathcal{P}]$. The above proofs can be carried out for each X in $\mathcal{N}[\mathcal{P}]$. Clearly $\mathcal{N}[\{\emptyset\}]$ is the class of normal spaces. In this way, natural generalizations of dim will result. Indeed, observe the following inclusions:

$$\mathcal{N}_H = \mathcal{N}[\mathcal{N}] \subset \mathcal{N}[\mathcal{P}] \subset \mathcal{N}[\mathcal{Q}] \subset \mathcal{N}[\{\emptyset\}] = \mathcal{N}, \qquad \text{for } \mathcal{P} \supset \mathcal{Q}.$$

The normal universe modulo $\mathcal{P}$ can be used in many other places also. We shall not concern ourselves with this more general approach.

6. $\mathcal{P}$-dim and mappings into spheres

Agreement. *The universe is contained in $\mathcal{N}_H$.*

The function $\mathcal{P}$-dim will be shown to be characterized by means of mappings into spheres just as dim is. The fact that such a characterization theorem exists is no surprise because of Theorem 5.9. Our discussion will begin with unstable maps. Unstable maps were already encountered in Section 4 where connections with $\mathcal{P}$-Ind were made. There the continuous maps f were defined on the whole space X. But for dim we shall see that condition (c) of Theorem 5.9 will introduce a closed $\mathcal{P}$-kernel G of X such that the maps f are continuous on $X \setminus G$. The following is a variation of Definition 4.6.

6.1. Definition. Let $\mathcal{P}$ be a class of spaces and let f be a map (not necessarily continuous) from a space X into a metric space Y. A point y of Y is called a *$\mathcal{P}$-unstable value* of f if for each positive number ε and for each closed $\mathcal{P}$-kernel G of X for which $f|(X \setminus G)$ is continuous there exists a closed $\mathcal{P}$-kernel H and there exists a continuous map g of $X \setminus H$ into Y such that $H \supset G$ and

$$d(f(x), g(x)) < \varepsilon \quad \text{and} \quad g(x) \neq y, \qquad x \in X \setminus H.$$

A point y of Y is called a *$\mathcal{P}$-stable value* of f if it is not a $\mathcal{P}$-unstable value of f.

Observe that the notion of $\mathcal{P}$-unstable is stronger than that of unstable modulo $\mathcal{P}$ of Definition 4.6.

Similar to $\mathcal{P}$-Ind we have the following unstable value theorem (compare with Theorem 4.7).

6.2. Theorem. *Let f be a map of X into $\mathbb{I}^{n+1}$. If $\mathcal{P}$-dim $X \leq n$, then each point of $\mathbb{I}^{n+1}$ is a $\mathcal{P}$-unstable value of f.*

Proof. It is clear that every point of the boundary $\partial\,\mathbb{I}^{n+1}$ of $\mathbb{I}^{n+1}$ is a $\mathcal{P}$-unstable value of f. So only the interior points of $\mathbb{I}^{n+1}$ need to be considered. Also it is easily seen that the assertion for interior points will follow from the special case of the origin 0 of $\mathbb{I}^{n+1}$. Let us prove this special case. The proof will follow the lines of that for Theorem 4.7. Let G be a closed $\mathcal{P}$-kernel of X such that $f|(X \setminus G)$ is continuous. Denote the coordinate maps of f by f_i. Then the maps $f_i|(X \setminus G)$ of $X \setminus G$ into $\mathbb{I}$, $i = 0, 1, \ldots, n$, are continuous. There is no loss in assuming that $6\varepsilon\sqrt{n+1} < 1$. For each i we define two closed sets $A_i = \mathrm{cl}\,(\{\, x : x \notin G,\ f_i(x) \geq \varepsilon \,\})$ and $B_i = \mathrm{cl}\,(\{\, x : x \notin G,\ f_i(x) \leq -\varepsilon \,\})$. The closed sets A_i and B_i are disjoint modulo G because the restriction of f to $X \setminus G$ is continuous. By condition (c) of Theorem 5.9 there is a closed $\mathcal{P}$-kernel H and there are partitions S_i between A_i and B_i modulo H, $i = 0, 1, \ldots, n$, such that $H = \bigcap\{\, S_i : i = 0, 1, \ldots, n \,\}$. The remainder of the proof is the same as that of Theorem 4.7 where X is replaced by $X \setminus G$ and H is replaced by $H \setminus G$.

There is the following interpolation lemma for $\mathcal{P}$-dim that is analogous to Lemma 4.8 for $\mathcal{P}$-Ind. Its proof is a simple modification of the one given in that lemma.

6.3. Lemma. *let f be a map (not necessarily continuous) of a space X into $\mathbb{I}^{n+1}$ and let G be a closed $\mathcal{P}$-kernel of X such that $f|(X \setminus G)$ is continuous. If the origin 0 of $\mathbb{I}^{n+1}$ is a $\mathcal{P}$-unstable value of f, then there is a closed $\mathcal{P}$-kernel H of X and there is a continuous map g of $X \setminus H$ into $\mathbb{I}^{n+1}$ such that $H \supset G$,*

$$\|g(x)\|_\infty = 1, \qquad x \in X \setminus H,$$
$$f(x) = g(x), \qquad x \in f^{-1}[\partial\,(\mathbb{I}^{n+1})] \setminus H.$$

The $\mathcal{P}$-dim analogue of Theorem 4.9 will be labeled as a lemma since its converse will be shown to be true also.

6.4. Lemma. *Let n be a natural number and let G be a closed $\mathcal{P}$-kernel of a space X. If $\mathcal{P}$-dim $X \leq n$ and C is a closed set of X, then for every continuous map f of $C \setminus G$ into $\mathbb{S}^n$ there exists a* *closed $\mathcal{P}$-kernel H with $G \subset H$ and $H \cap (C \setminus G) = \emptyset$ such that f has a continuous extension over $X \setminus H$.*

Proof. Consider a continuous map f of $C \setminus G$ into $\mathbb{S}^n$. We may extend f to all of C in an arbitrary manner since we are concerned with f being continuous only on $C \setminus G$. We shall identify $\mathbb{S}^n$ with the boundary $\partial(\mathbb{I}^{n+1})$. Since $\mathbb{S}^n$ is an absolute neighborhood retract, we can extend $f|(C \setminus G)$ to a continuous map of an open neighborhood U of $C \setminus G$ in the subspace $X \setminus G$. Clearly U is open in X. Let V be an open subset of $X \setminus G$ such that $\mathrm{cl}(V) \subset U \mod G$ and $C \setminus G \subset V$. Then $\mathrm{B}(V) \cup G$ is a closed subset of $X \setminus V$ and f restricted to $\mathrm{B}(V) \setminus G$ is continuous. We have $\mathcal{P}\text{-dim}(X \setminus V) \leq n$ by Theorem 5.6. Similar to the proof of Theorem 4.9, by Lemmas 6.2 and 6.3 there is a closed $\mathcal{P}$-kernel H of $X \setminus V$ and there is a continuous map g of $(X \setminus V) \setminus H$ into $\mathbb{S}^n$ such that $G \subset H$ and g and f agree on $\mathrm{B}(V) \setminus H$. Obviously H is a closed $\mathcal{P}$-kernel of X. The extension of $f|(C \setminus H)$ can now be constructed in a manner similar to that in Theorem 4.9.

6.5. Lemma. *Let n be a natural number. Suppose X is a space that satisfies the following property: For each closed set C, for each closed $\mathcal{P}$-kernel G and for each continuous map f of $C \setminus G$ into $\mathbb{S}^n$ there is a closed $\mathcal{P}$-kernel H with $G \subset H$ and $H \cap (C \setminus G) = \emptyset$ such that f has a continuous extension over $X \setminus H$. Then $\mathcal{P}\text{-dim}\, X \leq n$.*

Proof. We shall show that condition (c) of Theorem 5.9 holds for X. For $i = 0, 1, \ldots, n$ let (A_i, B_i) be a pair of closed sets and G be a closed $\mathcal{P}$-kernel such that A_i and B_i are disjoint modulo G and then select continuous maps $f_i \colon X \setminus G \to \mathbb{I}$ with $f_i^{-1}[-1] \supset A_i \setminus G$ and $f_i^{-1}[1] \supset B_i \setminus G$. Identify the boundary of $\mathbb{I}^{n+1}$ and $\mathbb{S}^n$. Since $f = (f_0, f_1, \ldots, f_n)$ is a continuous map of $X \setminus G$ into $\mathbb{I}^{n+1}$, we have that $C = f^{-1}[\mathbb{S}^n] \cup G$ is closed and $C \setminus G = f^{-1}[\mathbb{S}^n]$ holds. From the property given in the statement of the lemma there is a closed $\mathcal{P}$-kernel H with $H \supset G$ and $H \cap f^{-1}[\mathbb{S}^n] = \emptyset$ and there is a continuous extension $g = (g_0, g_1, \ldots, g_n)$ of $f|(C \setminus H)$ to $X \setminus H$. Clearly the set $S_i = g_i^{-1}[0] \cup H$ is a partition between A_i and B_i modulo H for every i and $\bigcap\{ S_i : i = 0, 1, \ldots, n \} = H$. We have shown that condition (c) of Theorem 5.9 holds. Consequently, $\mathcal{P}\text{-dim}\, X \leq n$.

Combining Lemmas 6.4 and 6.5, we have our characterization theorem.

6.6. Theorem. *Let n be a natural number. For every class $\mathcal{P}$ and every space X, $\mathcal{P}$-dim $X \leq n$ if and only if for each closed set C of X, for each closed $\mathcal{P}$-kernel G of X and for each continuous map f of $C \setminus G$ into $\mathbb{S}^n$ there exists a closed $\mathcal{P}$-kernel H of X with $G \subset H$ and $H \cap (C \setminus G) = \emptyset$ such that f can be continuously extended over $X \setminus H$.*

6.7. Example. In Example 4.11 the lower bound n for $\mathcal{K}$-Ind, $\mathcal{S}$-Ind and $\mathcal{C}$-Ind of the space $X \times \mathbb{I}^n$ was computed for spaces X that were noncompact, non-σ-compact and, in the case of metrizable spaces, noncomplete respectively. This lower bound was established by means of Theorem 4.9 which relates $\mathcal{P}$-Ind and extensions of mappings into $\mathbb{S}^n$. With the aid of Theorem 6.6, a straightforward modification of these proofs will yield the lower bounds

$$\mathcal{K}\text{-dim}\,(X \times \mathbb{I}^n) \geq n, \quad \mathcal{S}\text{-dim}\,(X \times \mathbb{I}^n) \geq n, \quad \mathcal{C}\text{-dim}\,(X \times \mathbb{I}^n) \geq n$$

when X is respectively noncompact, non-σ-compact and, in the case of metrizable spaces, noncomplete.

Let us now consider the universe $\mathcal{U} = \mathcal{M}_0$ of separable metrizable spaces and the class $\mathcal{K}$ of compact spaces. The next proposition will connect $\mathcal{K}$-Dim and $\mathcal{K}$-Sur. To do this we shall have need of the following lemma connecting Dim and Ind.

6.8. Lemma. *For every separable metrizable space X,*

$$\dim X = \operatorname{Dim} X = \operatorname{ind} X = \operatorname{Ind} X.$$

Proof. In view of Theorem I.8.10 and Proposition 5.5.A we only need to prove $\operatorname{Dim} X \leq \operatorname{Ind} X$. The proof is by induction on $\operatorname{Ind} X$. We shall give only the inductive step. Suppose $\operatorname{Ind} X \leq n$ and let $\boldsymbol{U}$ be locally finite open cover of X. As X is a separable metrizable space, we have that $\boldsymbol{U}$ is also a countable collection. Indexing $\boldsymbol{U}$ as $\{ U_i : i = 0, 1, \ldots \}$, we let $\boldsymbol{F} = \{ F_i : i = 0, 1, \ldots \}$ be a closed cover that shrinks $\boldsymbol{U}$. Then for each pair (F_i, U_i) there is a partition S_i between F_i and $X \setminus U_i$ such that $\operatorname{Ind} S_i \leq n-1$. Let V_i and W_i be disjoint open sets with $F_i \subset V_i$, $X \setminus U_i \subset W_i$ and $X \setminus S_i = V_i \cup W_i$. As $S_i \subset U_i$ holds for each i, $\{ S_i : i = 0, 1, \ldots \}$ is locally finite in X. So the set $S = \bigcup \{ S_i : 0, 1, \ldots \}$ is closed. By the sum theorem for Ind the inequailty $\operatorname{Ind} S \leq n-1$ holds. From the induction hypothesis we have a collection $\boldsymbol{Z}' = \{ Z'_i : i = 0, 1, \dots \}$ with $\operatorname{ord} \boldsymbol{Z}' \leq n$ such that each of its members Z'_i are open sets of S with $Z'_i \subset U_i$. There is an open collection $\boldsymbol{Z} = \{ Z_i : i = 0, 1, \dots \}$ of X such that $Z'_i = Z_i \cap S$ for each i and $\operatorname{ord} \boldsymbol{Z} = \operatorname{ord} \boldsymbol{Z}'$. Next let $V'_0 = V_0 \setminus S$ and for $i \geq 1$ let $V'_i = V_i \setminus \bigcup \{ (V_j \cup S) : j < i \}$. Then the open collection $\boldsymbol{V} = \{ V'_i : i = 0, 1, \dots \}$ satisfies $\operatorname{ord} \boldsymbol{V} \leq 1$ and $\bigcup \boldsymbol{V} = X \setminus S$. So the collection $\boldsymbol{V} \cup \boldsymbol{Z}$ satisfies $\operatorname{ord} (\boldsymbol{V} \cup \boldsymbol{Z}) \leq n + 1$ and refines $\boldsymbol{U}$. Thus we have $\operatorname{Dim} X \leq n$.

6.9. Proposition. *For every separable metrizable space X,*

$$\mathcal{K}\text{-dim}\, X \leq \mathcal{K}\text{-Dim}\, X \leq \mathcal{K}\text{-Sur}\, X.$$

Proof. Only the right inequality needs proof. Let $\mathcal{K}\text{-Sur}\, X \leq n$. Then there is a $\mathcal{K}$-kernel G of X such that $\operatorname{Ind} (X \setminus G) \leq n$. Let $\boldsymbol{U}$ be a $\mathcal{K}$-border cover of X with enclosure F such that $\boldsymbol{U}$ is locally finite in $X \setminus F$. Then $\boldsymbol{U}$ is locally finite in $X \setminus H$, where $H = F \cup G$, and $\operatorname{Ind} (X \setminus H) \leq n$. By Lemma 6.8 there is an open collection $\boldsymbol{V}$ whose union is $X \setminus H$ such that $\boldsymbol{V}$ refines $\boldsymbol{U}$ and $\operatorname{ord} \boldsymbol{V} \leq n + 1$. Since H is in $\mathcal{K}$, we have $\mathcal{K}\text{-Dim}\, X \leq n$.

The above proposition will be sharpened to equalities in Theorem 6.11. (See Proposition 7.4 for the inequalities corresponding to more general classes $\mathcal{P}$.) To do this we shall now use the characterization of $\dim X \leq n$ to calculate lower bounds for $\mathcal{K}\text{-Ind}\, X$ and $\mathcal{K}\text{-dim}\, X$. We first prove a lemma.

6.10. Lemma. *Let $n \geq 1$. Suppose that X is a separable metrizable space and $\{ X_k : k = 0, 1, \dots \}$ is a sequence of closed subsets of X such that*

(a) *$\{ X_k : k = 0, 1, \dots \}$ is a discrete collection in X,*
(b) *$\dim X_k \geq n$ for $k \geq 0$.*

Then there exists a closed set C and a continuous map f of C into $\mathbb{S}^{n-1}$ such that for any compact set H the set $C \setminus H$ is nonempty and the map $f|(C \setminus H)$ cannot be extended to a continuous map of $X \setminus H$ into $\mathbb{S}^{n-1}$.

Proof. For each k there is a closed subset C_k of X_k such that some continuous map f_k of C_k into $\mathbb{S}^{n-1}$ does not have a continuous extension over X_k. The collection $\{ C_k : k = 0, 1, \dots \}$ is discrete in X because of condition (a). So $C = \bigcup \{ C_k : k = 0, 1, \dots \}$

is closed in X. The sequence of functions f_k, $k = 0, 1, \dots$, will define a continuous function f on C in the obvious way. The collection $\{ X_k : k = 0, 1, \dots \}$ is an open cover of the closed subspace $F = \bigcup \{ X_k : k = 0, 1, \dots \}$ of X. So the compact set $H \cap F$ is contained in $\bigcup \{ X_k : 0 \le k < k_0 \}$ for some natural number k_0. Therefore if $f|(C \setminus H)$ has a continuous extension to $X \setminus H$ then f_{k_0} has a continuous extension to X_{k_0}. This is a contradiction.

The next theorem is intimately connected to the conjecture of de Groot (I.5.6). Many topological invariants were invented in the course of the development of the theory of dimension modulo a class of spaces. The theorem shows that several of these invariants for the class $\mathcal{K}$ actually coincide in the universe $\mathcal{M}_0$ of separable metrizable spaces, the universe that is most relevant to de Groot's conjecture. Indeed, the theorem rules out many candidates for an internal characterization of the compactness deficiency.

6.11. Theorem. *For spaces X in $\mathcal{M}_0$,*

$$\mathcal{K}\text{-Ind}\, X = \mathcal{K}\text{-Sur}\, X = \mathcal{K}\text{-dim}\, X = \mathcal{K}\text{-Dim}\, X.$$

Proof. The equalities are obvious for X in $\mathcal{K}$; and Theorem 3.6 and Proposition 6.9 yield the case where $\mathcal{K}\text{-Sur}\, X = 0$.

Because of Theorem 3.6 and Proposition 6.9, we only need to prove the inequalities $\mathcal{K}\text{-Sur}\, X \le \mathcal{K}\text{-Ind}\, X$ and $\mathcal{K}\text{-Sur}\, X \le \mathcal{K}\text{-dim}\, X$. So assume $n \ge 1$ and $\mathcal{K}\text{-Sur}\, X \ge n$. Then $\dim (X \setminus H) \ge n$ for every $\mathcal{K}$-kernel H of X. Let us use this property to establish the existence of a closed set C and a continuous map f from C into $\mathbb{S}^{n-1}$ such that for any compact set H the set $C \setminus H$ is nonempty and the map $f|(C \setminus H)$ cannot be extended to $X \setminus H$. Let A be the set of points p of X such that every neighborhood U of p has $\dim U \ge n$. Then $q \notin A$ implies that some neighborhood U of q has $\dim U < n$. So A is closed in X. Moreover, $\dim (X \setminus A) < n$. To verify this we observe that each point of $X \setminus A$ has a neighborhood U_q such that $\operatorname{cl}(U_q) \subset X \setminus A$ and $\dim \operatorname{cl}(U_q) < n$. From the Lindelöf property there is a countable collection of such sets $\operatorname{cl}(U_q)$ that covers $X \setminus A$. Thus $\dim (X \setminus A) < n$ holds by the countable sum theorem of dimension. Let us show next that $A \setminus H \ne \emptyset$ for every H in $\mathcal{K}$. Suppose that $H \supset A$ for some H in $\mathcal{K}$. Then $n \le \mathcal{K}\text{-Sur}\, X \le \dim (X \setminus H) \le \dim (X \setminus A) < n$, a contradiction. Hence the closed set A is not compact. So there is a sequence p_k, $k = 0, 1, \dots$, in A that has no limit point in X. For each k let X_k be the closure of a neighborhood U_{p_k} of p_k so that $\{ X_k : k = 0, 1, \dots \}$ is a discrete collection in X. Clearly $\dim X_k \geq n$ for all k. Lemma 6.10 yields the existence of the required closed set C and continuous map f.

The proof of the lemma will now be completed by means of contradictions. First assume $\mathcal{K}\text{-Ind}\, X \leq n - 1$. Since $n - 1 \geq 0$, by Theorem 4.9 there is a closed $\mathcal{K}$-kernel H of X such that the continuous map f of C into $\mathbb{S}^{n-1}$ has a continuous extension to $X \setminus H$ and $H \cap C = \emptyset$. This is a contradiction since such an extension cannot exist. Finally assume $\mathcal{K}\text{-dim}\, X \leq n - 1$. Let G be a closed $\mathcal{K}$-kernel of X. Then by Lemma 6.4 there exists a closed $\mathcal{K}$-kernel H with $H \cap (C \setminus G) = \emptyset$ and $G \subset H$ and there exists a continuous extension g of f to $X \setminus H$. Then, as in the first case, a contradiction will result.

In the example at the end of Section 3, a summary was made of various functions modulo the class $\mathcal{K}$. A sharpening of some of the inequalities presented there was promised to appear at the end of this section. We have the following examples.

6.12. Examples. A straightforward application of the above theorem yields the first example.

a. For each positive integer n,

$$0 = \operatorname{cmp} \mathbb{R}^n < n = \mathcal{K}\text{-Ind}\, \mathbb{R}^n = \mathcal{K}\text{-Sur}\, \mathbb{R}^n = \mathcal{K}\text{-dim}\, \mathbb{R}^n.$$

b. Suppose $1 \leq n < m$ and let X be the disjoint topological sum of $\mathbb{R}^n$ and $\mathbb{I}^m$. Then

$$0 = \operatorname{cmp} X < n = \mathcal{K}\text{-Ind}\, X = \mathcal{K}\text{-Sur}\, X < m = \operatorname{Ind} X.$$

7. Comparison of $\mathcal{P}$-Ind and $\mathcal{P}$-dim

Agreement. *The universe is contained in $\mathcal{N}_H$.*

The comparison of Ind and dim for normal spaces can be nicely made by means of mappings into spheres. The same comparison will be shown to be true for $\mathcal{P}$-Ind and $\mathcal{P}$-dim under suitable restrictions imposed on the space X. These restrictions occur because in general the function $\mathcal{P}$-Ind lacks certain properties. One can assure that $\mathcal{P}$-Ind will have these properties by imposing conditions that will be discussed in Chapter III; the conditions are related to additive and monotone properties of normal families $\mathcal{P}$ for spaces that are in the Dowker universe $\mathcal{D}$. So for now we shall procede by including these restrictions in the hypotheses of our propositions.

7.1. Proposition. *Let X be a space that satisfies the two conditions:*

(a) *For each closed $\mathcal{P}$-kernel G of X it is true that $X \setminus G$ is a member of the universe and $\mathcal{P}\text{-Ind}\,(X \setminus G) \leq \mathcal{P}\text{-Ind}\, X$.*
(b) *For each closed $\mathcal{P}$-kernel G of X and each $(\mathcal{P} \cap \mathcal{U})$-kernel H of X such that $G \cup H$ is closed it is true that $G \cup H$ is a $\mathcal{P}$-kernel of X.*

Then

$$\mathcal{P}\text{-dim}\, X \leq \mathcal{P}\text{-Ind}\, X.$$

In particular, for every X in $\mathcal{N}_H$

$$\dim X \leq \operatorname{Ind} X.$$

Proof. Suppose n is a natural number and $\mathcal{P}\text{-Ind}\, X \leq n$. Let G be a closed $\mathcal{P}$-kernel of X and C be a closed set. Then by (a) we have $\mathcal{P}\text{-Ind}\,(X \setminus G) \leq \mathcal{P}\text{-Ind}\, X \leq n$ and $X \setminus G$ is in $\mathcal{U}$. For each continuous map f of $C \setminus G$ into $\mathbb{S}^n$ there is by Theorem 4.9 a closed $\mathcal{P}$-kernel H_0 of the subspace $X \setminus G$ such that $H_0 \cap (C \setminus G) = \emptyset$ and f has a continuous extension to $(X \setminus G) \setminus H_0$. Clearly $H = G \cup H_0$ is closed in X. So we have by (b) that H is a closed $\mathcal{P}$-kernel of X. Since $G \subset H$, we have $\mathcal{P}\text{-dim}\, X \leq n$ by Theorem 6.6.

Observe that the condition (a) of the proposition fails for the universe $\mathcal{M}_0$ of separable metrizable spaces and the class $\mathcal{K}$ of compact spaces. Also when the universe is $\mathcal{N}_H$ and the class $\mathcal{P}$ is $\{\emptyset\}$, the inequality $\dim X \leq \operatorname{Ind} X$ will result. (The discussion of the coincidence of dim and Ind will be completed in Section V.3.) With this last observation in mind we make the following definition.

7.2. Definition. Let $\mathcal{P}$ be a class of spaces. For a space X in the universe the *$\mathcal{P}$-surplus of X* is the extended integer

$$\mathcal{P}\text{-sur}\, X = \min \{\dim (X \setminus G) : G \text{ is a } (\mathcal{P} \cap \mathcal{U})\text{-kernel of } X\}.$$

Note that the $\mathcal{P}$-surplus is undefined for spaces not in the universe. See also the additional remarks following Definition 3.2.

The definition has been made so as to yield

$$\mathcal{P}\text{-sur}\, X = (\mathcal{P} \cap \mathcal{U})\text{-sur}\, X$$

for each X in the universe. The next proposition explains the choice of notations in Definition 3.2 and the above definition.

7.3. Proposition. *For every space* X,

$$\mathcal{P}\text{-sur}\, X \leq \mathcal{P}\text{-Sur}\, X.$$

7.4. Proposition. *Let* X *be a space that satisfies the two conditions:*

(a) $\dim Z \leq \dim Y$ *whenever* $Z \subset Y \subset X$.

(b) *For each closed* $\mathcal{P}$*-kernel* G *of* X *and each* $(\mathcal{P} \cap \mathcal{U})$*-kernel* H *of* X *such that* $G \cup H$ *is closed it is true that* $G \cup H$ *is a* $\mathcal{P}$*-kernel of* X.

Then

$$\mathcal{P}\text{-dim}\, X \leq \mathcal{P}\text{-sur}\, X.$$

Proof. Suppose n is a natural number and $\mathcal{P}\text{-sur}\, X \leq n$. Then there is a $(\mathcal{P} \cap \mathcal{U})$-kernel H_0 such that $\dim (X \setminus H_0) \leq n$. Let $\boldsymbol{U} = \{\, U_i : i = 0, 1, \ldots, k \,\}$ be a finite $\mathcal{P}$-border cover of X with enclosure G_0 and let $Y = X \setminus (G_0 \cup H_0)$. Then $\dim Y \leq n$ holds in view of (a). Let Z be $X \setminus G_0 = \bigcup \boldsymbol{U}$ and let $\boldsymbol{U}' = \{\, U_i^! : i = 0, 1, \ldots, k \,\}$ be an open shrinking of $\boldsymbol{U}$ in the space Z such that $\mathrm{cl}_Z(U_i^!) \subset U_i$ for each i. By Propositions 4.3 and 5.5.C, there exists in the subspace Y a closed cover $\boldsymbol{F} = \{\, F_i : i = 0, 1, \ldots, k \,\}$ of Y such that $F_i \subset U_i^!$ for each i and $\mathrm{ord}\, \boldsymbol{F} \leq n + 1$. For each i, $\mathrm{cl}_Z(F_i) \subset \mathrm{cl}_Z(U_i^!)$. As $\boldsymbol{F}' = \{\, \mathrm{cl}_Z(F_i) : i = 0, 1, \ldots, k \,\}$ is finite, $E = \{\, x \in Z : \mathrm{ord}_x \boldsymbol{F}' \geq n + 2 \,\}$ is closed in Z. Also, $E \cap Y = \emptyset$. Let $Y' = \bigcup \boldsymbol{F}' \setminus E$. Then $Y \subset Y'$ holds and Y' is closed in $Z \setminus E$. By Proposition 4.4 there is an open swelling $\boldsymbol{W} = \{\, W_i : i = 0, 1, \ldots, k \,\}$ in $Z \setminus E$ of the collection $\boldsymbol{F}'$ with $\mathrm{ord}\, \boldsymbol{W} \leq n + 1$ and $W_i \subset U_i$ for all i. The set $L = X \setminus \bigcup \boldsymbol{W}$ is contained in $G_0 \cup H_0$. Since the sets $G = L \cap G_0$ and $H = L \cap H_0$ fulfill the requirements of (b), it follows that $\boldsymbol{W}$ is a $\mathcal{P}$-border cover of X. Therefore, $\mathcal{P}\text{-dim}\, X \leq n$.

8. Hulls and deficiency

Agreement. *The universe is contained* $\mathcal{N}_H$.

Early discoveries concerning rim-compact spaces lead to the concept of the deficiency of a space. This concept is associated with the structure of supersets, or hulls, of a space. Hulls and deficiency will be generalized from the class $\mathcal{K}$ to the the class $\mathcal{P}$. By working in the context of the universe $\mathcal{T}$ of all topological spaces, our notion of a hull will depend only on the class $\mathcal{P}$.

8.1. Definition. Let $\mathcal{P}$ be a class of spaces and X be a space in $\mathcal{T}$. A space Y is called a *$\mathcal{P}$-hull of X* if $X \subset Y$ and $Y \in \mathcal{P}$.

The definition of the next functions will be in the context of the universe of discourse.

8.2. Definition. Let $\mathcal{P}$ be a class of spaces. For a space X in the universe the *small $\mathcal{P}$-deficiency of X* is the extended integer

$$\mathcal{P}\text{-def}\, X = \min \{ \dim (G \setminus X) : G \text{ is a } (\mathcal{P} \cap \mathcal{U})\text{-hull of } X \}$$

and the *large $\mathcal{P}$-deficiency of X* is the extended integer

$$\mathcal{P}\text{-Def}\, X = \min \{ \operatorname{Ind} (G \setminus X) : G \text{ is a } (\mathcal{P} \cap \mathcal{U})\text{-hull of } X \}.$$

The functions $\mathcal{P}$-def and $\mathcal{P}$-Def are undefined for spaces that are not in the universe in spite of the fact that the defining formulas may be well defined for spaces not in the universe. But the function values are ∞ when no $(\mathcal{P} \cap \mathcal{U})$-hull of X exists. Obviously

$$\mathcal{P}\text{-def}\, X = (\mathcal{P} \cap \mathcal{U})\text{-def}\, X \quad \text{and} \quad \mathcal{P}\text{-Def}\, X = (\mathcal{P} \cap \mathcal{U})\text{-Def}\, X.$$

Similar to $\mathcal{P}$-sur and $\mathcal{P}$-Sur is the following.

8.3. Proposition. *For every space* X,

$$\mathcal{P}\text{-def}\, X \le \mathcal{P}\text{-Def}\, X.$$

8.4. Proposition. *If* $\mathcal{P}$ *and* $\mathcal{Q}$ *are classes with* $\mathcal{P} \supset \mathcal{Q}$, *then* $\mathcal{P}$-def $\le \mathcal{Q}$-def *and* $\mathcal{P}$-Def $\le \mathcal{Q}$-Def.

The next proposition connects $\mathcal{P}$-dim and $\mathcal{P}$-def. The statement uses the notion of open monotonicity which will be defined first.

8.5. Definition. The universe is said to be *open-monotone* if every open subspace of each space in the universe is also a member of the universe. A class $\mathcal{P}$ of spaces is said to be *open-monotone in* $\mathcal{U}$ if $\mathcal{P}$ satisfies the condition:

If X is a space in $\mathcal{P} \cap \mathcal{U}$, then every open subspace of X is in $\mathcal{P} \cap \mathcal{U}$.

When $\mathcal{P}$ is open-monotone in the the universe $\mathcal{T}$ of all topological spaces, the class $\mathcal{P}$ will be called *absolutely open-monotone.*

8.6. Proposition. *Let $\mathcal{P}$ be a class that is open-monotone in $\mathcal{U}$ and satisfies the condition*

(a) *for every Y in $\mathcal{P}$ the inequality* $\dim Z_1 \leq \dim Z_2$ *holds whenever $Z_1 \subset Z_2 \subset Y$.*

If a space X has the property

(b) *for each closed $\mathcal{P}$-kernel G of X and each $(\mathcal{P} \cap \mathcal{U})$-kernel H of X such that $G \cup H$ is closed it is true that $G \cup H$ is a $\mathcal{P}$-kernel of X,*

then

$$\mathcal{P}\text{-dim}\, X \leq \mathcal{P}\text{-def}\, X.$$

Proof. Assume $\mathcal{P}\text{-def}\, X \leq n$. Let us show that $\mathcal{P}\text{-dim}\, X \leq n$ holds by verifying the condition (c) in Theorem 5.9. Consider a closed $\mathcal{P}$-kernel G of X and pairs (A_i, B_i), $i = 0, 1, \dots, n$, of closed subsets A_i and B_i of X that are disjoint modulo G. For each i there are closed sets A'_i and B'_i with $A_i \subset A'_i$, $B_i \subset B'_i$ and $A'_i \cap B'_i = G$ such that $A'_i \setminus G$ and $B'_i \setminus G$ are neighborhoods of $A_i \setminus G$ and $B_i \setminus G$ respectively in the subspace $X \setminus G$. There exists a $(\mathcal{P} \cap \mathcal{U})$-hull Y of X with $\dim (Y \setminus X) \leq n$. For each i we denote by C_i the subset $\mathrm{cl}_Y(A'_i) \cap \mathrm{cl}_Y(B'_i)$ of Y. Let $Z = Y \setminus \bigcup\{ C_i : i = 0, 1, \dots, n \}$. A simple calculation will yield $X \cap Z = X \setminus G$. Also $Z \setminus X \subset Y \setminus X$ and $Z \cap \mathrm{cl}_Y(A'_i) \cap \mathrm{cl}_Y(B'_i) = \emptyset$ for each i. As $\dim (Z \setminus X) \leq n$ by (a), condition (c) of Theorem 5.9 will produce in the space $Z \setminus X$ partitions S'_i between $(Z \setminus X) \cap \mathrm{cl}_Y(A'_i)$ and $(Z \setminus X) \cap \mathrm{cl}_Y(B'_i)$ such that $\bigcap\{ S'_i : i = 0, 1, \dots, n \} = \emptyset$. Observe that S'_i is a partition between $A_i \setminus G$ and $B_i \setminus G$ in the subspace $(A_i \cup (Z \setminus X) \cup B_i) \setminus G$ for each i. By Lemma I.4.5 there is a partition T'_i between $A_i \setminus G$ and $B_i \setminus G$ in Z with $T'_i \cap (Z \setminus X) = S'_i$. It easily follows that the set $H_0 = \bigcap\{ T'_i : i = 0, 1, \dots, n \}$ is closed in Z and is contained in $X \setminus G$

because $X \cap Z = X \setminus G$. Therefore H_0 is closed in $X \setminus G$. We define $S_i = (X \cap T_i') \cup G$ for every i and $H = G \cup H_0$. Then the sets H and $S_0, S_1, \ldots, S_n$ are closed in X. One easily shows that S_i is a partition between A_i and B_i modulo H for each i. It is clear that $H = \bigcap \{ S_i : i = 0, 1, \ldots, n \}$ holds. Because Z is an open set of Y and Y is a space in $\mathcal{P} \cap \mathcal{U}$, we have that Z is in $\mathcal{P} \cap \mathcal{U}$. So H_0 is also in $\mathcal{P} \cap \mathcal{U}$. We have by (b) that H is in $\mathcal{P} \cap \mathcal{U}$.

There is the following connection between $\mathcal{P}$-Ind and $\mathcal{P}$-Def.

8.7. Proposition. *Let $\mathcal{P}$ be a class that is open-monotone in $\mathcal{U}$ and satisfies the condition*

(a) *for each Y in $\mathcal{P} \cap \mathcal{U}$ the inequality $\operatorname{Ind} Z \leq \operatorname{Ind} Y$ holds for every open set Z of Y.*

Then

$$\mathcal{P}\text{-}\operatorname{Ind} X \leq \mathcal{P}\text{-}\operatorname{Def} X$$

for every space X.

Proof. The proof will be by induction on $n = \mathcal{P}\text{-}\operatorname{Def} X$. We shall provide only the induction step. Assume $\mathcal{P}\text{-}\operatorname{Def} X \leq n$. Let A and B be disjoint closed sets of X and let Y be a $(\mathcal{P} \cap \mathcal{U})$-hull of X with $\operatorname{Ind}(Y \setminus X) \leq n$. We may assume Y satisfies $\operatorname{cl}_Y(A) \cap \operatorname{cl}_Y(B) = \emptyset$. Indeed, from condition (a) the space $Y \setminus \big(\operatorname{cl}_Y(A) \cap \operatorname{cl}_Y(B)\big)$ can be used in place of Y. Theorem 2.21 applied to Ind will give a partition S between $\operatorname{cl}_Y(A)$ and $\operatorname{cl}_Y(B)$ with $\operatorname{Ind}\big(S \cap (Y \setminus X)\big) \leq n - 1$. As $(S \cap X) \in \mathcal{U}$ and $S \in \mathcal{P} \cap \mathcal{U}$ and as $S \setminus (S \cap X) = S \cap (Y \setminus X)$, we have $\mathcal{P}\text{-}\operatorname{Def}(S \cap X) \leq n - 1$. The induction hypothesis gives $\mathcal{P}\text{-}\operatorname{Ind}(S \cap X) \leq n - 1$. Therefore, $\mathcal{P}\text{-}\operatorname{Ind} X \leq n$.

Let us end the section with a few remarks. The dimension functions dim and Ind do not satisfy the conditions (a) of Propositions 8.6 and 8.7 in the universe $\mathcal{N}_H$ of hereditarily normal spaces. But they will for spaces in the Dowker universe which will be defined in Chapter III. And the property (b) of Proposition 8.6 is related to the notion of strongly closed-additive classes defined in Chapter III. Observe that the class $\mathcal{K}$ of compact spaces fails to be open-monotone. But the class $\mathcal{C}$ of complete metric spaces is open-monotone. So the propositions apply to $\mathcal{C}$ but not to $\mathcal{K}$ in the universe $\mathcal{M}_0$ of separable metrizable spaces. Finally we have

$$\mathcal{K}\text{-}\operatorname{Def} \mathbb{R}^2 = \mathcal{K}\text{-}\operatorname{def} \mathbb{R}^2 = 0 < 2 = \mathcal{K}\text{-}\operatorname{Ind} \mathbb{R}^2 = \mathcal{K}\text{-}\operatorname{dim} \mathbb{R}^2.$$

9. Absolute Borel classes in metric spaces

Agreement. *The universe of discourse is the class $\mathcal{M}$ of metrizable spaces.*

The examples up to this point have been the classes $\mathcal{K}$, $\mathcal{C}$ and $\mathcal{S}$ of compact, complete and σ-compact metrizable spaces respectively. This section concerns other classes of this type, namely the absolute Borel classes. These are the next obvious examples and will be used in the next section. The Borel sets in metrizable spaces are well-known in mathematics today, but the Borel classes of sets are not as well known. Since it is the Borel classes of sets that is needed in the next series of examples, a brief discussion of them will be given here.

Let X be a space. The family of *Borel sets of* X is the smallest family $\mathbf{B}$ of subsets of X that satisfies the following conditions.

BF1: Every closed set is a member of $\mathbf{B}$.
BF2: If H is a member of $\mathbf{B}$, then $X \setminus H$ is a member of $\mathbf{B}$.
BF3: If H_n, $n = 0, 1, 2, \ldots$, is a sequence of members of $\mathbf{B}$, then the set $\bigcup\{ H_n : n = 0, 1, 2, \ldots\}$ is a member of $\mathbf{B}$.

Equivalently, the family of Borel sets of X is the smallest family $\mathbf{B}$ of subsets of X that satisfies the following conditions.

BG1: Every open set is a member of $\mathbf{B}$.
BG2: BG2 = BF2.
BG3: BG3 = BF3.

Because of the De Morgan's rules, conditions BF2 and BF3 together are equivalent to condition BF2 and the condition BF4 given below.

BF4: If H_n, $n = 0, 1, 2, \ldots$, is a sequence of members of $\mathbf{B}$, then the set $\bigcap\{ H_n : n = 0, 1, 2, \ldots\}$ is a member of $\mathbf{B}$.

Consequently the family of Borel sets of X is the smallest family $\mathbf{B}$ of subsets of X that satisfies the conditions BF1, BF2 and BF4, or equivalently the conditions BG1, BG2 and BG4 = BF4.

The family of Borel sets can be generated by an inductive transfinite process. To describe this, the notions of even and odd ordinal numbers must be explained. To this end let ω_0 denote the first infinite ordinal number. If α is an infinite ordinal number, then α can be uniquely written as $\alpha = \xi + n$ where ξ is a limit ordinal number and $n < \omega_0$; so every infinite ordinal number is called even or odd as n is even or odd. (This makes limit ordinal numbers even.) Clearly no discussion of even or odd is needed for ordinal numbers n less than ω_0. For every ordinal number α the Borel classes $\mathbf{F}_\alpha$ are defined transfinitely as follows.

(i-F) $\mathbf{F}_0$ is the family of closed subsets of X.

(ij-F) If α is odd, then $\mathbf{F}_\alpha$ is the smallest family of subsets of X that is closed under countable unions and that contains the family $\bigcup\{\, \mathbf{F}_\xi : \xi < \alpha \,\}$.

(iij-F) If α is even, then $\mathbf{F}_\alpha$ is the smallest family of subsets of X that is closed under countable intersections and that contains the family $\bigcup\{\, \mathbf{F}_\xi : \xi < \alpha \,\}$.

The complementary process gives the Borel classes $\mathbf{G}_\alpha$ defined transfinitely for every ordinal number α as follows.

(i-G) $\mathbf{G}_0$ is the family of open subsets of X.

(ij-G) If α is odd, then $\mathbf{G}_\alpha$ is the smallest family of subsets of X that is closed under countable intersections and that contains the family $\bigcup\{\, \mathbf{G}_\xi : \xi < \alpha \,\}$.

(iij-G) If α is even, then $\mathbf{G}_\alpha$ is the smallest family of subsets of X that is closed under countable unions and that contains the family $\bigcup\{\, \mathbf{G}_\xi : \xi < \alpha \,\}$.

Let ω_1 denote the first uncountable ordinal number. Then the following is a basic fact about the family of Borel sets and the classes $\mathbf{F}_\alpha$ and $\mathbf{G}_\alpha$.

9.1. Theorem. *For any space X,*

$$\mathbf{B} = \bigcup\{\, \mathbf{F}_\alpha : \alpha < \omega_1 \,\} \quad \text{and} \quad \mathbf{B} = \bigcup\{\, \mathbf{G}_\alpha : \alpha < \omega_1 \,\}.$$

Moreover,

(a) *$M \in \mathbf{F}_\alpha$ if and only if $X \setminus M \in \mathbf{G}_\alpha$,*

(b) *both $\mathbf{F}_\alpha$ and $\mathbf{G}_\alpha$ are closed under finite unions and under finite intersections,*

(c) *if $Z \subset Y \subset X$, then Z is a member of the family $\mathbf{F}_\alpha$ (or $\mathbf{G}_\alpha$) for Y if and only if there exists a subset Z' of X such that $Z' \cap Y = Z$ and Z' is a member of the family $\mathbf{F}_\alpha$ (or $\mathbf{G}_\alpha$) for X.*

Next the classes $\mathbf{F}_\alpha$ and $\mathbf{G}_\alpha$ will be identified by their properties of closure under countable intersections and closure under countable unions.

9.2. Definition. For $\alpha < \omega_1$, the *multiplicative Borel class* α is the family $\mathbf{F}_\alpha$ or $\mathbf{G}_\alpha$ according as α is even or odd respectively, and the *additive Borel class* α is the family $\mathbf{F}_\alpha$ or $\mathbf{G}_\alpha$ according as α is odd or even respectively.

Note that the multiplicative Borel class α is closed under countable intersections and that the additive Borel class α is closed under countable unions. Clearly for $\alpha < \beta < \omega_1$ we have that both the multiplicative class α and the additive class α are contained in the multiplicative class β and contained in the additive class β.

The next theorem has a topological flavor and the proof found in Engelking [1967] will be given.

9.3. Theorem. *The union of a locally finite collection of members of an additive or a multiplicative Borel class α is a member of the same class.*

Proof. The proof is by transfinite induction on α. The statement is obviously true for the classes $\mathbf{F}_0$ and $\mathbf{G}_0$. Let $\alpha > 0$ and assume that the statement is true when $\xi < \alpha$. Consider a locally finite collection $\{ A_s : s \in S \}$ of members of the additive Borel class α. Then for each s in S we have $A_s = \bigcup\{ A_{s,n} : n = 0, 1, 2, \dots \}$, where $A_{s,n}$ is a member of $\mathbf{F}_{\zeta(s,n)} \cup \mathbf{G}_{\zeta(s,n)}$ for some $\zeta(s,n)$ less than α. For each n and each ξ less than α we consider the collection $\{ A_{s,n} : \zeta(s,n) < \xi \}$. For each fixed n the induction hypothesis gives $\bigcup\{ A_{s,n} : \zeta(s,n) < \xi \}$ is a member of $\mathbf{F}_\xi \cup \mathbf{G}_\xi$. As $\xi < \alpha$, we have $\mathbf{F}_\xi \cup \mathbf{G}_\xi \subset \mathbf{F}_\alpha \cap \mathbf{G}_\alpha$. Also one or the other of $\mathbf{F}_\alpha$ and $\mathbf{G}_\alpha$ is the additive Borel class α. Because $\alpha < \omega_1$, there is a nondecreasing sequence ξ_m, $m = 0, 1, 2, \dots$, of ordinal numbers converging to α. For each pair of natural numbers (m, n) define $B_{m,n}$ as the set $\bigcup\{ A_{s,n} : \zeta(s,n) < \xi_m \}$. Then $\bigcup\{ B_{m,n} : m \in \mathbb{N},\, n \in \mathbb{N} \}$ is a member of the additive Borel class α. Since $\bigcup\{ A_s : s \in S \} = \bigcup\{ B_{m,n} : m \in \mathbb{N},\, n \in \mathbb{N} \}$, the case of the additive Borel class α is proved. Suppose next that $\{ A_s : s \in S \}$ is a locally finite collection of sets from the multiplicative Borel class α where the indexing by S is injective. As X is metrizable, there exists a locally finite open cover $\{ V_t : t \in T \}$ of X such that for each t in T there is a finite subset S_t of S with $V_t \cap \left(\bigcup\{ A_s : s \notin S_t \}\right) = \emptyset$. So $V_t \setminus \bigcup\{ A_s : s \in S \}$ is equal to $V_t \setminus \bigcup\{ A_s : s \in S_t \}$ and the last set is a member of the additive Borel class α. Consequently the set $\bigcup\{ (V_t \setminus \bigcup\{ A_s : s \in S_t \}) : t \in T \}$ is a member of the additive Borel class α. Therefore $\bigcup\{ A_s : s \in S \}$ is a member of the multiplicative Borel class α because $\bigcup\{\,(V_t \setminus \bigcup\{\,A_s : s \in S\,\}) : t \in T\,\}$ is its complement. The transfinite induction is completed.

The following corollary is an easy consequence of the fact that the additive classes are closed under countable unions.

9.4. Corollary. *The union of a σ-locally finite collection of sets in the additive Borel class α is a member of the same class.*

Although (c) of Theorem 9.1 states that the additive Borel class α and the multiplicative Borel class α both have subspace and superspace relationships, it is not strong enough for the purposes of dimension theory modulo a class $\mathcal{P}$. The property defined by membership in $\mathcal{P}$ is topologically invariant and hence is independent of any particular ambient space. Therefore a set A in a Borel class α for a space X is suitable for dimension theory modulo a class $\mathcal{P}$ if and only if for every space Y and every subset B of Y such that A and B are homeomorphic the set B is a member of the same Borel class α for Y. This motivates the following definition.

9.5. Definition. Let $\alpha < \omega_1$. A metrizable space X is said to be of *absolute multiplicative (additive) Borel class* α provided that X is a member of the multiplicative (additive) Borel class α in Y whenever X is a subspace of a metrizable space Y. The respective absolute Borel classes α will be denoted by $\mathsf{M}(\alpha)$ and $\mathsf{A}(\alpha)$. In a similar way one can define the *absolute Borel classes* $\mathbf{F}_\alpha$ and $\mathbf{G}_\alpha$, which will be denoted by $|\mathsf{F}_\alpha|$ and $|\mathsf{G}_\alpha|$ respectively.

There are natural inclusion relationships among the various absolute Borel classes. These inclusions are diagrammed on page 72.

The following is a characterization of $\mathsf{M}(\alpha)$ and $\mathsf{A}(\alpha)$.

9.6. Theorem. (a) *$X \in \mathsf{A}(0)$ if and only if $X = \emptyset$.*

(b) *$X \in \mathsf{M}(0)$ if and only if X is a compact space.*

(c) *$X \in \mathsf{A}(1)$ if and only if X is a σ-locally compact space.*

(d) *For $1 \le \alpha < \omega_1$, $X \in \mathsf{M}(\alpha)$ if and only if X belongs to the multiplicative Borel class α in some complete space Y.*

(e) *For $2 \le \alpha < \omega_1$, $X \in \mathsf{A}(\alpha)$ if and only if X belongs to the additive Borel class α in some complete space Y.*

Proof. Statements (a) and (b) are easily proved. The proof of the statement (c), which is due to Stone [1962], will not be given here because it is rather involved and lengthy. The proofs of (d) and (e) will be given to illustrate the techniques employed in the universe of metrizable spaces, namely the use of the Lavrentieff theorem.

Let us prove the sufficiency parts of (d) and (e). Suppose that X_1 is a member of the Borel class α of a complete space Y_1, that X_2 is a subset of a space Y_2 and that $f\colon X_1 \to X_2$ is a homeomorphism. Since every space has a complete extension, let Y be such an extension for Y_2. By virtue of the theorem of Lavrentieff (Theorem I.7.3) the function f can be extended to a homeomorphism $\widetilde{f}\colon Z_1 \to Z_2$, where Z_1 and Z_2 are G_δ-sets of Y_1 and Y respectively (containing X_1 and X_2). Because X_1 is a member of the Borel class α of Z_1 (Theorem 9.1), the set X_2 is a member of the Borel class α of Z_2. By Theorem 9.1 there is a subset $\widetilde{X}_2$ of Y such that $\widetilde{X}_2 \cap Z_2 = X_2$ and $\widetilde{X}_2$ belongs to the Borel class α of Y. Now observe the equality $\widetilde{X}_2 \cap Z_2 = (\widetilde{X}_2 \cap Z_2) \cap Y_2 = X_2$. With the assumption on α in statements (d) and (e) of the theorem, $\widetilde{X}_2 \cap Z_2$ is a member of the Borel class α in the space Y because G_δ-sets are members of the additive Borel classes α when $\alpha \geq 2$ and are members of the multiplicative Borel classes α when $\alpha \geq 1$. Thus the sufficiency of the conditions have been established. The necessity parts are trivial. So statements (d) and (e) are proved.

9.7. Corollary. *In the universe $\mathcal{M}_0$ of separable metrizable spaces the following hold.*

$$\mathsf{A}(0) = \{\emptyset\}, \quad \mathsf{M}(0) = \mathcal{K} \cap \mathcal{M}, \quad \mathsf{A}(1) = \mathcal{S} \cap \mathcal{M}_0, \quad \mathsf{M}(1) = \mathcal{C}.$$

Proof. Only $\mathsf{A}(1) = \mathcal{S} \cap \mathcal{M}_0$ requires proof. The proof will be left to the reader.

Another easy consequence of the last theorem is the following.

9.8. Corollary. *For $\alpha \geq 2$ the classes $\mathsf{A}(\alpha)$ and $\mathsf{M}(\alpha)$ are F_σ-monotone and G_δ-monotone in $\mathcal{M}$.*

(The meaning of the two monotonicities should be evident.)

The next theorem establishes closure properties under unions and intersections for the absolute additive and multiplicative classes.

9.9. Theorem. *Let $X \subset Y$. Then the following statements are true.*

(a) *If X is the intersection or the union of a finite subcollection $\boldsymbol{H}$ of $\mathsf{A}(\alpha)$ $(\mathsf{M}(\alpha))$ where each member of $\boldsymbol{H}$ is a subset of Y, then X is a member of $\mathsf{A}(\alpha)$ $(\mathsf{M}(\alpha))$.*

(b) *If X is the union of a σ-locally finite subcollection of $\mathsf{A}(\alpha)$, then X is a member of $\mathsf{A}(\alpha)$.*

(c) *For $1 \leq \alpha < \omega_1$, if X is the union of a locally finite subcollection of $\mathsf{M}(\alpha)$, then X is a member of $\mathsf{M}(\alpha)$.*

(d) *If X is the union of a σ-locally finite subcollection of $\mathsf{M}(0)$, then X is a member of $\mathsf{A}(1)$.*

(e) *If X is the intersection of a countable subcollection $\boldsymbol{H}$ of $\mathsf{M}(\alpha)$ where each member of $\boldsymbol{H}$ is a subset of Y, then X is a member of $\mathsf{M}(\alpha)$.*

Proof. The proofs for the smaller values 0 and 1 of α are easily handled separately. The proof for each of the remaining α's uses Theorem 9.6 and the Lavrentieff theorem. Here one uses a complete extension of Y together with Theorem 9.3 and Corollary 9.4.

A property possessed by nonempty Borel subsets of a complete separable metrizable space is that of being the continuous image of the space $\mathbb{P}$ of irrational numbers. This property yields the further property that such Borel subsets which are uncountable must contain a copy of the Cantor set. We shall use the remainder of the section to give a proof of these properties. The space X in the discussion will always be a nonempty *Polish space*, that is, a complete separable metrizable space.

9.10. Theorem. *Suppose X is a Polish space. Then there exists a continuous map f of $\mathbb{P}$ onto X.*

Proof. Assume that complete metrics have been chosen for the spaces X and $\mathbb{P}$. As X is separable, there is a closed cover $\boldsymbol{F}_{\langle 0\rangle} = \{ F_{\langle 0j\rangle} : j \in \mathrm{N} \}$ of X such that $\operatorname{diam} F_{\langle 0j\rangle} \leq 1$ for j in N. For each pair $\langle 0n\rangle$ there is a closed cover $\boldsymbol{F}_{\langle 0n\rangle} = \{ F_{\langle 0nj\rangle} : j \in \mathrm{N} \}$ of $F_{\langle 0n\rangle}$ such that $F_{\langle 0nj\rangle} \subset F_{\langle 0n\rangle}$ and $\operatorname{diam} F_{\langle 0nj\rangle} \leq 2^{-1}$ for j in N. In general, for each finite sequence $\langle 0n_1n_2\ldots n_k\rangle$ there is a closed cover

$$\boldsymbol{F}_{\langle 0n_1n_2\ldots n_k\rangle} = \{ F_{\langle 0n_1n_2\ldots n_k j\rangle} : j \in \mathrm{N} \}$$

of $F_{\langle 0n_1n_2\dots n_k\rangle}$ such that

$$F_{\langle 0n_1n_2\dots n_k j\rangle} \subset F_{\langle 0n_1n_2\dots n_k\rangle}, \qquad j \in \mathbb{N}$$

and

$$\operatorname{diam} F_{\langle 0n_1n_2\dots n_k j\rangle} \leq 2^{-k}, \qquad j \in \mathbb{N}.$$

Similarly we can construct corresponding closed covers

$$\mathbf{G}_{\langle 0n_1n_2\dots n_k\rangle} = \{\, G_{\langle 0n_1n_2\dots n_k j\rangle} : j \in \mathbb{N}\,\}$$

for the space $\mathbb{P}$ with the additional feature that the covers consist of mutually disjoint sets that are simultaneously open and closed in $\mathbb{P}$. Now observe that each point p in $\mathbb{P}$ is associated with a unique sequence $\langle 0n_1n_2\dots n_k\dots\rangle$ such that $\bigcap\{\, G_{\langle 0n_1n_2\dots n_k\rangle} : k \in \mathbb{N}\,\}$ is the singleton set corresponding to p. As the metric on X is complete, the corresponding set $\bigcap\{\, F_{\langle 0n_1n_2\dots n_k\rangle} : k \in \mathbb{N}\,\}$ is a singleton subset of X. In this way we have defined a map $f\colon \mathbb{P} \to X$. Since the metric on $\mathbb{P}$ is complete, the map f is onto X. The map f is obviously continuous at each point p of $\mathbb{P}$ because the sets $G_{\langle 0n_1n_2\dots n_k\rangle}$ are open in $\mathbb{P}$ and are mapped into sets of diameter less than or equal to 2^{-k}.

9.11. Theorem. *Suppose that X is a Polish space and let $f_n\colon \mathbb{P} \to X$, $n = 0,1,2,\dots$, be a sequence of continuous maps. Then the following statements hold.*

(a) *The set $S = \bigcup\{\, f_n[\mathbb{P}] : n = 0,1,2,\dots\}$ is the image of a continuous map f on $\mathbb{P}$.*

(b) *The set $M = \bigcap\{\, f_n[\mathbb{P}] : n = 0,1,2,\dots\}$ is either empty or the image of a continuous map f on $\mathbb{P}$.*

Proof. To prove statement (a) we observe that $\mathbb{P}$ is homeomorphic to the disjoint topological sum of a countably infinite number of copies of $\mathbb{P}$. Define f to be the union of the continuous function f_n.

For the proof of statement (b), assume $M \neq \emptyset$. It is well-known that the infinite product space $\mathbb{P}^{\mathbb{N}}$ is a complete separable metrizable space. For the continuous map $F\colon \mathbb{P}^{\mathbb{N}} \to X^{\mathbb{N}}$ defined by

$$F(p_0, p_1, \dots, p_k, \dots) = \big(f_0(p_0), f_1(p_1), \dots, f_k(p_k), \dots\big),$$

the set

$$Z = \{\, (p_0, p_1, \dots, p_k, \dots) \in \mathbb{P}^{\mathbb{N}} : f_0(p_0) = f_i(p_i),\ i \in \mathbb{N}\,\}$$

is closed because Z is the inverse image under F of the diagonal Δ of $X^{\mathbb{N}}$. As Δ is homeomorphic to X, we have M is homeomorphic to $F[Z]$. And as the subspace Z is complete, there is by Theorem 9.10 a continuous map g of $\mathbb{P}$ onto Z. Define f to be the composition of F and g.

9.12. Corollary. *Let X be a Polish space. If B is a nonempty Borel subset of a space X, then there is a continuous map f of $\mathbb{P}$ onto B.*

Proof. Let $\mathbf{A}$ be the collection of subsets of X consisting of the empty set and the continuous images of $\mathbb{P}$. Clearly every G_δ-set of X is a member of $\mathbf{A}$ (Theorem I.7.1), in particular the closed sets and the open sets are members of $\mathbf{A}$. Since the collection $\mathbf{A}$ is closed under countable unions and under countable intersections, every Borel subset of X is a member of $\mathbf{A}$.

9.13. Theorem. *Every uncountable Borel subset of a Polish space X contains a copy of the Cantor set.*

Proof. Let $f\colon \mathbb{P} \to X$ be a continuous map such that $f[\mathbb{P}]$ is uncountable. Observe that $\{ x \in X : f^{-1}[x] \text{ has nonempty interior} \}$ is countable because $\mathbb{P}$ is separable. Consequently the set

$$D = \{ p \in \mathbb{P} : f^{-1}[f(p)] \text{ is nowhere dense} \}$$

is an uncountable G_δ-set of $\mathbb{P}$. It is now a simple matter to construct a copy C of the Cantor set contained in D such that $f|C$ is injective.

10. Dimension modulo Borel classes

The agreement of the last section will be continued.

Agreement. *The universe is the class $\mathcal{M}$ of metrizable spaces.*

Let us make some easily verified observations. From the agreement we have $\mathsf{A}(\alpha) = \mathsf{A}(\alpha) \cap \mathcal{M}$ and $\mathsf{M}(\alpha) = \mathsf{M}(\alpha) \cap \mathcal{M}$. Both of these classes are closed-monotone, and they are, except for $\mathsf{M}(0)$, also open-monotone. Finally if X is a space that is the union of two members of one of these classes, then X is also a member of the same class.

In order to apply the theorems and propositions of the chapter it will be convenient to state the following theorem.

10.1. Theorem. *Suppose that the class $\mathcal{P}$ satisfies the conditions:*

(a) *If $X = F \cup G$ is such that F is closed and F and G are in $\mathcal{P}$, then X is in $\mathcal{P}$.*
(b) *If $X = \bigcup\{ X_i : i = 0, 1, 2, \ldots\}$ where $\{ X_i : i = 0, 1, 2, \ldots\}$ is a countable locally finite closed collection in X with X_i in $\mathcal{P}$ for all i, then X is in $\mathcal{P}$.*

Then $\mathcal{P}$-Ind is simultaneously closed-monotone and open-monotone and satisfies the conditions:

(c) *If X is the union of two subsets F and G, one of which is closed, then $\max\{ \mathcal{P}\text{-Ind}\, F, \mathcal{P}\text{-Ind}\, G\} \leq \mathcal{P}\text{-Ind}\, X$.*
(d) *If $\{ X_i : i = 0, 1, 2, \ldots\}$ is a countable locally finite closed cover of X, then*

$$\sup\{ \mathcal{P}\text{-Ind}\, X_i : i = 0, 1, 2, \ldots\} = \mathcal{P}\text{-Ind}\, X.$$

The proof of the theorem will be given later in subsection 10.3. In addition to the theorems of this chapter we shall need to refer to several theorems from Chapter V which is devoted to basic dimension functions, that is, dimension functions defined by the existence of special bases for a space.

The first two consequences of Theorem 10.1 are that $\mathsf{A}(\alpha)$-Ind is open-monotone for $0 \leq \alpha < \Omega$ and that $\mathsf{M}(\alpha)$-Ind is open-monotone for $1 \leq \alpha < \Omega$. Therefore, on applying Proposition 7.1 to all absolute Borel classes except for $\mathsf{M}(0)$, we have

$$\begin{aligned} &\mathsf{A}(\alpha)\text{-dim} \leq \mathsf{A}(\alpha)\text{-Ind}, && \alpha \geq 0, \\ &\mathsf{M}(\alpha)\text{-dim} \leq \mathsf{M}(\alpha)\text{-Ind}, && \alpha \geq 1. \end{aligned} \tag{1}$$

Theorems 3.6 and 8.7 give the inequalities

$$\begin{aligned} &\mathsf{A}(\alpha)\text{-Ind} \leq \mathsf{A}(\alpha)\text{-Sur}, && \alpha \geq 0, \\ &\mathsf{M}(\alpha)\text{-Ind} \leq \mathsf{M}(\alpha)\text{-Def}, && \alpha \geq 1. \end{aligned} \tag{2}$$

The proof of the next inequality will make use of a consequence of Tumarkin's theorem (Theorem V.2.12).

$$\begin{aligned} &\mathsf{A}(\alpha)\text{-Sur} \leq \mathsf{A}(\alpha)\text{-Def}, && \alpha \geq 1, \\ &\mathsf{M}(\alpha)\text{-Sur} \leq \mathsf{M}(\alpha)\text{-Def}, && \alpha \geq 2. \end{aligned} \tag{3}$$

Proof. We shall only prove the second inequality since the other inequality is proved in the same way. Suppose that $n = \mathsf{M}(\alpha)\text{-Def}\, X$ is finite and let Y be a $\mathsf{M}(\alpha)$-hull of X with $n = \text{Ind}\,(Y \setminus X)$. By Tumarkin's theorem there is a G_δ-set Z of Y such that $Y \setminus X \subset Z$ and $n = \text{Ind}\, Z$. As $X \setminus Z = Y \setminus Z$ and $\alpha \geq 2$, we have the F_σ-set $X \setminus Z$ is a $\mathsf{M}(\alpha)$-kernel of X by Corollary 9.8. By the subspace theorem for Ind we have $\text{Ind}\,(X \setminus Z) \leq n$. Thus we have shown $\mathsf{M}(\alpha)\text{-Sur}\, X \leq n$.

The next proposition takes advantage of the complementary relationships between the additive and multiplicative Borel classes. Further discussions on complementary relations and ambiguous classes can be found in Section V.2.

10.2. Proposition. *Suppose that Z is in $\mathsf{M}(1)$ and that X and Y are disjoint subsets such that $Z = X \cup Y$. Then, for $\alpha \geq 1$,*

$$\mathsf{A}(\alpha)\text{-Sur}\, X \geq \mathsf{M}(\alpha)\text{-Def}\, Y,$$
$$\mathsf{M}(\alpha)\text{-Sur}\, X \geq \mathsf{A}(\alpha)\text{-Def}\, Y.$$

Proof. Suppose that $n = \mathsf{A}(\alpha)\text{-Sur}\, Y$ is finite and let S be an $\mathsf{A}(\alpha)$-kernel of Y such that $n = \text{Ind}\,(Y \setminus S)$. As $\alpha \geq 1$ and Z is in $\mathsf{M}(1)$, the set $T = Z \setminus S$ is a member of $\mathsf{M}(\alpha)$ that contains X and $n = \text{Ind}\,(T \setminus X)$ holds. From this we infer $\mathsf{M}(\alpha)\text{-Def}\, X \leq n$.

We leave the proof of the second inequality to the reader.

We can now prove the following formulas.

(4) $\mathsf{A}(\alpha)\text{-Sur} = \mathsf{A}(\alpha)\text{-Def}$ and $\mathsf{M}(\alpha)\text{-Sur} = \mathsf{M}(\alpha)\text{-Def}$, $\alpha \geq 2$.

Proof. From (3) and Proposition 10.2 we have

$$\mathsf{A}(\alpha)\text{-Def}\, X \geq \mathsf{A}(\alpha)\text{-Sur}\, X \geq \mathsf{M}(\alpha)\text{-Def}\, Y$$
$$\geq \mathsf{M}(\alpha)\text{-Sur}\, Y \geq \mathsf{A}(\alpha)\text{-Def}\, X$$

when Z is a disjoint union of X and Y and Z is in $\mathsf{M}(1)$. We complete the proof by letting Z be any completion of X for the first equation of (4) and by letting Z be any completion of Y for the second equation.

The reader is referred to Examples I.10.12 and I.10.13 for $\alpha = 1$ in formula (4) above.

Let us now use some more material from Chapter V. The coincidence theorem $\dim X = \operatorname{Ind} X$ for metrizable spaces X (originally proved by Morita in [1954] and in Katětov [1952]) will be established in Theorem V.3.14. This coincidence leads immediately to the following.

$$\tag{5} \begin{array}{lll} \mathsf{A}(\alpha)\text{-sur} = \mathsf{A}(\alpha)\text{-Sur} & \text{and} & \mathsf{A}(\alpha)\text{-def} = \mathsf{A}(\alpha)\text{-Def}, \\ \mathsf{M}(\alpha)\text{-sur} = \mathsf{M}(\alpha)\text{-Sur} & \text{and} & \mathsf{M}(\alpha)\text{-def} = \mathsf{M}(\alpha)\text{-Def}. \end{array}$$

The coincidence theorem for dimension is a special case of the sharpening of the inequalities (1), namely for $\mathsf{A}(0)$. The special role of the universe $\mathcal{M}$ allows us to take advantage of the basic dimension functions discussed in Chapter V. Indeed, Theorems V.2.4, V.2.21 and V.3.12 give the following sharpening of (1) and (2).

$$\tag{6} \begin{array}{ll} \mathsf{A}(\alpha)\text{-dim} = \mathsf{A}(\alpha)\text{-Ind}, & \alpha \geq 0, \\ \mathsf{M}(\alpha)\text{-dim} = \mathsf{M}(\alpha)\text{-Ind}, & \alpha \geq 1. \end{array}$$

$$\tag{7} \begin{array}{ll} \mathsf{A}(\alpha)\text{-Sur} = \mathsf{A}(\alpha)\text{-Ind}, & \alpha \geq 0, \\ \mathsf{M}(\alpha)\text{-Def} = \mathsf{M}(\alpha)\text{-Ind}, & \alpha \geq 1. \end{array}$$

Results in Section IV.2 will yield the following two formulas.

$$\tag{8} \begin{array}{ll} \mathsf{A}(\alpha)\text{-dim} = \mathsf{A}(\alpha)\text{-Dim}, & \alpha \geq 0, \\ \mathsf{M}(\alpha)\text{-dim} = \mathsf{M}(\alpha)\text{-Dim}, & \alpha \geq 1. \end{array}$$

Let us turn to the universe $\mathcal{M}_0$ of separable metrizable spaces. The connections between the small and large inductive dimensions modulo the absolute Borel classes can be established in the same manner as that in Chapter I for the special cases of $\mathcal{C}$ and $\mathcal{S}$. (Also see Corollary V.3.13.) The argument presented there fails only for the absolute Borel class $\mathcal{M}_0 \cap \mathsf{M}(0)$. Theorem 10.1 will be needed in the proof.

$$\tag{9} \begin{array}{ll} \mathsf{A}(\alpha)\text{-ind}\, X = \mathsf{A}(\alpha)\text{-Ind}\, X, & X \in \mathcal{M}_0, \\ \mathsf{M}(\alpha)\text{-ind}\, X = \mathsf{M}(\alpha)\text{-Ind}\, X, & \alpha \geq 1,\ X \in \mathcal{M}_0. \end{array}$$

Returning to arbitrary metrizable spaces, we have that the inclusion relationships between the various absolute Borel classes will result in inequalities for the functions under discussion. For example,

$$\text{(10)} \qquad \mathsf{A}(\alpha)\text{-sur} \le \mathsf{M}(\beta)\text{-sur} \le \mathsf{A}(\gamma)\text{-sur}, \qquad \alpha > \beta > \gamma.$$

Finally we remark that

$$\text{(11)} \qquad \mathsf{A}(\alpha)\text{-ind} \ne \mathsf{A}(\alpha)\text{-Ind} \quad \text{and} \quad \mathsf{M}(\alpha)\text{-ind} \ne \mathsf{M}(\alpha)\text{-Ind}.$$

This will be shown in Example III.1.16.

10.3. Let us now turn to the proof of Theorem 10.1. We have need of a general construction that will be used twice in the course of our proof. Let Y be an open subset of a metrizable space X. Then, since Y is a cozero-set of X, there is a continuous function $f\colon X \to [0,1]$ such that $f^{-1}\big[(0,1]\big] = Y$. For i in $\mathbb{N}$ we define the closed sets

$$H_i = \{\, x : 2^{-(i+1)} \le f(x) \le 2^{-i} \,\}$$

and the open sets

$$U_i = \{\, x : 2^{-(i+2)} < f(x) < 2^{-(i-1)} \,\}.$$

Then $H_i \subset U_i \subset \operatorname{cl}_X(U_i) \subset Y$ holds for every i and the collection $\{\, U_i : i = 0, 1, 2, \ldots \}$ is locally finite in Y.

Proof of Theorem 10.1. The proof is by induction and has the flavor of the normal family argument that is exploited in Chapter III. For each class $\mathcal{P}$ we define a collection $\mathcal{P}'$ by

$$\mathcal{P}' = \{\, X \in \mathcal{U} : \mathcal{P} \text{ contains a partition between each pair of disjoint closed subsets of } X \,\}.$$

With $\mathcal{P}^{(-1)} = \mathcal{P} \cap \mathcal{U}$ we further define $\mathcal{P}^{(n)} = (\mathcal{P}^{(n-1)})'$ for n in $\mathbb{N}$. As $\mathcal{U} = \mathcal{M}$, we have $\mathcal{P}^{(n)} = \{\, X \in \mathcal{M} : \mathcal{P}\text{-Ind}\, X \le n \,\}$ for $n \ge -1$. The theorem will follow easily if it can be shown that $\mathcal{P}'$ satisfies conditions (a) and (b) whenever $\mathcal{P}$ does. Assume $\mathcal{P}$ satisfies the conditions (a) and (b).

We shall first prove $\mathcal{P}$ and $\mathcal{P}'$ are open-monotone (in $\mathcal{U} = \mathcal{M}$ of course). Then we shall address the proof that $\mathcal{P}'$ satisfies conditions (a) and (b). Let us show that $\mathcal{P}$ is open-monotone. Suppose X

is in $\mathcal{P}$ and Y is an open set of X. Then from the construction immediately preceding the proof we have Y is the union of the locally finite collection $\{ H_i : i = 0, 1, 2, \ldots \}$ in Y where each set H_i is closed in X. Since $\mathcal{P}$ is closed-monotone, each H_i is in $\mathcal{P}$. The set Y is in $\mathcal{P}$ by condition (b). We infer from Theorem 2.16 that $\mathcal{P}'$ is closed-monotone. Now we can show $\mathcal{P}'$ is also open-monotone. To this end we suppose X is in $\mathcal{P}'$ and Y is an open set of X. Consider disjoint subsets A and B of Y that are closed in Y. Let H_i and U_i be as in the construction above. Then $A_i = A \cap H_i$ and $B_i = B \cup (X \setminus U_i)$ are disjoint closed sets of X. From the inclusions $\mathrm{cl}_X(U_i) \subset Y \subset X$ we have $\mathrm{cl}_X(U_i) \in \mathcal{P}'$. And from the definition of $\mathcal{P}'$ there is a partition S_i between A_i and B_i in X such that $S_i \cap \mathrm{cl}_X(U_i)$ is in $\mathcal{P}$. Since $S_i \subset U_i$, we have $S_i \in \mathcal{P}$. Let V_i and W_i be disjoint open sets in X with $A_i \subset V_i$ and $B_i \subset W_i$ such that $X \setminus S_i = V_i \cup W_i$. Then the set $V = \bigcup\{ V_i : i = 0, 1, 2, \ldots \}$ is open in Y and $A \subset V$. Also the subset $\bigcup\{ S_i : i = 0, 1, 2, \ldots \}$ of Y is in $\mathcal{P}$ and

$$W = \bigcap\{ W_i \cap Y : i = 0, 1, 2, \ldots \} = Y \setminus \bigcup\{ S_i \cup V_i : i = 0, 1, 2, \ldots \}$$

is an open set of Y containing B because $S_i \cup V_i \subset U_i \setminus B$ for each i and the collection $\{ U_i : i = 0, 1, 2, \ldots \}$ is locally finite in Y. Finally let $S = \bigcup\{ S_i : i = 0, 1, 2, \ldots \} \setminus V$. Then

$$\begin{aligned} Y \setminus S &= Y \setminus \bigcup\{ S_i \setminus V : i = 0, 1, 2, \ldots \} \\ &= \bigcap\{ (V_i \cup (W_i \cap Y) \cup V) : i = 0, 1, 2, \ldots \} \\ &= \left(\bigcap\{ W_i \cap Y : i = 0, 1, 2, \ldots \}\right) \cup V \\ &= W \cup V. \end{aligned}$$

So S is a partition between A and B in Y and S is in $\mathcal{P}$. Thus we have shown $Y \in \mathcal{P}'$ and $\mathcal{P}'$ is open-monotone.

Let us show $\mathcal{P}'$ satisfies the condition (a). Suppose F and G are in $\mathcal{P}'$ and F is closed in $X = F \cup G$. Since $\mathcal{P}'$ is open-monotone, we have $G \setminus F$ is in $\mathcal{P}'$. So we may further assume $F \cap G = \emptyset$. Let A and B be disjoint closed sets. Then there is a partition S_0 between $A \cap F$ and $B \cap F$ in F with $S_0 \in \mathcal{P}$. Clearly S_0 is also a partition between A and B in $Z = F \cup A \cup B$. Let A_0 and B_0 be disjoint open sets of Z such that $Z \setminus S_0 = A_0 \cup B_0$, $A \subset A_0$ and $B \subset B_0$. As Z is closed in X, the sets A_0 and B_0 are also closed in $X \setminus S_0$. Let A_1 and B_1 be disjoint closed neighborhoods of A_0 and B_0 in

$X \setminus S_0$ respectively. In the space G there is a partition S_1 between the disjoint closed sets $A_1 \cap G$ and $B_1 \cap G$ with $S_1 \in \mathcal{P}$. As A_1 and B_1 are disjoint closed neighborhoods in $X \setminus S_0$ of A_0 and B_0, from $F \cap G = \emptyset$ we have $\mathrm{cl}_X(S_1) \subset S_0 \cup S_1$. Hence $S = S_0 \cup S_1$ is a partition between A and B in X. Since $\mathcal{P}$ satisfies condition (a), we have S is in $\mathcal{P}$. Thus we find $\mathcal{P}'$ also satisfies condition (a).

Finally we shall show that $\mathcal{P}'$ satisfies the condition (b). Suppose $\{ X_i : i = 0, 1, 2, \dots \}$ is a locally finite closed cover of X such that each X_i is in $\mathcal{P}'$. From (a), $F_i = \bigcup \{ X_j : j = 0, 1, \dots, i \}$ is in $\mathcal{P}'$ for each i. Moreover, from the open monotonicity of $\mathcal{P}'$ we have $G_{i+1} = F_{i+1} \setminus F_i$ is also in $\mathcal{P}'$. Let A and B be disjoint closed sets. Employing the construction of the preceding paragraph, we can inductively construct subsets S_i of G_i that are in $\mathcal{P}$ and disjoint sets V_i and W_i that are open in the space F_i satisfying $A \cap F_i \subset V_i$, $B \cap F_i \subset W_i$ and $F_i \setminus \bigcup \{ S_j : 0 \leq j \leq i \} = V_i \cup W_i$. The local finiteness of $\{ X_i : i = 0, 1, 2, \dots \}$ yields $V = \bigcup \{ V_i : i = 0, 1, 2, \dots \}$ and $W = \bigcup \{ W_i : i = 0, 1, 2, \dots \}$ are open, $S = \bigcup \{ \mathrm{cl}\,(S_i) : i = 0, 1, \dots \}$ is the set $\mathrm{cl}\,(\bigcup \{ S_i : i = 0, 1, 2, \dots \})$ and $X \setminus S = V \cup W$ holds. Clearly, $A \subset V$ and $B \subset W$. Because $\mathcal{P}$ satisfies condition (a), we have $\mathrm{cl}\,(S_i)$ is in $\mathcal{P}$; and, because $\mathcal{P}$ satisfies condition (b), we have S is in $\mathcal{P}$. We have shown $X \in \mathcal{P}'$. So $\mathcal{P}'$ satisfies condition (b). The theorem is now proved.

Our final task of the section will be to exhibit for the functions $\mathsf{A}(\alpha)$-Ind and $\mathsf{M}(\alpha)$-Ind a space for each possible value. The Example I.7.13 used the existence of totally imperfect sets (= contains no nonempty perfect set) to achieve this for the class $\mathcal{C}$. The discussion of that example contains a proof of the fact that every uncountable complete separable metrizable space X contains a subset Y such that neither Y nor $X \setminus Y$ contains an uncountable F_σ-set of X. Moreover, it is proved there that if $\dim X = n$, then $\dim Y \geq n - 1$ and $\dim (X \setminus Y) \geq n - 1$. These facts will be summarized in the following proposition.

10.4. Proposition. *Let Z be an uncountable complete separable metrizable space. Then Z can be written as the the union of two disjoint subsets X and Y such that both X and Y are totally imperfect. Moreover, if $\dim Z = n$, then $\dim X \geq n - 1$ and $\dim Y \geq n - 1$.*

10.5. Example. Let X_n and Y_n be disjoint totally imperfect subsets of $\mathbb{I}^{n+1}$ such that $\mathbb{I}^{n+1} = X_n \cup Y_n$. We shall prove

$$\mathsf{A}(\alpha)\text{-sur}\, X_n = n \quad \text{and} \quad \mathsf{M}(\alpha)\text{-sur}\, X_n = n, \qquad \alpha \geq 0.$$

As both X_n and Y_n are dense in $\mathbb{I}^{n+1}$, we have by induction on n that $\dim X_n = n$ and $\dim Y_n = n$. Let B be an absolute Borel set contained in X_n. We infer from Theorem 9.13 that B is a countable set. The two disjoint sets $X_n \setminus B$ and $Y_n \cup B$ are totally imperfect sets whose union is $\mathbb{I}^{n+1}$. Therefore we have $\dim(X_n \setminus B) = n$ for every absolute Borel set B contained in X_n. Our assertions are now easily proved.

From formulas (5), (6), (7) and (9) we have

$$\begin{aligned}\mathsf{A}(\alpha)\text{-ind}\, X_n &= \mathsf{A}(\alpha)\text{-Ind}\, X_n \\ &= \mathsf{A}(\alpha)\text{-dim}\, X_n = \mathsf{A}(\alpha)\text{-sur}\, X_n = n, \qquad \alpha \geq 0.\end{aligned}$$

Formulas (3), (4) and (5) yield

$$\begin{gathered}\mathsf{A}(\alpha)\text{-sur}\, X_n = \mathsf{A}(\alpha)\text{-def}\, X_n, \qquad \alpha \geq 2, \\ n = \mathsf{A}(1)\text{-sur}\, X_n \leq \mathsf{A}(1)\text{-def}\, X_n \leq n, \\ \mathsf{A}(0)\text{-def}\, X_n = +\infty.\end{gathered}$$

For the multiplicative absolute Borel classes we find with the help of Proposition 10.2 and Theorem 6.11

$$\begin{aligned}\mathsf{M}(\alpha)\text{-ind}\, X_n &= \mathsf{M}(\alpha)\text{-Ind}\, X_n = \mathsf{M}(\alpha)\text{-dim}\, X_n \\ &= \mathsf{M}(\alpha)\text{-sur}\, X_n = \mathsf{M}(\alpha)\text{-def}\, X_n = n, \qquad \alpha \geq 0.\end{aligned}$$

11. Historical comments and unsolved problems

The Definitions 2.1 and 2.11 are essentially those of Lelek [1964] who introduced them under the name of *inductive invariants.*

The systematic introduction of a universe of discourse into the general development of dimension modulo a class of spaces appears here for the first time. Earlier papers implicitly used a universe. But also implicit in these papers is the assumption that the class $\mathcal{P}$ is contained in the universe. This last assumption does not make the dimensions $\mathcal{P}$-ind, $\mathcal{P}$-Ind and $\mathcal{P}$-dim independent of the universe. The introduction of the universe into the development also has resulted in cleaner statements of the theorems and propositions.

The partition definitions of the inductive dimensions modulo a class $\mathcal{P}$ has been chosen over the more commonly used boundaries of open sets. In proofs that use boundaries of open sets, it is the partitioning property of the boundaries that is used. By using the partition definitions, we find that many proofs become much shorter because they arrive more quickly to the essential part of the argument.

The early papers on surplus and deficiency (and hence kernels and hulls) used implicitly some universe. Indeed, it was always assumed that the kernels and hulls were to be taken from $\mathcal{P} \cap \mathcal{U}$. This has been formally stated in this chapter. The domain of each of the functions $\mathcal{P}$-sur, $\mathcal{P}$-def, etc., is the universe.

The discussion of Section 4 essentially appeared first in Aarts and Nishiura [1972]. The unstable value theorems appear here for the first time.

The Theorems 5.7 and 5.9 characterizing $\mathcal{P}$-dim are new. The concept of the normal universe modulo $\mathcal{P}$ is new and perhaps can be exploited further. The Theorem 6.6 characterizing $\mathcal{P}$-dim by means of mappings into spheres is also new. In this connection, reference is made to Baladze [1982]. For $\mathcal{P}$-dim in the metrizable setting, the reader is directed to the papers Aarts [1972] and Aarts and Nishiura [1973a]. The equalities $\mathcal{K}$-Ind $=$ $\mathcal{K}$-dim $=$ $\mathcal{K}$-Dim $=$ $\mathcal{K}$-Sur in Theorem 6.11 are new. The comparison of $\mathcal{P}$-Ind and $\mathcal{P}$-dim by means of mappings into spheres in Proposition 7.1 appears here for the first time. In this chapter the surpluses $\mathcal{P}$-sur and $\mathcal{P}$-Sur and the deficiencies $\mathcal{P}$-def and $\mathcal{P}$-Def are defined in terms of dim and Ind respectively. The need for these distinctions will become apparent in the next two chapters.

The question of determining a reasonable universe of discourse in which all of the agreements and special technical hypotheses imposed in the various propositions of this chapter are valid will be addressed in the subsequent chapters. Several classes of spaces have been proposed in the literature in response to this question for the functions Ind and dim. A historical discussion of these classes will be given in Chapters III and IV. The Dowker universe $\mathcal{D}$, which includes all of the previously given classes, is defined In Chapter III.

The examples of the absolute Borel classes first appeared in the papers of Aarts [1972] and Aarts and Nishiura [1973a]. For each α the intersection $\mathsf{A}(\alpha) \cap \mathsf{M}(\alpha)$ is called the ambiguous absolute Borel class α. The computations found in the proof of Proposition 10.2 are special cases of those related to the concept of complementary dimension functions discussed in these papers. The reader is referred to Sections V.2 for a discussion of this concept.

The class $\mathcal{C}$ ($= \mathsf{M}(1)$) of complete metrizable spaces can be generalized to a subclass of completely regular spaces by means of the notion of Čech completeness of a space X (= a G_δ-subspace of a Hausdorff compactification). This will permit the generalization of the large inductive completeness degree Icd and the completeness deficiency $\mathcal{C}$-def to the universe of completely regular spaces. It was shown by van Mill in [1982] that there is a completely regular space X such that $\operatorname{Icd} X = 0$ and $\mathcal{C}\text{-def}\, X = \infty$. Thus it is not possible to generalize the completeness degree results in the universe $\mathcal{M}$ of metrizable spaces to the larger universe $\mathcal{R}_c$ of completely regular spaces.

Unsolved problems

1. What are the values of the gaps that are possible for the inequalities (10) of Section 10 and the corresponding inequalities of the other functions that have been defined in this chapter?

2. Extend in a meaningful way the results of Section 10 to a universe that properly contains the universe $\mathcal{M}$. In particular, find a universe strictly between the class $\mathcal{M}$ and the class $\mathcal{R}_c$ of completely regular spaces in which the pathology found by van Mill in [1982] fails to exist.

CHAPTER III

FUNCTIONS OF INDUCTIVE DIMENSIONAL TYPE

Introduced in Chapter II were various dimension functions modulo a class $\mathcal{P}$. Just as in dimension theory, there were two distinct types; one used an inductive approach and the other used a covering approach. The discussion of Chapter I indicated that these two approaches gave rise to a dichotomy for certain classes $\mathcal{P}$ in the universe $\mathcal{M}_0$ of separable metrizable spaces. It was found that the combination of kernels, surplus and the inductive approach was a natural one for the class $\mathcal{S}$ of σ-compact spaces and that the combination of hulls, deficiency and the covering approach was a natural one for the class $\mathcal{C}$ of complete spaces. This dichotomy reflected the fact that, on the one hand, the class $\mathcal{S}$ has closure under countable unions of closed sets and the class $\mathcal{C}$ does not and, on the other hand, the class $\mathcal{C}$ has closure under countable intersections of open sets and the class $\mathcal{S}$ does not. To distinguish the two approaches, the terminology *functions of inductive dimensional type* and *functions of covering dimensional type* will be used.

In this chapter the functions of inductive dimensional type will be studied. These functions include $\mathcal{P}$-ind, $\mathcal{P}$-Ind, $\mathcal{P}$-Sur and $\mathcal{P}$-Def. The exposition will begin with addition theorems. The results, in particular that of point addition, will be used to study connections between $\mathcal{P}$-ind and $\mathcal{P}$-Ind. The investigation will move next to the normal families of dimension theory. The functions $\mathcal{P}$-ind and $\mathcal{P}$-Ind lead naturally to normal families. Normal families were introduced by Hurewicz in [1927] and by Morita in [1954] in the universes of separable and general metrizable spaces respectively. A general investigation of normal and related families will be carried out in Sections 2 and 3. These investigations are applied in a manner similar to that of Dowker [1953] to create what will be called the Dowker universe. The chapter will end with axiomatics. Menger proposed in [1929] an axiom scheme for the dimension function. His axioms were strongly influenced by the work of Hurewicz [1927] on normal families. A discussion of Menger-type axioms and other axioms will be given in Section 5.

Agreement. *The universe of discourse is closed-monotone and is contained in $\mathcal{N}_H$. Every class has the empty space as a member and is closed-monotone in $\mathcal{U}$. Unless indicated otherwise, X will always be a space in the universe.*

1. Additivity

Generally speaking, we shall reserve the notion of additivity for discussions concerning finite unions of spaces. Obviously, addition theorems involve relationships between spaces situated in an ambient space. It is important that the ambient spaces as well as the spaces be in the universe of discourse. (This arrangement is similar to that of surplus and deficiency found in Chapter II.) We shall now define four types of additivity for classes $\mathcal{P}$.

1.1. Definition. A class $\mathcal{P}$ is said to be *additive in $\mathcal{U}$* if each space X in the universe is a member of $\mathcal{P}$ whenever

(i) $X = Y \cup Z$,
(ij) Y and Z are in $\mathcal{P} \cap \mathcal{U}$.

When the additional condition

(iij-c) Y and Z are closed

is imposed, the class $\mathcal{P}$ is said to be *closed-additive in $\mathcal{U}$*.
When the additional condition is

(iij-s) Y is closed,

the class $\mathcal{P}$ is said to be *strongly closed-additive in $\mathcal{U}$*.
And finally when the additional condition is

(iij-p) Y is a singleton,

the class $\mathcal{P}$ is said to be *point-additive in $\mathcal{U}$*.

The definitions of the various additivities have emphasized the membership of the spaces in the universe. In a sense, this emphasis is redundant. We have included this redundancy because comparisons between results for the universe and results for subcollections of the universe will be made.

In the universe $\mathcal{M}_0$ we have already seen three main addition theorems in Chapter I. The first of these is the inequality:

A. *For X and Y in $\mathcal{M}_0$,* $\operatorname{ind}(X \cup Y) \leq \operatorname{ind} X + \operatorname{ind} Y + 1$.

The second and third theorems concern closed-additive and strongly closed-additive classes. Both are consequences of the countable sum theorem.

B. *For each n, the class $\{X : \operatorname{ind} X \leq n\}$ is closed-additive in the universe $\mathcal{M}_0$.*

C. *For each n, the class $\{X : \operatorname{ind} X \leq n\}$ is strongly closed-additive in the universe $\mathcal{M}_0$.*

Of course, the strongly closed-additive statement **C** implies the statement **B**.

With respect to point additivity, recall that there exists a space X in $\mathcal{M}_0$ with $\operatorname{cmp} X = 1$ such that $\operatorname{cmp}(X \setminus \{p\}) = 0$ for some point p in X. Thus we have that the class $\mathcal{K}' = \{X : \operatorname{cmp} X \leq 0\}$ fails to be point-additive in the universe $\mathcal{M}_0$. Obviously, from **C** above we know that the small inductive dimension ind cannot be raised by the addition of a point to spaces X in the universe $\mathcal{M}_0$. In other words, the class $\{X : \operatorname{ind} X \leq 0\}$ is point-additive in the universe $\mathcal{M}_0$. The next example will show that this class is not point-additive in the universe $\mathcal{M}$ of all metrizable spaces.

1.2. Example. There exists a space X in $\mathcal{M}$ such that $\operatorname{ind} X = 1$ and $\operatorname{ind}(X \setminus \{p\}) = 0$ for some point p in X. To see this, let Δ be the famous example of Roy [1962] which has the properties that Δ is a complete metric space, $\operatorname{ind} \Delta = 0$ and $\operatorname{Ind} \Delta = 1$ (see Section I.4 for other references). There are disjoint closed sets F and G of Δ such that each partition S between F and G in Δ is nonempty. The set X is formed from Δ by identifying the set F to be the point p. With ρ as a metric on Δ for which the sets F and G have a distance 1 between them, we shall define a metric d on X by means of a formula. In the formula, $\rho(x, F)$ will denote the usual distance between the point x and the set F.

$$d(x,z) = \begin{cases} \min\{\rho(x,F) + \rho(z,F), \rho(x,z)\}, & \text{for } x \neq p,\ z \neq p \\ \rho(x,F), & \text{for } x \neq p,\ z = p \\ \rho(z,F), & \text{for } x = p,\ z \neq p \\ 0, & \text{for } x = p,\ z = p. \end{cases}$$

A straightforward calculation will yield that d is a metric for X. Moreover, when $x \neq p$ and $z \neq p$, the inequality $d(x,z) \leq \rho(x,z)$

holds. To see that $X \setminus \{p\}$ and $\Delta \setminus F$ have the same topology we let x and ε be such that $x \neq p$ and $0 < 2\varepsilon \leq \rho(x, F)$. If $\rho(x, z) < \varepsilon$, then $\rho(z, F) \geq \rho(x, F) - \rho(x, z) \geq 2\varepsilon - \rho(x, z) \geq \varepsilon > \rho(x, z)$. Thus the definition of d gives $d(x, z) = \rho(x, z)$ whenever $\rho(x, z) < \varepsilon$. Consequently, the open balls $\{ z \in \Delta : \rho(x, z) < \varepsilon \}$ and $\{ z \in X : d(x, z) < \varepsilon \}$ coincide when $2\varepsilon \leq \rho(x, F)$. Clearly the space X has the required properties.

Let us return to inequalities like those in **A** above. We shall begin with $\mathcal{P}$-ind. The following proposition, which is an immediate consequence of the definition, will be needed.

1.3. Proposition. *Let q be an isolated point of a space X. Then*

$$\mathcal{P}\text{-ind}\, X = \mathcal{P}\text{-ind}\, (X \setminus \{ q \})$$

whenever $\mathcal{P}$-ind $(X \setminus \{ q \}) \geq 0$.

The proposition with $\mathcal{P} = \{ \emptyset \}$ is used in the next theorem.

1.4. Theorem. *Let A and B be subspaces of X such that B is closed and $X = A \cup B$. Then*

$$\mathcal{P}\text{-ind}\, X \leq \mathcal{P}\text{-ind}\, A + \text{ind}\, B + 1.$$

It is to be observed that the statement of the theorem is in the context of the universe of discourse. Consequently the spaces A and B are in the universe; arbitrary subsets of X need not be in the universe because the universe is only closed-monotone by agreement.

Proof. The proof is by induction on ind B. Since the inequality is obvious for ind $B = -1$, we need prove only the inductive step. Suppose that the inequality holds when ind $B < n$ and assume that B is such that ind $B = n$. Let x be a point of X and F be a closed set of X with $x \notin F$. Since B is closed in X, we have $B \cup \{ x \}$ is closed and ind $(B \cup \{ x \}) =$ ind B. Therefore no generality is lost in assuming x is a member of B. Then by Theorem II.2.20 there is a partition S between x and F in X with ind $(B \cap S) \leq n - 1$. As $\mathcal{P}$-ind $(A \cap S) \leq \mathcal{P}$-ind A, we have $\mathcal{P}$-ind $S \leq \mathcal{P}$-ind $A + n$ by the induction hypothesis. Thereby the inequality holds when ind $B = n$.

Even though point addition theorems fail for $\mathcal{K}$-ind ($=$ cmp) in the universe $\mathcal{M}_0$ and for ind in the universe $\mathcal{M}$, the last theorem does provide an upper bound on how badly it may fail. That is, the difference becomes at most 1 by letting B be a singleton subset of X.

The large inductive dimension analogue of the above theorem permits the set B to be other than closed in X.

1.5. Theorem. *Suppose $X = A \cup B$. Then*

$$\mathcal{P}\text{-Ind}\, X \leq \mathcal{P}\text{-Ind}\, A + \text{Ind}\, B + 1.$$

Proof. The proof is by induction on $\text{Ind}\, B = n$. We shall indicate only the inductive step. Let F and G be disjoint closed sets of X. Then by Theorem II.2.21 there is a partition S between F and G in X with $\text{Ind}\,(B \cap S) \leq n - 1$. The induction hypothesis yields $\mathcal{P}\text{-Ind}\, S \leq \mathcal{P}\text{-Ind}\,(A \cap S) + n$. The proof is now easily completed.

Concerning the replacement of ind and Ind with $\mathcal{P}$-ind and $\mathcal{P}$-Ind in the above two theorems, we have the following results. The additivity conditions defined in the section will be used.

1.6. Theorem. *Suppose that $\mathcal{P}$ is closed-additive in $\mathcal{U}$. Let A and B be closed sets of a space X with $X = A \cup B$. Then*

$$\mathcal{P}\text{-ind}\, X \leq \mathcal{P}\text{-ind}\, A + \mathcal{P}\text{-ind}\, B + 1.$$

Proof. Since both A and B are closed, we need to check only the points of $A \cap B$. An induction on $n = \mathcal{P}\text{-ind}\, A + \mathcal{P}\text{-ind}\, B + 1$ will be made. The case $n = -1$ is obvious. For the inductive step, observe that $n \geq 0$ implies $\mathcal{P}\text{-ind}\, A \geq 0$ or $\mathcal{P}\text{-ind}\, B \geq 0$. Suppose $\mathcal{P}\text{-ind}\, A \geq \mathcal{P}\text{-ind}\, B$ and let $x \in A \cap B$. Then for any closed set F of X not containing x there is by Proposition II.2.20 a partition S between x and F in X such that $\mathcal{P}\text{-ind}\,(S \cap A) \leq \mathcal{P}\text{-ind}\, A - 1$. Applying the induction hypothesis to the sets S, $S \cap A$ and $S \cap B$, we have $\mathcal{P}\text{-ind}\, S \leq n - 1$. That is, $\mathcal{P}\text{-ind}\, X \leq n$.

1.7. Theorem. *Suppose that $\mathcal{P}$ is additive in $\mathcal{U}$. Let A and B be subspaces of X such that $X = A \cup B$. Then*

$$\mathcal{P}\text{-Ind}\, X \leq \mathcal{P}\text{-Ind}\, A + \mathcal{P}\text{-Ind}\, B + 1.$$

Proof. We shall induct on the sum $n = \mathcal{P}\text{-Ind}\, A + \mathcal{P}\text{-Ind}\, B + 1$. As the case $n = -1$ is obvious, we shall pass to the inductive step.

Let F and G be disjoint closed sets of X and suppose that $n \geq 0$. Then we have $\mathcal{P}$-Ind $A \geq 0$ or $\mathcal{P}$-Ind $B \geq 0$. The inductive step is completed in a manner similar to that of the proof of the previous theorem. Instead of using Proposition II.2.20, one uses Theorem II.2.21.

The following analogue for strongly closed-additive classes is established in a manner similar to the last theorem.

1.8. Theorem. *Suppose that $\mathcal{P}$ is strongly closed-additive in $\mathcal{U}$. Let A and B be subspaces of X such that A or B is closed and $X = A \cup B$. Then*

$$\mathcal{P}\text{-Ind}\, X \leq \mathcal{P}\text{-Ind}\, A + \mathcal{P}\text{-Ind}\, B + 1.$$

With $\mathcal{U} = \mathcal{N}_H$ and $\mathcal{P} = \{\emptyset\}$, Theorems 1.5 and 1.7 both yield the second inequality of the following important theorem of dimension theory. The first one is proved by observing that $\operatorname{ind} X \neq -1 \neq \operatorname{ind} Y$ may be assumed.

1.9. Theorem (Addition theorem). *For any hereditarily normal space $X \cup Y$,*

$$\operatorname{ind}(X \cup Y) \leq \operatorname{ind} X + \operatorname{ind} Y + 1,$$
$$\operatorname{Ind}(X \cup Y) \leq \operatorname{Ind} X + \operatorname{Ind} Y + 1.$$

We shall conclude the section with a discussion on the interrelationships of the finite sum theorem, the addition theorem and the point addition theorem for $\mathcal{P}$-ind in the universe $\mathcal{M}$ of metrizable spaces.

Agreement. *In the remainder of the section the universe is the class $\mathcal{M}$ of metrizable spaces.*

It is already known that the finite sum theorem, the addition theorem and the point addition theorems do not have analogues for the class $\mathcal{K}$ in the universe $\mathcal{M}_0$. For the reader's convenience we shall list the three inequalities corresponding to these theorems. (A = addition; P = point addition; S = finite sum.)

Theorem A. $\mathcal{P}\text{-ind}\,(X \cup Y) \leq \mathcal{P}\text{-ind}\, X + \mathcal{P}\text{-ind}\, Y + 1$.

Theorem P. $\mathcal{P}\text{-ind}\,(X \cup \{p\}) \leq \mathcal{P}\text{-ind}\, X$ *when* $X \notin \mathcal{P}$.

Theorem S. $\mathcal{P}\text{-ind}\,(X \cup Y) \leq \max\{\mathcal{P}\text{-ind}\, X, \mathcal{P}\text{-ind}\, Y\}$ *whenever X and Y are closed in $X \cup Y$.*

The idea of the proof of the next proposition stems from Example I.5.10.f.

1.10. Proposition. *Theorem S implies Theorem P.*

Proof. Assume $X \notin \mathcal{P}$ and let ρ be a metric for $X \cup \{p\}$. Then the set $X \cup \{p\}$ is the union of the two closed sets

$$F = \{p\} \cup \{x \in X : 2^k \leq \rho(p,x) \leq 2^{k+1} \text{ for some odd integer } k\},$$

$$G = \{p\} \cup \{x \in X : 2^k \leq \rho(p,x) \leq 2^{k+1} \text{ for some even integer } k\}.$$

Since $\mathcal{P}\text{-ind}\, X \geq 0$, from the definition of $\mathcal{P}$-ind one easily sees that $\mathcal{P}\text{-ind}\, F$ and $\mathcal{P}\text{-ind}\, G$ are both less than or equal to $\mathcal{P}\text{-ind}\, X$. Theorem S now completes the proof.

1.11. Proposition. *If $\mathcal{P} \cap \mathcal{M} \neq \{\emptyset\}$, then Theorem A implies Theorem P.*

Proof. The proposition follows easily because $\{p\} \in \mathcal{P}$ holds as a result of the closed-monotonicity in $\mathcal{M}$ of the class $\mathcal{P}$ and the existence of a nonempty space in the class $\mathcal{P} \cap \mathcal{M}$.

1.12. Proposition. *If $\mathcal{P}$ is an additive class, then Theorem P implies Theorem A.*

(Recall that the present discussion has assumed $\mathcal{M}$ as the universe. Consequently, additive refers to this universe.)

Proof. The proof is similar to that of Theorem 1.6. Again we shall induct on $n = \mathcal{P}\text{-ind}\, X + \mathcal{P}\text{-ind}\, Y + 1$. The case $n = -1$ is the statement that $\mathcal{P}$ is additive. For the inductive step assume $n \geq 0$ and let $Z = X \cup Y$. Consider a point p and a closed set F in Z with $p \notin F$. Without loss of generality we may assume $\mathcal{P}\text{-ind}\, X \geq 0$. Then by Theorem P we have $\mathcal{P}\text{-ind}\,(X \cup \{p\}) \leq \mathcal{P}\text{-ind}\, X$. There is a partition S between p and F with $\mathcal{P}\text{-ind}\,(S \cap X) < \mathcal{P}\text{-ind}\,(X \cup \{p\})$ by Proposition II.2.20. Consequently, as in the proof of Theorem 1.6, we have $\mathcal{P}\text{-ind}\, S \leq n - 1$. Thereby we have $\mathcal{P}\text{-ind}\,(X \cup Y) \leq n$ and the induction is completed.

In the next proposition we shall use the point addition theorem for $\mathcal{P}$-Ind. Its proof is a simple modification of the one for Ind in Theorem I.4.12 and is left to the reader.

1.13. Proposition. *$\mathcal{P}$-ind $=$ $\mathcal{P}$-Ind if and only if Theorem P holds.*

Proof. Assume $\mathcal{P}$-ind $=$ $\mathcal{P}$-Ind. Then Theorem P will follow from the point addition theorem for $\mathcal{P}$-Ind (Theorem I.4.12).

Conversely, assume that Theorem P holds. We already know that $\mathcal{P}$-ind $\leq$ $\mathcal{P}$-Ind holds. The reverse inequality will be proved by induction. Assume that the reverse inequality holds for spaces X with $\mathcal{P}\text{-ind}\, X < n$. Let X be such that $\mathcal{P}\text{-ind}\, X = n$ and consider disjoint closed sets F and G. We may assume that both F and G are nonempty for the sets can be partitioned by $S = \emptyset$ in the contrary case. If $\mathcal{P}\text{-ind}\,(X \setminus F) = -1$, then there is partition S between F and G in X such that $S \in \mathcal{P}$. So assume $\mathcal{P}\text{-ind}\,(X \setminus F) \geq 0$. Let $Y = X \setminus F$ and identify the closed set F to be p. By Theorem II.2.17 we have $\mathcal{P}\text{-ind}\, Y \leq n$. Then, using the metric d defined in Example 1.2, we have that $Y \cup \{p\}$ is a metric space such that Y is homeomorphic to $X \setminus F$. By Theorem P, $\mathcal{P}\text{-ind}\,(Y \cup \{p\}) \leq n$. There exists a partition S between p and G in $Y \cup \{p\}$ such that $\mathcal{P}\text{-ind}\, S \leq n - 1$. The induction hypothesis gives $\mathcal{P}\text{-Ind}\, S \leq n - 1$. Since $S \subset Y$, we have that S is also a partition between F and G in X. Thus we have proved $\mathcal{P}\text{-Ind}\, X \leq n$ and the induction is now completed.

At the beginning of our discussion of Theorems A, P and S we observed that ind and cmp failed to satisfy Theorem P. That is, the absolute Borel classes $\{\emptyset\}$ and $\mathcal{K}$ are examples of classes $\mathcal{P}$ for which Proposition 1.13 applies. We shall show next that the other absolute Borel classes are also examples. To this end, we shall construct a subspace of Roy's space Δ for which the point addition inequality fails for each absolute Borel class α. This will be achieved by showing that $\mathcal{P}$-ind $\neq$ $\mathcal{P}$-Ind whenever $\mathcal{P}$ is an absolute Borel class. A preliminary discussion of the space Δ will be needed. The space Δ is complete and dense in itself and has cardinality $\mathfrak{c} = 2^{\aleph_0}$. So it has exactly $\mathfrak{c}$ countable subsets and hence exactly $\mathfrak{c}$ separable closed subspaces. Using the classical Bernstein construction referred to earlier in Example I.7.13, we can decompose Δ into two disjoint subsets A and B so that every Cantor set contained in Δ intersects both A and B.

1.14. Proposition. *Let A and B be disjoint subsets of Δ such that $\Delta = A \cup B$ and such that every Cantor set contained in Δ inter-*

sects both A and B. Let E be a Borel subset of Δ that contains B. Then the sets $\widetilde{A} = E \cap A$ and $\widetilde{B} = \Delta \setminus \widetilde{A}$ also have the property that every Cantor set contained in Δ intersects both $\widetilde{A}$ and $\widetilde{B}$.

Proof. Let K be a Cantor set contained in Δ and E be a Borel set that contains B. Then $K \setminus E$ is an absolute Borel space in $\mathcal{M}_0$. Consequently, as in the discussion of Example II.10.5, the set $K \setminus E$ must be countable since it is contained in A. So the uncountable absolute Borel space $K \cap E$ in $\mathcal{M}_0$ contains a Cantor set L. As $L \cap A \neq \emptyset$, we have $K \cap \widetilde{A} = K \cap E \cap A \supset L \cap A \neq \emptyset$. And as $L \cap B \neq \emptyset$, we also have $K \cap \widetilde{B} \supset K \cap B \neq \emptyset$.

The space Δ is complete. So if the Borel subset E in the above proposition is a set of Borel class α in Δ, then it is also a space of absolute Borel class α when $\alpha \geq 2$. And if E is a G_δ-set in Δ, then it is in $\mathsf{M}(1)$.

1.15. Proposition. *Let A and B be disjoint subsets of Δ such that $\Delta = A \cup B$ and such that every Cantor set contained in Δ intersects both A and B. Then*

$$\operatorname{Ind} A = 1 \quad \textit{and} \quad \operatorname{Ind} B = 1.$$

Proof. By way of a contradiction we shall show $\operatorname{Ind} A \geq 1$. Assume $\operatorname{Ind} A = 0$. Let F and G be disjoint closed subsets of Δ such that every partition between F and G is nonempty. By Theorem II.2.21 there exists a partition S between F and G such that $S \cap A = \emptyset$. Since S is closed and $S \subset B$, we have that S contains no Cantor sets. Observe that every dense in itself complete metric space contains a Cantor set. So no closed subset of S is dense in itself. That is, S is a scattered set.

Let us show that the closed set S (indeed, any scattered metrizable space) is σ-discrete. To this end, let d be a metric on the space Δ. For each ordinal number α we shall denote by D_α the derived set of order α of the set S. As S is scattered, there exists for each point x in S an ordinal number α_x such that $x \in D_{\alpha_x}$ and $x \notin D_{\alpha_x+1}$. Let X_n be the set $\{\, x \in S : d(x, D_{\alpha_x+1}) > \frac{1}{n+1} \,\}$ for n in $\mathbb{N}$. Clearly we have $S = \bigcup\{\, X_n : n \in \mathbb{N} \,\}$. It remains to be shown that each set X_n is σ-discrete. Let Y_n be a maximal subset of X_n such that no two points of Y_n have distance smaller than $\frac{1}{2(n+1)}$. If $d(y,x) < \frac{1}{n+1}$ for a point y in Y_n and a point x in X_n, then from the definition of the set X_n we have $\alpha_y = \alpha_x$. Consequently the sets $X_{n\alpha} = \{ x \in X_n : \alpha_x = \alpha \}$, where α is an ordinal number, form a collection of open subsets of the subspace X_n such that the distance between two distinct sets $X_{n\alpha}$ and $X_{n\alpha'}$ exceeds $\frac{1}{2(n+1)}$. Since the set $X_{n\alpha}$ is a subset of $D_\alpha \setminus D_{\alpha+1}$, the set $X_{n\alpha}$ is σ-discrete. It follows that each set X_n is σ-discrete. By reindexing the sets we have a countable collection $\{ S_i : i < \omega_0 \}$ of discrete subsets of Δ such that $S = \bigcup\{ S_i : i < \omega_0 \}$. We may also assume that the S_i's are disjoint.

We shall construct a neighborhood W of S that is open-and-closed and such that $W \cap (F \cup G) = \emptyset$. We may assume that the metric for Δ is complete. Using the facts that Δ is collectionwise normal and $\operatorname{ind} \Delta = 0$, we can find for each i less than ω_0 a discrete collection $\{ W_s : s \in S_i \}$ of open-and-closed sets such that

(1) $s \in W_s$ and $W_s \cap (F \cup G) = \emptyset$ for each s in S_i,
(2) diameter $W_s < 2^{-i}$ for each s in S_i.

Let $W = \bigcup\{ W_s : s \in S_i,\ i < \omega_0 \}$. Then the set W is open and disjoint from $F \cup G$. To show that W is closed, let x be in $\operatorname{cl}(W)$ and let $\{ x_k : k < \omega_0 \}$ be a sequence in W converging to x. Pick an point s_k in S with $x_k \in W_{s_k}$ for each k. On the one hand, if the set $\{ s_k : k < \omega_0 \} \cap S_i$ is infinite for some i, then x is a member of $\operatorname{cl}(\bigcup\{ W_s : s \in S_i \})$ and thereby, from $\operatorname{cl}(\bigcup\{ W_s : s \in S_i \}) = \bigcup\{ W_s : s \in S_i \} \subset W$, is a member of W. And on the other hand, if the sets $\{ s_k : k < \omega_0 \} \cap S_i$ are finite for every i, then it follows that $\lim_{k\to\infty} d(x_k, s_k) = 0$ and consequently $\lim_{k\to\infty} s_k = x$ because $\lim_{k\to\infty} x_k = x$. Since S is closed, we have $x \in S \subset W$. Thus W is closed.

Let U and V be disjoint open sets in Δ with $F \subset U$ and $G \subset V$ such that $S = \Delta \setminus (U \cup V)$. Then $U' = U \cup W$ and $V' = V \setminus W$ are disjoint open sets of Δ such that $\Delta = U' \cup V'$ with $F \subset U'$ and $G \subset V'$. This is contrary to the choice of F and G. So $\operatorname{Ind} A = 1$. The proposition easily follows.

It was remarked in Section II.10 that $\mathcal{P}$-ind $\neq$ $\mathcal{P}$-Ind held when $\mathcal{P}$ was an absolute Borel class α in the universe $\mathcal{M}$. The next example shows this to be the case.

1.16. Example. For every countable ordinal number α there exists a space X in $\mathcal{M}$ such that

$$\mathsf{A}(\alpha)\text{-ind}\, X < \mathsf{A}(\alpha)\text{-Ind}\, X \quad \text{and} \quad \mathsf{M}(\alpha)\text{-ind}\, X < \mathsf{M}(\alpha)\text{-Ind}\, X.$$

Such spaces have been already exhibited for $\alpha = 0$. We shall construct the others by employing decompostions of Roy's space Δ that were discussed above. The space Δ can be decomposed into disjoint subsets A and B with the property that each of these sets intersects every Cantor set contained in Δ. Let X be the subspace A. Suppose that G is an $\mathsf{A}(\alpha)$-kernel of X. Then, with $\widetilde{A}$ being the set in Proposition 1.14 that corresponds to $E = \Delta \setminus G$, we have $X \setminus G = \widetilde{A}$ and hence $\operatorname{Ind}(X \setminus G) = 1$. Therefore, $\mathsf{A}(\alpha)\text{-Sur}\, X = 1$. From formula (7) of Section II.10 we have $\mathsf{A}(\alpha)\text{-Ind}\, X = 1$. For $\mathsf{M}(\alpha)$, $\alpha \geq 1$, let $X = B$. We infer from Lemma I.7.5 that $\mathsf{M}(\alpha)\text{-Def}\, X = 1$. Formula (7) of Section II.10 yields $\mathsf{M}(\alpha)\text{-Ind}\, X = 1$. Finally we have $\mathsf{A}(\alpha)\text{-ind}\, X = 0$ and $\mathsf{M}(\alpha)\text{-ind}\, X = 0$ from Proposition II.2.3.

In view of the above example, it now follows from Proposition 1.13 that when the universe is $\mathcal{M}$ the point addition theorem for the small inductive dimension modulo a class (Theorem P) fails for every absolute Borel class α. Consequently, Theorem S fails for every absolute Borel class α by Proposition 1.10, and Theorem A fails for all absolute Borel classes except for the absolute Borel class $\mathsf{A}(0) = \{\emptyset\}$ by Proposition 1.11 and Theorem 1.9.

2. Normal families

The inductive theory of dimension naturally derives from $\{\emptyset\}$ two sequences of new classes, namely the classes $\{X : \operatorname{ind} X \leq n\}$ and $\{X : \operatorname{Ind} X \leq n\}$ for n in $\mathbb{N}$. Each of the two partitioning operations of inductive dimension creates a new class from the previous class in the sequence. The study of normal families investigates those properties of the original class that are inherited by the new class, especially the properties of closures under subsets and under sums, that is, unions. In fact, normal families are natural for modelling inductive proofs of these properties. (See the proof given in II.10.3 of Theorem II.10.1 for an example of such an argument.) A very successful theory exists when the universe is $\mathcal{M}_0$ or $\mathcal{M}$. The aim of this section is to develop a parallel theory for functions of inductive dimensional type in the universe $\mathcal{M}_0$. The proofs of the sum and other theorems promised in Chapter I will be provided by means of normal family techniques. The definitions of this section will be given in the general universe $\mathcal{U}$ since they will be used in the next section where an optimal universe called the Dowker universe will be constructed. (Of course, we mean optimal in the sense that the subspace theorem and the sum theorems will hold.)

The two partitioning operations of the inductive approach to dimension theory will be formally defined next. But before giving the definitions we note that for each class of spaces $\mathcal{P}$ the definitions of $\mathcal{P}$-ind X and $\mathcal{P}$-Ind X apply to any space X in the class $\mathcal{T}$. The same will be true for the definition of the partitioning operations.

2.1. Definition. Let $\mathcal{P}$ be a class.

(i) The *operation* H assigns to $\mathcal{P}$ the class $\mathsf{H}[\mathcal{P}]$ consisting of all spaces X in $\mathcal{T}$ with the property that for each point p of X and for each closed set F with $p \notin F$ there is a partition S between p and F such that S is in $\mathcal{P}$.

(ij) The *operation* M assigns to $\mathcal{P}$ the class $\mathsf{M}[\mathcal{P}]$ consisting of all spaces X in $\mathcal{T}$ with the property that for each pair of disjoint closed sets F and G there is a partition S between F and G such that S is in $\mathcal{P}$.

Obviously one can iterate these operations. Indeed, one sees immediately that

$$\mathsf{H}^n[\mathcal{P}] = \{ X : \mathcal{P}\text{-ind}\, X \leq n-1 \}, \qquad n \in \mathbb{N},$$
$$\mathsf{M}^n[\mathcal{P}] = \{ X : \mathcal{P}\text{-Ind}\, X \leq n-1 \}, \qquad n \in \mathbb{N},$$

where, by definition, $\mathsf{H}^0[\mathcal{P}] = \mathsf{M}^0[\mathcal{P}] = \mathcal{P}$.

The next proposition has a straightforward proof.

2.2. Proposition. *The following are true.*

(a) $\mathsf{H}[\mathcal{P}] \supset \mathsf{M}[\mathcal{P}]$.

(b) *If* $\mathcal{P} \subset \mathcal{Q}$, *then* $\mathsf{H}[\mathcal{P}] \subset \mathsf{H}[\mathcal{Q}]$ *and* $\mathsf{M}[\mathcal{P}] \subset \mathsf{M}[\mathcal{Q}]$.

(c) $\mathsf{H}[\mathcal{P} \cap \mathcal{U}] \cap \mathcal{U} = \mathsf{H}[\mathcal{P}] \cap \mathcal{U}$ *and* $\mathsf{M}[\mathcal{P} \cap \mathcal{U}] \cap \mathcal{U} = \mathsf{M}[\mathcal{P}] \cap \mathcal{U}$.

Observe that only (c) required the universe to be closed-monotone.

The proof of the next theorem is related to those found in Proposition II.2.7 and Theorem II.2.16.

2.3. Theorem. *The classes* $\mathbf{H}[\mathcal{P}]$ *and* $\mathbf{M}[\mathcal{P}]$ *are closed-monotone in* $\mathcal{U}$.

Proof. We shall give only the proof for the $\mathbf{M}$ operation since the proof for the $\mathbf{H}$ operation is no harder. Suppose that the space X is in $\mathbf{M}[\mathcal{P}]$ and Y is a closed set of X. Let F and G be disjoint closed subsets of Y. Since Y is closed in X, the sets F and G are also closed in X. From the definition of the $\mathbf{M}$ operation there is a partition S in X between F and G such that S is in $\mathcal{P}$. Because the universe is closed-monotone, the space S is also in the universe. Since $\mathcal{P}$ is closed-monotone in $\mathcal{U}$, the space $S \cap Y$ is in $\mathcal{P}$. Thus we have shown that Y is in $\mathbf{M}[\mathcal{P}]$.

The corresponding monotonicity theorem requires more than the obvious change of hypothesis in the case of $\mathbf{M}[\mathcal{P}]$. But the following is true for the $\mathbf{H}$ operation.

2.4. Theorem. *For a monotone or open-monotone universe, let* $\mathcal{P}$ *be respectively a monotone or an open-monotone class in* $\mathcal{U}$. *Then the class* $\mathbf{H}[\mathcal{P}]$ *is also respectively monotone or open-monotone in* $\mathcal{U}$.

Proof. Suppose the universe is open-monotone and the class $\mathcal{P}$ is open-monotone $\mathcal{U}$. Let X be in $\mathbf{H}[\mathcal{P}]$ and Y be an open set of X. Consider a point y of Y and a set F that is closed in the subspace Y with $y \notin F$. The point y is not in $\mathrm{cl}(F)$. From the definition of the $\mathbf{H}$ operation there is a partition S in X between y and $\mathrm{cl}(F)$ such that S is in $\mathcal{P}$. We have that $S \cap Y$ is in $\mathcal{P} \cap \mathcal{U}$. Consequently we have Y is in $\mathbf{H}[\mathcal{P}]$. The proof for the monotone case is an easy modification of the above.

In the same vein as the proof of Theorem II.2.21 we have the following theorem concerning ambient spaces. (Recall that the universe is contained in the class $\mathcal{N}_H$ of hereditarily normal spaces.)

2.5. Theorem. *Suppose that* Y *is a subspace of* X. *If* Y *is in* $\mathbf{M}[\mathcal{P}]$ *then for each pair of separated subsets* A *and* B *of* X *such that* $\mathrm{cl}_X(A) \cap \mathrm{cl}_X(B) \cap Y = \emptyset$ *there exists a partition* S *between* A *and* B *in* X *such that* $S \cap Y$ *is in* $\mathcal{P}$.

It has already been shown in Section 1 that closed-additivity of the class $\mathcal{K}$ is not preserved by the $\mathbf{H}$ operation in the universe $\mathcal{M}_0$. Also in [1949] Lokucievskiĭ gave an example of a compact space X

with $\operatorname{Ind} X = 2$ such that X is the union of two closed subspaces each with Ind less than 2 (see also Engelking [1978]). That is, closed-additivity is not preserved by the **M** operation as well. Consequently one must impose more conditions on the class $\mathcal{P}$ to assure that these partitioning operations will preserve the various additivity properties defined in Section 1. We begin by defining other forms of closed-additivity for a class $\mathcal{P}$.

2.6. Definition. A class $\mathcal{P}$ is said to be *countably* (*locally finitely*, *σ-locally finitely*) *closed-additive in $\mathcal{U}$* when the following statement (respectively) is true: If $\mathbf{V}$ is a countable (locally finite, σ-locally finite) closed cover of a space X in the universe with $\mathbf{V} \subset \mathcal{P}$, then X is in $\mathcal{P}$.

Using these newly defined forms of closed-additivity, we now define normal and related families in the universe.

2.7. Definition. A class $\mathcal{P}$ is said to be a *normal family in $\mathcal{U}$* if $\mathcal{P}$ satisfies

N1: $\mathcal{P}$ is monotone in $\mathcal{U}$,
N2: $\mathcal{P}$ is σ-locally finitely closed-additive in $\mathcal{U}$.

A class $\mathcal{P}$ is said to be a *semi-normal family in $\mathcal{U}$* if $\mathcal{P}$ satisfies

S1: $\mathcal{P}$ is closed-monotone in $\mathcal{U}$,
S2: $\mathcal{P}$ is σ-locally finitely closed-additive in $\mathcal{U}$.

A class $\mathcal{P}$ is said to be a *regular family in $\mathcal{U}$* if $\mathcal{P}$ satisfies

R1: $\mathcal{P}$ is closed-monotone in $\mathcal{U}$,
R2: $\mathcal{P}$ is countably closed-additive in $\mathcal{U}$.

A class $\mathcal{P}$ is said to be a *cosmic family in $\mathcal{U}$* if $\mathcal{P}$ satisfies

C1: $\mathcal{P}$ is closed-monotone in $\mathcal{U}$,
C2: $\mathcal{P}$ is locally finitely closed-additive in $\mathcal{U}$.

The conditions S1, R1 and C1 are redundant in view of the agreement made at the beginning of the chapter. We have repeated them for emphasis.

Remark. We have come to the last of the definitions that are dependent on the universe of discourse. From here on we shall suppress the "in $\mathcal{U}$" whenever the context of the discussion will permit it.

There are obvious relationships among the various families. In particular, a class that is both regular and cosmic is necessarily semi-normal. And a semi-normal family is both regular and cosmic. In the universe $\mathcal{M}$ one can easily verify that the absolute Borel classes $\mathsf{A}(\alpha)$ are examples of semi-normal families for every α and that the absolute Borel classes $\mathsf{M}(\alpha)$ are examples of cosmic families for $\alpha \geq 1$.

Historically, normal families were defined to be subclasses of $\mathcal{M}_0$ that are monotone and countably closed-additive. Later, normal families were defined precisely as above for the universe $\mathcal{M}$ of metrizable spaces. One can easily show for spaces in the universe $\mathcal{M}_0$ that locally finite collections are countable ones. Therefore regular families are semi-normal in this universe. Indeed, in the universe $\mathcal{M}_0$ a class $\mathcal{P}$ is a normal family if and only if $\mathcal{P}$ is monotone and countably closed-additive.

The operations $\mathbf{H}$ and $\mathbf{M}$ and the notion of a normal family were introduced to give an elegant proof of the sum and decomposition theorems in the universe $\mathcal{M}$. For now we shall concentrate on the separable metrizable case to provide the proofs that were promised in Section I.3. (The decomposition theorem for general metrizable spaces will be discussed in Section V.2.) The universe will be $\mathcal{M}_0$ for the remainder of the section.

2.8. Theorem. *In the universe $\mathcal{M}_0$ the class $\mathbf{M}[\mathcal{P}]$ is a normal family whenever $\mathcal{P}$ is a normal family.*

Proof. (Compare the proof to that in Section II.10.3.) Let us show that $\mathbf{M}[\mathcal{P}]$ is monotone. To this end let us first prove that it is open-monotone. Suppose that X is in $\mathbf{M}[\mathcal{P}]$ and Y is an open set of X. Choose a metric d on X that is bounded by 1. With $f(x) = d(x, X \setminus Y)$, for each k in $\mathbb{N}$ let $H_k = f^{-1}[(2^{-k-2}, 2^{-k+2})]$ and $I_k = f^{-1}[[2^{-k-1}, 2^{-k+1}]]$. Then the collections $\{H_k : k \in \mathbb{N}\}$ and $\{I_k : k \in \mathbb{N}\}$ are respectively open and closed in the space X, are locally finite in Y and are covers of Y. Clearly, $I_k \subset H_k \subset Y$. Now let F and G be disjoint sets that are closed in Y. For each natural number k let $F_k = F \cap I_k$ and $G_k = (X \setminus H_k) \cup G$. The sets F_k and G_k are disjoint closed subsets of X. So there is a partition S_k between F_k and G_k in X such that S_k is in $\mathcal{P}$. We have that $M = \bigcup\{S_k : k \in \mathbb{N}\}$ is in $\mathcal{P}$ because $\mathcal{P}$ is a regular family. Let U_k and V_k be disjoint open sets with $F_k \subset U_k$ and $G_k \subset V_k$

such that $X \setminus S_k = U_k \cup V_k$. Then denote by S the set $Y \setminus (U \cup V)$ where $U = \bigcup\{ U_k : k \in \mathbb{N} \}$ and $V = \bigcap\{ V_k : k \in \mathbb{N} \}$. As the collection $\{ H_k : k \in \mathbb{N} \}$ is locally finite in Y, the set $V \cap Y$ is open. One easily sees that S is a closed subspace of M and hence S is in $\mathcal{P}$ because $\mathcal{P}$ is closed-monotone. As S is a partition between F and G in Y, we have shown that $\mathbf{M}[\mathcal{P}]$ is open-monotone.

Let us now show that $\mathbf{M}[\mathcal{P}]$ is monotone. Suppose that X is in $\mathbf{M}[\mathcal{P}]$ and Y is a subset of X. Let F and G be disjoint sets that are closed in Y. Clearly, F and G are separated sets in the open subspace $Z = X \setminus (\mathrm{cl}_X(F) \cap \mathrm{cl}_X(G))$. Since Z is in $\mathbf{M}[\mathcal{P}]$ by the first part of the proof, we have by Theorem 2.5 that there is a partition S' between F and G in Z such that S' is in $\mathcal{P}$. Consequently $S = S' \cap Y$ is a partition between F and G in Y. Finally S is in $\mathcal{P}$ because $\mathcal{P}$ is monotone.

It remains to be shown that $\mathbf{M}[\mathcal{P}]$ is countably closed-additive. Suppose that $X = \bigcup\{ X_k : k \in \mathbb{N} \}$, where each set X_k is closed in X and is a member of $\mathbf{M}[\mathcal{P}]$. There is no loss in generality in assuming $X_0 = \emptyset$. We have from the open monotonicity of $\mathbf{M}[\mathcal{P}]$ that the set $K_k = X_k \setminus \bigcup\{ X_j : j < k \}$ is also in $\mathbf{M}[\mathcal{P}]$. Consider disjoint closed sets F and G of X. Let A_0 and B_0 be open sets with disjoint closures such that $F \subset A_0$ and $G \subset B_0$. By Theorem 2.5 there is a partition S_1 between A_0 and B_0 such that $L_1 = S_1 \cap K_1$ is in $\mathcal{P}$. Let U_1 and V_1 be open sets in X_1 such that $X \setminus S_1 = U_1 \cup V_1$ with $A_0 \cap X_1 \subset U_1$ and $B_0 \cap X_1 \subset V_1$. Then the sets $F_1 = A_0 \cup U_1$ and $G_1 = B_0 \cup V_1$ are separated and $\mathrm{cl}_X(F_1) \cap \mathrm{cl}_X(G_1) \subset L_1$. Now select disjoint open sets A_1 and B_1 such that $F_1 \subset A_1$, $G_1 \subset B_1$ and $\mathrm{cl}_X(A_1) \cap \mathrm{cl}_X(B_1) \subset L_1$. Obviously, $X_1 \setminus (A_1 \cup B_1) \subset L_1$. By Theorem 2.5 there is a partition S_2 between A_1 and B_1 in X such that $L_2 = S_2 \cap K_2$ is in $\mathcal{P}$. We have that $L_1 \cup L_2$ is in $\mathcal{P}$ because L_2 is a countable union of closed subsets and $\mathcal{P}$ is countably closed-additive. Repeating the above construction, we have disjoint open sets A_2 and B_2 satisfying the inclusions $A_1 \subset A_2$ and $B_1 \subset B_2$ and such that $(X_1 \cup X_2) \setminus (A_2 \cup B_2) \subset L_1 \cup L_2$. Continuing inductively, we have a sequence of disjoint open sets A_k and B_k and F_σ-sets L_k in $\mathcal{P}$ such that

$$F \subset A_k \subset A_{k+1} \quad \text{and} \quad G \subset B_k \subset B_{k+1}, \qquad k \in \mathbb{N},$$

$$\bigcup\{ X_j : j \leq k \} \setminus (A_k \cup B_k) \subset \bigcup\{ L_i : i \leq k \}, \qquad k \in \mathbb{N}.$$

The set $L = \bigcup\{ L_k : k \in \mathbb{N} \}$ is in $\mathcal{P}$ because $\mathcal{P}$ is countably closed-additive. Let $S = X \setminus \bigcup\{ (A_k \cup B_k) : k \in \mathbb{N} \}$. Then S is partition between F and G in X and $S \subset L$. By the closed-monotonicity of $\mathcal{P}$ we have $S \in \mathcal{P}$. Thus we have shown that X is in $\mathsf{M}[\mathcal{P}]$.

2.9. Theorem. *Suppose that $\mathcal{P}$ is a normal family in the universe $\mathcal{M}_0$. The following are equivalent for spaces X in the universe.*

(a) $X \in \mathsf{M}[\mathcal{P}]$.
(b) $X = P \cup N$, *where* $P \in \mathcal{P}$ *and* $\operatorname{ind} N \leq 0$.
(c) $X = P \cup Z$, *where* $P \in \mathcal{P}$ *and* $\operatorname{Ind} Z \leq 0$.

Proof. Let us show that (a) implies (b). Let $\mathcal{B} = \{ U_i : i \in \mathbb{N} \}$ be a countable base for the open sets of X. Consider the countable family $\{ (C_j, D_j) : j \in \mathbb{N} \}$ of all pairs of elements of $\mathcal{B}$ such that $\operatorname{cl}(C_j) \subset D_j$. For each j there is a partition S_j between $\operatorname{cl}(C_j)$ and $X \setminus D_j$ in X with $S_j \in \mathcal{P}$. The set $\bigcup\{ S_j : j \in \mathbb{N} \}$ is in $\mathcal{P}$ because $\mathcal{P}$ is a normal family. Obviously, $\operatorname{ind}(X \setminus \bigcup\{ S_j : j \in \mathbb{N} \}) \leq 0$.

That (b) implies (c) will follow from the identity $\mathsf{H}[\{\emptyset\}] \cap \mathcal{M}_0 = \mathsf{M}[\{\emptyset\}] \cap \mathcal{M}_0$. The proof of this identity is the same as the one for the inductive step of Theorem I.4.4 and will not be repeated here. Indeed, the proof of this identity is much less complicated than that of the inductive step because the open sets will have empty boundaries.

Finally, that (c) implies (a) follows easily from Proposition I.4.6.

2.10. Theorem. *In the universe $\mathcal{M}_0$ of separable metrizable spaces the dimension* Ind *satisfies the countable sum theorem and the decomposition theorem.*

Proof. As the class $\{\emptyset\}$ is a normal family, the classes $\mathsf{M}^n[\mathcal{P}]$ are also normal families by Theorem 2.8. We have that $\operatorname{Ind} X \leq n - 1$ if and only if $X \in \mathsf{M}^n[\{\emptyset\}]$. So the countable sum theorem for Ind follows. Also from the equivalence of (a) and (c) of Theorem 2.9 we have the statement:

$X \in \mathsf{M}^n[\{\emptyset\}]$ *if and only if* $X = Y \cup Z$ *where* $Y \in \mathsf{M}^{n-1}[\{\emptyset\}]$ *and* $\operatorname{Ind} Z \leq 0$.

The decomposition theorem for Ind now follows.

It has been shown already in Theorem I.4.4 that ind and Ind coincide for separable metrizable spaces. The proof given there was based on the countable sum theorem (which was assumed to have been proved) together with the construction of a specific locally finite cover. We shall give here a somewhat more elegant proof of the same theorem by using the decomposition theorem. The proof is a simple consequence of normal family arguments.

2.11. Theorem. *For each separable metrizable space X,*

$$\operatorname{ind} X = \operatorname{Ind} X.$$

Proof. We already know that $\mathbf{H}^n[\{\emptyset\}] \supset \mathbf{M}^n[\{\emptyset\}]$. We shall prove the reverse inclusion. The reverse inclusion is true for $n = 0$ by definition. For the inductive step we let $X \in \mathbf{H}^n[\{\emptyset\}]$. Then we infer the existence of a countable base $\mathcal{B} = \{U_i : i \in \mathbb{N}\}$ for the open sets of X such that $\mathrm{B}(U_i) \in \mathbf{H}^{n-1}[\{\emptyset\}]$ for every i. From the induction hypothesis we have $\{\mathrm{B}(U_i) : i \in \mathbb{N}\} \subset \mathbf{M}^{n-1}[\{\emptyset\}]$. As $\mathbf{M}^{n-1}[\{\emptyset\}]$ is a normal family, we have $P = \bigcup\{\mathrm{B}(U_i) : i \in \mathbb{N}\}$ is in $\mathbf{M}^{n-1}[\{\emptyset\}]$. Moreover, $\operatorname{ind}(X \setminus P) \leq 0$. So by Theorem 2.9 we have $X \in \mathbf{M}^n[\{\emptyset\}]$.

Combining Theorems 2.10 and 2.11, we now have the promised proofs of Theorems I.3.7 and I.3.8.

3. Optimal universe

The following agreement will be in force for this section. Its content will facilitate the investigations of the properties of the Dowker universe which will be defined later (see Definition 3.22). The focus will be the "optimal" property of the Dowker universe.

Agreement. *In addition to the agreement made in the introduction to the chapter, the universe is open-monotone.*

The utility of normal families was illustrated in the last section for the universe $\mathcal{M}_0$ of separable metrizable spaces. The separability reduced the considerations to monotone regular families. For the universe $\mathcal{M}$ of metrizable spaces the full force of normal families is needed to derive the sum theorems of dimension theory. As we shall discover in this section, the development of a sum theorem when the universe is larger than $\mathcal{M}$ will require much more than the techniques found in the previous section for separable metrizable spaces. The main task is to find a universe in which the class $\mathbf{M}[\mathcal{P}]$ will be open-monotone in $\mathcal{U}$ (which is contained in the class $\mathcal{N}_H$ of hereditarily normal spaces by agreement) whenever $\mathcal{P}$ is. The operation $\mathbf{S}$

is introduced to determine a necessary and sufficient condition for the open monotonicity of $\mathsf{M}[\mathcal{P}]$. The techniques of this section are essentially those used by Dowker [1953] in his successful study of the class $\mathcal{N}_T$ of totally normal spaces. (A space X is *totally normal* if it is a normal space with the property that every open set U is the union of a locally finite in U collection of open F_σ-sets in X.) The class $\mathcal{N}_T$ of totally normal spaces includes the class $\mathcal{N}_P$ of perfectly normal spaces and hence the class $\mathcal{M}$ of metrizable spaces. Here we shall use normal families techniques instead of induction which was used by Dowker. The advantage of this approach is that one can isolate those properties of a family which are needed for the inductive step to preserve the sum and subspace theorems on applying the M operation to a normal family. Among these properties are open monotonicity and strong closed additivity. (Both properties are free for the universe $\mathcal{M}$.) The end result of the investigation is the creation of an optimal universe, called the Dowker universe, in which a satisfactory theory of $\mathcal{P}$-Ind can be developed.

The notions of normal and related families and the two operations H and M were defined in the previous section. A class $\mathcal{P}$ need not be a normal family in the universe. An obvious procedure would be the expansion of the class to a normal family. This is our first definition. The operation defined here will be used later to create the Dowker universe.

3.1. Definition. For classes $\mathcal{P}$ and $\mathcal{Q}$ the *normal family extension of $\mathcal{P}$ in $\mathcal{Q}$* is the class of spaces X in $\mathcal{Q}$ for which $X = \bigcup \boldsymbol{F}$ for some subcollection $\boldsymbol{F}$ of $\mathcal{P}$ such that $\boldsymbol{F}$ is a σ-locally finite closed cover of X. This class of spaces is denoted by $\mathsf{N}[\mathcal{P} : \mathcal{Q}]$.

Obviously one has the inclusions

$$\mathcal{P} \cap \mathcal{Q} \subset \mathsf{N}[\mathcal{P} : \mathcal{Q}] \subset \mathcal{Q},$$

and, when $\mathcal{P}$ is a normal family in the universe,

$$\mathcal{P} \cap \mathcal{U} = \mathsf{N}[\mathcal{P} : \mathcal{U}].$$

Moreover,

$$\mathsf{N}[\mathcal{P} : \mathcal{U}] \subset \mathsf{N}[\mathcal{Q} : \mathcal{U}] \quad \text{when} \quad \mathcal{P} \subset \mathcal{Q}.$$

Let us now prove a key relationship between cosmic families and cozero-sets of a space X. The following proof is reminiscent of the proof that cozero-sets of a normal space are normal subspaces. Its proof has been foreshadowed in the first part of the proof of Theorem 2.8.

3.2. Theorem. *Suppose that $\mathcal{P}$ is a cosmic family. Then every cozero-set of a space X in $\mathbf{M}[\mathcal{P}]$ is also in $\mathbf{M}[\mathcal{P}]$.*

Proof. Let Y be a cozero set of X. There is a continuous function f on X into [0,1] such that $Y = f^{-1}\big[(0,1]\big]$. For each natural number k, let H_k be the open set $f^{-1}\big[(2^{-k-2}, 2^{-k+2})\big]$ and I_k be the closed set $f^{-1}\big[[2^{-k-1}, 2^{-k+1}]\big]$. The collections $\{ H_k : k \in \mathbb{N} \}$ and $\{ I_k : k \in \mathbb{N} \}$ are covers of Y that are locally finite in the subspace Y. Clearly, $I_k \subset H_k \subset Y$. Let F and G be disjoint sets that are closed in Y. For each natural number k, let $F_k = F \cap I_k$ and $G_k = (X \setminus H_k) \cup G$. The sets F_k and G_k are disjoint closed sets of X. So there are partitions S_k between F_k and G_k in X such that S_k is in $\mathcal{P}$. Let U_k and V_k be disjoint open sets such that $X \setminus S_k = U_k \cup V_k$, $F_k \subset U_k$ and $G_k \subset V_k$. Clearly, $S_k \subset H_k$. So $\bigcup\{ S_k : k \in \mathbb{N} \}$ is in $\mathcal{P}$ because $\mathcal{P}$ is a cosmic family. It can be shown that $U = \bigcup\{ U_k : k \in \mathbb{N} \}$ and $V = Y \cap \bigcap\{ V_k : k \in \mathbb{N} \}$ are open sets. Then $S = Y \setminus (U \cup V)$ is a partition between F and G in Y such that S is in $\mathcal{P}$.

We shall generate special open sets of a space X, to be called Dowker-open sets, by employing unions of point-finite collections of cozero-sets of X. Before proceeding along this direction we shall give a lemma on point-finite covers of a general topological space. (The collections that are being considered are indexed ones.)

3.3. Lemma. *Let $\{ U_s : s \in S \}$ be a point-finite open cover of a topological space Y. For each positive integer i, denote by K_i the set of all points of the space Y which belong to exactly i members of the cover $\{ U_s : s \in S \}$ and by $\mathbf{T}_i$ the family of all subsets of S that have exactly i elements. Then*

(a) $Y = \bigcup\{ K_i : i \geq 1 \}$,
(b) $K_i \cap K_j = \emptyset$ *for* $i \neq j$,
(c) $F_i = \bigcup\{ K_j : j \leq i \}$ *is closed for* $i \geq 1$,
(d) $K_i = \bigcup\{ K_T : T \in \mathbf{T}_i \}$, *where the sets K_T, defined by letting $K_T = \bigcap\{ K_i \cap U_s : s \in T \}$ for $T \in \mathbf{T}_i$, are open in K_i and pairwise disjoint.*

Proof. The equalities (a) and (b) follow directly from the definition of the sets K_i. For statement (c), observe that if $x \notin F_i$, then $x \in U_{s_1} \cap U_{s_2} \cap \cdots \cap U_{s_{i+1}} \subset X \setminus F_i$, where $s_1, s_2, \ldots, s_{i+1}$ are distinct elements of S. To establish (d) it will suffice to note that $K_T \subset K_i$ for T in $\mathbf{T}_i$ and that whenever T and T' are distinct members of $\mathbf{T}_i$ the set $K_T \cap K_{T'}$ is the empty set since the union $T \cup T'$ contains at least $i+1$ elements of S.

We shall now resume our development of normal families by defining Dowker-open sets of a space X.

3.4. Definition. A subset U of a topological space X is called *Dowker-open in* X if it is the union of a point-finite collection of cozero-sets of X.

Obviously, if Y is a subspace of X and U is Dowker-open in X, then the trace of U on Y is Dowker-open in Y.

3.5. Lemma. *Suppose that $\mathcal{P}$ be a strongly closed-additive semi-normal family. Let Y be Dowker-open set of a space X in $\mathsf{M}[\mathcal{P}]$. Then Y is in $\mathsf{M}[\mathcal{P}]$.*

Proof. Suppose that $\{U_s : s \in S\}$ is a point-finite collection of cozero-sets of X whose union is Y. The notation of Lemma 3.3 will be used. Let us first show that K_T is in $\mathsf{M}[\mathcal{P}]$ for each T in $\mathbf{T}_i$. The set $X_i = (X \setminus Y) \cup F_i$ is closed in X. From the equalities $K_i = F_i \setminus F_{i-1} = X_i \cap (Y \setminus F_{i-1})$ and the property (d) of Lemma 3.3 we infer that the set K_T is open in X_i. Clearly we have $\bigcap\{U_s : s \in T\} \cap F_{i-1} = \emptyset$. Also the set $X_i \cap \bigcap\{U_s : s \in T\} = F_i \cap \bigcap\{U_s : s \in T\}$ is an F_σ-set of X_i. So K_T is a co-zero set of X_i. As X_i is a closed set, it is in $\mathsf{M}[\mathcal{P}]$. By Theorems 2.3 and 3.2, we have that K_T is in $\mathsf{M}[\mathcal{P}]$. As a consequence of this fact, we have from property (d) of Lemma 3.3 that K_i is an open set of X_i that is in $\mathsf{M}[\mathcal{P}]$ because a semi-normal family is a cosmic family .

We are ready to use the strong closed additivity of the class $\mathcal{P}$. At this juncture, only the properties of regular families are used. Let F and G be disjoint sets that are closed in Y. By Theorem 2.5 there is a partition S_0 between F and G in Y such that $L_0 = K_0 \cap S_0$ is in $\mathcal{P}$. There are disjoint open subsets U_0' and V_0' of Y such that $F \subset U_0'$, $G \subset V_0'$ and $Y \setminus S_0 = U_0' \cup V_0'$. Since Y is in $\mathcal{N}_H$, there are open sets U_0 and V_0 in Y such that $F_0 = (U_0' \cap K_0) \cup F \subset U_0$, $G_0 = (V_0' \cap K_0) \cup G \subset V_0$ and $\mathrm{cl}_Y(U_0) \cap \mathrm{cl}_Y(V_0) \subset L_0$. Theorem 2.5

yields a partition S_1 between U_0 and V_0 in Y with $L_1' = K_1 \cap S_1$ in $\mathcal{P}$. Obviously, $K_0 \cap S_1 \subset L_0$. Since $\mathcal{P}$ is strongly closed-additive, we have that $L_1 = L_0 \cup L_1'$ is in $\mathcal{P}$. Inductively, one constructs sequences of disjoint open sets U_i, V_i and closed sets L_i that are in $\mathcal{P}$ such that

$$U_i \subset U_{i+1}, \quad V_i \subset V_{i+1}, \quad L_i \subset L_{i+1},$$
$$F \subset U = \bigcup\{ U_i : i \in \mathbb{N}\}, \quad G \subset V = \bigcup\{ V_i : i \in \mathbb{N}\},$$
$$Y \setminus \bigcup\{ L_i : i \in \mathbb{N}\} = U \cup V.$$

Since $S = \bigcup\{ L_i : i \in \mathbb{N}\}$ is closed in Y, we have that S is in $\mathcal{P}$. Thereby we have shown that Y is in $\mathbf{M}[\mathcal{P}]$.

In [1953] Dowker introduced the class of totally normal spaces. Subsequently several generalizations of totally normal spaces have appeared. (See Section 6 for further references.) In [1978] Engelking followed the methods of Dowker to define the class of strongly hereditarily normal spaces. We shall give the definition of this class next. The definition finds its origin in the following characterization of hereditarily normal spaces:

A Hausdorff space X is hereditarily normal if and only if for each pair of separated sets F and G in X there is a pair of disjoint open sets U and V such that $F \subset U$ and $G \subset V$.

This last condition can be restated in terms of partitions as:

For each pair of separated sets F and G in X there is a partition S between F and G in X.

3.6. Definition. A space X is called *strongly hereditarily normal* if it is a Hausdorff space such that for each pair of separated sets F and G in X there is a pair of disjoint Dowker-open sets U and V such that $F \subset U$, and $G \subset V$. The class of strongly hereditarily normal spaces will be denoted by $\mathcal{E}$.

The class $\mathcal{E}$ is easily shown to be absolutely monotone. Clearly each strongly hereditarily normal space is hereditarily normal. The following equivalence is easily proved.

3.7. Theorem. *A space X is strongly hereditarily normal if and only if X is a Hausdorff space such that for each pair of separated* *sets F and G in X there is a partition S between F and G in X such that $X \setminus S$ is a Dowker-open set of X.*

The last theorem suggests another operation on classes of spaces based on the concept of partitions that is complementary to that of the operation $\mathbf{M}$. More precisely, we have the following definition.

3.8. Definition. Let $\mathcal{P}$ be a class of spaces. The *operation* $\mathbf{S}$ assigns to $\mathcal{P}$ the class $\mathbf{S}[\mathcal{P}]$ consisting of all topological spaces X such that for each pair of separated sets F and G in X there is a partition S between F and G in X with $X \setminus S$ in $\mathcal{P}$.

Suppose that a space X is in $\mathbf{S}[\mathcal{P}]$. Since $F = \emptyset$ and $G = X$ are separated sets in X, we have that X is in $\mathcal{P}$. Thus we have

$$\mathbf{S}[\mathcal{P}] \subset \mathcal{P}.$$

3.9. Theorem. *For each class $\mathcal{P}$, the class $\mathbf{S}[\mathcal{P}]$ is closed-monotone in $\mathcal{U}$.*

Proof. Let Y be a closed subspace of a space X in $\mathbf{S}[\mathcal{P}] \cap \mathcal{U}$. Let F and G be separated subsets of Y. As they are also separated in X, there is a partition S between F and G such that $X \setminus S$ is in $\mathcal{P}$. As the universe is open-monotone, $X \setminus S$ is in $\mathcal{U}$. Then the set $S \cap Y$ is partition between F and G in Y such that $Y \setminus S$ is in $\mathcal{P} \cap \mathcal{U}$ because the class $\mathcal{P}$ is closed-monotone in $\mathcal{U}$.

There is a simple property one can derive for the $\mathbf{S}$ operation applied to the universe. Clearly, $\mathbf{S}[\mathcal{U}] \subset \mathcal{U}$. Even more, as the universe is open-monotone, we have

$$\mathbf{S}[\mathcal{U}] = \mathcal{U} \subset \mathcal{N}_H.$$

We shall create a new universe from the universe $\mathcal{U}$. Obviously the universe is a class. In this context, consider the normal family extension $\mathbf{N}[\mathcal{U} : \mathcal{N}_H]$ of $\mathcal{U}$ in the class $\mathcal{N}_H$.

3.10. Theorem. *The class $\mathbf{N}[\mathcal{U} : \mathcal{N}_H]$ is absolutely open-monotone and absolutely closed-monotone and*

$$\mathbf{S}[\mathbf{N}[\mathcal{U} : \mathcal{N}_H]] = \mathbf{N}[\mathcal{U} : \mathcal{N}_H].$$

Proof. Let Y be an open subset of a space X in $\mathbf{N}[\mathcal{U} : \mathcal{N}_H]$. There is a σ-locally finite closed subcollection $\boldsymbol{F}$ of $\mathcal{U}$ such that $X = \bigcup \boldsymbol{F}$.

Since $\mathcal{U}$ is open-monotone, the collection $\{F \cap Y : F \in \mathbf{F}\}$ is also a subcollection of $\mathcal{U}$ that is a σ-locally finite closed cover of the subspace Y. That is, Y is in $\mathsf{N}[\mathcal{U}: \mathcal{N}_H]$. The proof is the same for closed-monotone. The remainder of the proof follows from the properties of the operation S listed above.

3.11. Theorem. *For any class* $\mathcal{Q}$,

$$\mathsf{S}[\mathcal{Q} \cap \mathcal{U}] = \mathsf{S}[\mathcal{Q}] \cap \mathcal{U}.$$

Proof. Clearly, $\mathsf{S}[\mathcal{Q} \cap \mathcal{U}] \subset \mathsf{S}[\mathcal{Q}] \cap \mathsf{S}[\mathcal{U}] \subset \mathsf{S}[\mathcal{Q}] \cap \mathcal{U}$. For the other inclusion suppose that X is in $\mathsf{S}[\mathcal{Q}] \cap \mathcal{U}$ and let F and G be separated subsets of X. Then there is a partition S between F and G in X such that $X \setminus S$ is in $\mathcal{Q}$. Also $X \setminus S$ is in $\mathcal{U}$ because $\mathcal{U}$ is open-monotone. Therefore X is in $\mathsf{S}[\mathcal{Q} \cap \mathcal{U}]$.

The last two theorems show that N and S operations behave nicely. The next theorem shows the connection between open monotonicity and the operations M and S.

3.12. Theorem. *Suppose that the class* $\mathcal{P}$ *is open-monotone. Then* $\mathsf{S}\big[\mathsf{M}[\mathcal{P}]\big]$ *is open-monotone. Moreover,*

$$\mathsf{S}\big[\mathsf{M}[\mathcal{P}]\big] \cap \mathcal{U} = \{X : Y \in \mathsf{M}[\mathcal{P}] \cap \mathcal{U} \text{ for each open set } Y \text{ of } X\}.$$

Proof. In view of the last theorem the inclusion of the right-hand collection of the equation in the left-hand one is easily established. We shall prove the opposite inclusion. Let Y be an open subset of a space X in $\mathsf{S}\big[\mathsf{M}[\mathcal{P}] \cap \mathcal{U}\big]$. We must show $Y \in \mathsf{M}[\mathcal{P}] \cap \mathcal{U}$. Let F and G be disjoint subsets of Y that are closed in Y. Then F and G are separated in X. Since X is in $\mathsf{S}\big[\mathsf{M}[\mathcal{P}]\big]$, there is a partition S between F and G in X such that $X \setminus S$ is in $\mathsf{M}[\mathcal{P}]$. Let U and V be disjoint open sets with $F \subset U, G \subset V$ and $X \setminus S = U \cup V$. Since $F \subset U$ and F is closed in the space Y, we have that F and $Y \setminus U$ are separated sets in X and that $\mathrm{cl}_X(F) \cap \mathrm{cl}_X(Y \setminus U) \cap (X \setminus S) = \emptyset$. By Theorem 2.5 there is a partition L between F and $Y \setminus U$ in X such that $L \cap (X \setminus S)$ is in $\mathcal{P}$. Clearly L is a partition between F and G in X. Since $L \cap (Y \setminus U) = \emptyset$, it follows that $L \cap Y = L \cap Y \cap U$. And so we have $Y \cap L \cap (X \setminus S) = (L \cap Y \cap U) \cup (L \cap Y \cap V) = L \cap Y \cap U = Y \cap L$. Because $\mathcal{P}$ is open-monotone, we have $Y \cap L$ is in $\mathcal{P}$. Thus we have shown that Y is in $\mathsf{M}[\mathcal{P}]$ and the opposite inclusion is now established. That $\mathsf{S}\big[\mathsf{M}[\mathcal{P}] \cap \mathcal{U}\big]$ is open-monotone follows immediately from the equality of the two collections.

3.13. Theorem. *For each $\mathcal{P}$ that is open-monotone, the class $\mathbf{M}[\mathcal{P}]$ is open-monotone if and only if $\mathbf{S}[\mathbf{M}[\mathcal{P}]] \cap \mathcal{U} = \mathbf{M}[\mathcal{P}] \cap \mathcal{U}$.*

Proof. If $\mathbf{M}[\mathcal{P}]$ is open-monotone, then the class $\mathbf{M}[\mathcal{P}] \cap \mathcal{U}$ is absolutely open-monotone by definition. Consequently the equality $\mathbf{S}[\mathbf{M}[\mathcal{P}] \cap \mathcal{U}] = \mathbf{M}[\mathcal{P}] \cap \mathcal{U}$ holds. The converse statement follows from Theorem 3.12.

From open monotonicity we go next to monotonicity.

3.14. Theorem. *Suppose that the universe is monotone and the class $\mathcal{P}$ is monotone. Then the following statements are equivalent for every space X.*

(a) *If Y is a subspace of X, then Y is in $\mathbf{M}[\mathcal{P}] \cap \mathcal{U}$.*

(b) *If Y is an open subspace of X, then Y is in $\mathbf{M}[\mathcal{P}] \cap \mathcal{U}$.*

Consequently,

$$\mathbf{S}[\mathbf{M}[\mathcal{P}] \cap \mathcal{U}] = \{ X : Y \in \mathbf{M}[\mathcal{P}] \cap \mathcal{U} \text{ for each subspace } Y \text{ of } X \}$$

and hence the class $\mathbf{S}[\mathbf{M}[\mathcal{P}]]$ is monotone.

Proof. Clearly (a) implies (b).

Assume (b) holds and suppose that Y is a subspace of a space X. We must show that Y is in $\mathbf{M}[\mathcal{P}] \cap \mathcal{U}$. Let F and G be disjoint subsets of Y that are closed in Y. The sets F and G are then separated in X and $Z = X \setminus (\mathrm{cl}_X(F) \cap \mathrm{cl}_X(G))$ is in $\mathbf{M}[\mathcal{P}] \cap \mathcal{U}$. By Theorem 2.5 there is a partition S between F and G in Z such that S is in $\mathcal{P}$. From the absolute monotonicity of $\mathcal{P} \cap \mathcal{U}$ and the inclusion $Y \subset Z$ we have that $S \cap Y$ is a partition between F and G in Y such that $S \cap Y$ is in $\mathcal{P} \cap \mathcal{U}$. Thus Y is in $\mathbf{M}[\mathcal{P}] \cap \mathcal{U}$ and (a) holds.

The last statement of the theorem is a consequence of Theorems 3.11 and 3.12.

The next theorem concerns the strongly closed-additive property in the universe.

3.15. Theorem. *Suppose that $\mathcal{P}$ is open-monotone and strongly closed-additive. Then $\mathbf{S}[\mathbf{M}[\mathcal{P}]]$ is open-monotone and strongly closed-additive.*

Proof. We have from Theorems 3.11 and 3.12 that $\mathbf{S}[\mathbf{M}[\mathcal{P}]]$ is open-monotone. Let us show that $\mathbf{S}[\mathbf{M}[\mathcal{P}]]$ is strongly closed-additive. Assume that $X = Y \cup Z$ with Y and Z in $\mathbf{S}[\mathbf{M}[\mathcal{P}]] \cap \mathcal{U}$ and Y

closed in X. Since $\mathbf{S}[\mathbf{M}[\mathcal{P}]]$ is open-monotone, we may assume further that Y and Z are disjoint. Let F and G be disjoint closed sets in X. By Theorem 2.5 there is a partition S_0 between F and G such that $S_0 \cap Y$ is in $\mathcal{P}$. Let U_0 and V_0 be disjoint open sets in X such that $F \subset U_0$, $G \subset V_0$ and $X \setminus S_0 = U_0 \cup V_0$. With $F_1 = (Y \cap U_0) \cup F$ and $G_1 = (Y \cap V_0) \cup G$ we have from Theorem 2.5 a partition S_1 between F_1 and G_1 in X such that $S_1 \cap Z$ is in $\mathcal{P}$. Clearly $S_1 \cap Y$ is contained in $S_0 \cap Y$. So $S = S_1 \cup (S_0 \cap Y)$ is a partition between F and G in X. By the strongly closed-additivity of $\mathcal{P}$ we have that S is in $\mathcal{P}$. Therefore X is in $\mathbf{M}[\mathcal{P}] \cap \mathcal{U}$. Moreover, if $\widetilde{X}$ is an open subspace of X, then $\widetilde{Y} = \widetilde{X} \cap Y$ and $\widetilde{Z} = \widetilde{X} \cap Z$ are in $\mathbf{S}[\mathbf{M}[\mathcal{P}]]$ and $\widetilde{Y}$ is closed in $\widetilde{X}$. Since $\mathcal{U}$ is open-monotone, we have that $\widetilde{X}$ is also in $\mathbf{M}[\mathcal{P}]$. By Theorem 3.12 we have that X is in $\mathbf{S}[\mathbf{M}[\mathcal{P}]]$.

To prepare for the theorems concerning normal families and its variants, the concept of strong closed additivity will be generalized to a well-ordered collection $\{ X_\beta : \beta < \alpha \}$ of subsets of a space X in $\mathcal{U}$. The natural requirement on this collection is that $\bigcup\{ X_\gamma : \gamma < \beta \}$ be closed in X whenever $\beta \leq \alpha$. In general, some further condition must be imposed. For the study of normal families there are two such conditions. The first is that $\alpha \leq \omega_0$, where ω_0 is the first infinite ordinal number. And the second is that $\{ X_\beta : \beta < \alpha \}$ be a locally finite collection in X. The first condition is related to regular families and the second is related to cosmic families. Clearly the cases for the ordinal numbers $\alpha = 0$, 1 and 2 are either trivial or the case of strong closed additivity.

We shall begin with the regular families in $\mathcal{U}$.

3.16. Theorem. *Suppose that $\mathcal{P}$ is an open-monotone, strongly closed-additive regular family. Then $\mathbf{S}[\mathbf{M}[\mathcal{P}]]$ is an open-monotone, strongly closed-additive regular family.*

Only the condition R2 for regular families remains to be verified. Due to Theorem 3.12, this is easily achieved from the next lemma.

3.17. Lemma. *Let $\mathcal{P}$ be a strongly closed-additive regular family. Suppose that X is a space and $\mathbf{K} = \{ K_i : i < \omega_0 \}$ is a collection of subsets of X such that*

(a) *$K_i \in \mathbf{M}[\mathcal{P}] \cap \mathcal{U}$ for $i < \omega_0$,*
(b) *$F_i = \bigcup\{ K_j : j < i \}$ is closed for $i < \omega_0$,*
(c) *$K_i \cap K_j = \emptyset$ for $i \neq j$,*

(d) $X = \bigcup \boldsymbol{K}$.

Then X is in $\mathbf{M}[\mathcal{P}]$.

Proof. The proof of the lemma is found in the last paragraph of the proof of Lemma 3.5.

Next we shall prove the analogous statements for cosmic families in $\mathcal{U}$.

3.18. Theorem. *Suppose that $\mathcal{P}$ is an open-monotone, strongly closed-additive cosmic family. Then $\mathbf{S}\big[\mathbf{M}[\mathcal{P}]\big]$ is an open-monotone, strongly closed-additive cosmic family.*

Only the condition C2 for cosmic families remains to be verified. Again due to Theorem 3.12, this is easily achieved from the next lemma.

3.19. Lemma. *Let $\mathcal{P}$ be a strongly closed-additive cosmic family. Suppose that X is a space and $\boldsymbol{K} = \{ K_\beta : \beta < \alpha \}$ is a locally finite collection of subsets of X such that*

(a) $K_\beta \in \mathbf{M}[\mathcal{P}] \cap \mathcal{U}$ *for* $\beta < \alpha$,

(b) $X_\beta = \bigcup \{ K_\gamma : \gamma < \beta \}$ *is closed for* $\beta < \alpha$,

(c) $K_\beta \cap K_\gamma = \emptyset$ *for* $\beta \neq \gamma$,

(d) $X = \bigcup \boldsymbol{K}$.

Then X is in $\mathbf{M}[\mathcal{P}]$.

Proof. The proof is by transfinite induction on α. Suppose for each β with $\beta < \alpha$ that the closed subspace X_β is in $\mathbf{M}[\mathcal{P}]$. Let F and G be disjoint closed sets of X. One can inductively construct sets U_γ, V_γ and L_γ such that

(1) L_γ is a partition between $F_\gamma = F \cap K_\gamma$ and $G_\gamma = G \cap K_\gamma$ in K_γ, and U_γ and V_γ are disjoint open sets of K_γ with $F_\gamma \subset U_\gamma$, $G_\gamma \subset V_\gamma$ and $K_\gamma \setminus L_\gamma = U_\gamma \cup V_\gamma$,

(2) for $\beta < \alpha$ the set $S_\beta = \bigcup \{ L_\gamma : \gamma < \beta \}$ is a partition between $F \cap X_\beta$ and $G \cap X_\beta$ in X_β and the sets $\bigcup \{ U_\gamma : \gamma < \beta \}$ and $\bigcup \{ V_\gamma : \gamma < \beta \}$ are open in X_β.

The construction is straightforward because the collection $\boldsymbol{K}$ is locally finite in X. Let S be the union $\bigcup \{ S_\beta : \beta < \alpha \}$. From (1) and (2) and the local finiteness of the collection $\boldsymbol{K}$ we have that S is partition between F and G in X. To verify that S is in $\mathcal{P}$ let us first verify inductively that each S_β is in $\mathcal{P}$. Suppose that S_δ is in $\mathcal{P}$ when δ is less than β and let γ be any ordinal number less than β. Then $L_\gamma \cup S_\gamma$ is in $\mathcal{P}$ because $\mathcal{P}$ is strongly closed-additive. So $\mathrm{cl}_X(L_\gamma)$ is in $\mathcal{P}$. Since the collection $\boldsymbol{K}$ is locally finite and X_β is closed, the equality $S_\beta = \bigcup\{\, \mathrm{cl}_X(L_\gamma) : \gamma < \beta \,\}$ holds. So S_β is in $\mathcal{P}$ by the fact that $\mathcal{P}$ is a cosmic family. It is now a simple matter to show that S is in $\mathcal{P}$. The induction step for the ordinal number α is now completed and the lemma is proved.

Combining Theorems 3.16 and 3.18, we have the following theorem for semi-normal families in $\mathcal{U}$.

3.20. Theorem. *Suppose that $\mathcal{P}$ is an open-monotone, strongly closed-additive semi-normal family. Then $\mathbf{S}[\,\mathbf{M}[\,\mathcal{P}\,]\,]$ is an open-monotone, strongly closed-additive semi-normal family.*

From Theorems 3.14 and 3.20 the following normal family theorem will result.

3.21. Theorem. *Suppose that the universe is monotone and that $\mathcal{P}$ is a strongly closed-additive normal family. Then $\mathbf{S}[\,\mathbf{M}[\,\mathcal{P}\,]\,]$ is a strongly closed-additive normal family.*

It is to be noted that a semi-normal family $\mathcal{P}$ is necessarily open-monotone and strongly closed-additive whenever the universe is contained in the class $\mathcal{M}$ of metrizable spaces. And when $\mathcal{M}$ is replaced by the class $\mathcal{M}_0$ of separable metrizable spaces, the semi-normal families are necessarily regular families.

The operation $\mathbf{S}$ was used to single out the part of the class $\mathbf{M}[\,\mathcal{P}\,]$ which is open-monotone (see Theorem 3.12). Indeed, in the development of normal families and its variants it was the class $\mathbf{S}[\,\mathbf{M}[\,\mathcal{P}\,]\,]$ that inherited open monotonicity. The class $\mathcal{E}$ of strongly hereditarily normal spaces uses the Dowker-open sets in conjunction with the operation $\mathbf{S}$. From Lemma 3.5 we see that the Dowker-open sets are, in a sense, the largest collection of open sets that contains the cozero-sets and are related to σ-locally finite unions of closed sets, a property of semi-normal families. Unfortunately the class $\mathcal{E}$ is not closed-additive in the universe $\mathcal{N}_H$ of hereditarily normal spaces. The next definition will correct this deficiency, thereby creating a universe which is, in a sense, optimal with respect to cozero-sets and normal families.

3.22. Definition. The *Dowker universe* $\mathcal{D}$ is the normal family extension of the class $\mathcal{E}$ in $\mathcal{N}_H$. That is, $\mathcal{D} = \mathbf{N}[\,\mathcal{E} : \mathcal{N}_H\,]$.

Clearly the inclusions $\mathcal{E} \subset \mathcal{D} \subset \mathcal{N}_H$ hold, and by the proof of Theorem 3.10 the universe $\mathcal{D}$ is absolutely monotone.

3.23. Theorem. *Suppose that the universe is contained in the Dowker universe. Let $\mathcal{P}$ be an open-monotone, strongly closed-additive semi-normal family. Then $\mathbf{M}[\mathcal{P}]$ is also an open-monotone, strongly closed-additive semi-normal family.*

Proof. From Lemma 3.5 and the proof of Theorem 3.12 we have that $\mathbf{M}[\mathcal{P}] \cap \mathcal{E} \cap \mathcal{U}$ is absolutely open-monotone. Let us show that $\mathbf{M}[\mathcal{P}] \cap \mathcal{U}$ is absolutely open-monotone. Assume $X \in \mathbf{M}[\mathcal{P}] \cap \mathcal{U}$. Then $X = \bigcup \mathbf{X}$, where $\mathbf{X}$ is a σ-locally finite closed cover of X such that $\mathbf{X} \subset \mathcal{E} \cap \mathcal{U}$. Let $\mathbf{X} = \bigcup\{\, \mathbf{X}_n : n = 1, 2, \dots\}$ where $\mathbf{X}_n$ is locally finite in X. Because $\mathbf{M}[\mathcal{P}] \cap \mathcal{E} \cap \mathcal{U}$ is absolutely open-monotone, we have from Lemma 3.19 that $X_n = \bigcup \mathbf{X}_n$ is in $\mathbf{M}[\mathcal{P}] \cap \mathcal{U}$. Indeed, for each open subspace Y of X the set $Y \cap X_n$ is in $\mathbf{M}[\mathcal{P}] \cap \mathcal{U}$. So X is in $\mathbf{M}[\mathcal{P}]$ by Lemma 3.17. Again each open subspace Y of X is in $\mathbf{M}[\mathcal{P}] \cap \mathcal{U}$. Thus we have that $\mathbf{M}[\mathcal{P}] \cap \mathcal{U}$ is absolutely open-monotone. Theorem 3.13 gives $\mathbf{S}\big[\mathbf{M}[\mathcal{P}]\big] \cap \mathcal{U} = \mathbf{M}[\mathcal{P}] \cap \mathcal{U}$; and, Theorem 3.20 completes the proof.

Observe at this point that the open-monotone and strongly closed-additive hypotheses on the class $\mathcal{P}$ are redundant for the universe $\mathcal{N}_P$ of perfectly normal spaces since each open set of a perfectly normal space is a cozero-set.

The last theorem and Theorem 3.14 give the following one for normal families.

3.24. Theorem. *Suppose that the universe is monotone and is contained in the Dowker universe. Let $\mathcal{P}$ be a strongly closed-additive normal family. Then $\mathbf{M}[\mathcal{P}]$ is also a strongly closed-additive normal family.*

With $\mathcal{P} = \{\emptyset\}$ we now have the extension of the theorem of Dowker [1953] from the universe $\mathcal{N}_T$ of totally normal spaces to the Dowker universe $\mathcal{D}$.

3.25. Theorem. *In the Dowker universe $\mathcal{D}$, the strong inductive dimension* Ind *has the following properties:*

(a) Monotone: *If $Y \subset X$, then* $\operatorname{Ind} Y \leq \operatorname{Ind} X$.

(b) Countably closed-additive: *If $X = \bigcup\{\, X_k : k \geq 1\}$ where each X_k is closed, then* $\operatorname{Ind} X$ *is the supremum of the set* $\{\operatorname{Ind} X_k : k \geq 1\}$.

(c) Locally finitely closed-additive: *If* $X = \bigcup \boldsymbol{F}$ *for a locally finite closed collection* $\boldsymbol{F}$ *in* X, *then* $\operatorname{Ind} X$ *is the supremum of the set* $\{ \operatorname{Ind} F : F \in \boldsymbol{F} \}$.
(d) Strongly closed-additive: *If* $X = F \cup G$ *and* F *is closed, then* $\operatorname{Ind} X$ *is the maximum of* $\operatorname{Ind} F$ *and* $\operatorname{Ind} G$.
(e) Additive: $\operatorname{Ind}(X \cup Y) \leq \operatorname{Ind} X + \operatorname{Ind} Y + 1$.

A consequence of the last theorem is that if a space X in the Dowker universe is such that it is the union of $n+1$ subspaces with Ind not exceeding 0 then $\operatorname{Ind} X \leq n$. This is one-half of the decomposition theorem in the universe $\mathcal{M}$ of metrizable spaces. The decomposition theorem with the additional condition of separability has been shown in Theorem 2.10. The theory of Ind in the universe $\mathcal{M}$ will be completed in Section V.1.

The condition that $\mathcal{P}$ be a normal family can be replaced by cosmic family when the universe is contained in the class $\mathcal{N}_P$ of perfectly normal spaces.

3.26. Theorem. *Suppose that the universe is contained in the class* $\mathcal{N}_P$ *of perfectly normal spaces. Let* $\mathcal{P}$ *be a strongly closed-additive cosmic family. Then* $\mathcal{P}$ *is open-monotone and* $\mathbf{M}[\mathcal{P}]$ *is also an open-monotone, strongly closed-additive cosmic family.*

Proof. Each open set of a perfectly normal space is a cozero-set. Consequently $\mathcal{P}$ is open-monotone. The remaining part of the theorem follows from Theorems 3.2, 3.12 and 3.18.

3.27. Example. Consider the universe $\mathcal{M}$ of metrizable spaces. The class $\mathcal{C}$ of complete metrizable spaces is a strongly closed-additive cosmic family that is not a semi-normal family. There are cosmic families that are not strongly closed-additive. In particular, the class $\mathcal{L}$ of locally compact spaces is one.

We shall conclude the section with a theorem showing the coincidence of $\mathcal{P}$-ind X and $\mathcal{P}$-Ind X for σ-totally paracompact spaces X.

3.28. Definition. A space X is said to be *σ-totally paracompact* if for every basis $\mathcal{B}$ of X there exists a σ-locally finite open cover $\boldsymbol{U}$ of X such that for each U in $\boldsymbol{U}$ there is a V in $\mathcal{B}$ such that $U \subset V$ and $\mathrm{B}_X(U) \subset \mathrm{B}_X(V)$.

3.29. Theorem. *Suppose that the universe is contained in the Dowker universe. Let $\mathcal{P}$ be an open-monotone, strongly closed-additive semi-normal family. If X is a σ-totally paracompact space, then*

$$\mathcal{P}\text{-ind}\, X = \mathcal{P}\text{-Ind}\, X.$$

Proof. Only the inequality $\mathcal{P}\text{-ind}\, X \geq \mathcal{P}\text{-Ind}\, X$ requires a proof. The proof is by induction on $\mathcal{P}\text{-ind}\, X$. We shall provide the inductive step of the proof. Assume $\mathcal{P}\text{-ind}\, X \leq n$ and let F and G be disjoint closed sets of X. Let F_0 and G_0 be disjoint closed neighborhoods of F and G respectively. Then the collection $\mathcal{B}$ of all open sets V such that $\mathcal{P}\text{-ind}\, \mathrm{B}_X(V) < n$ holds and such that $V \cap F_0 = \emptyset$ or $V \cap G_0 = \emptyset$ holds is a basis for the open sets of X. Since X is σ-totally paracompact, there is a σ-locally finite open cover $\boldsymbol{U}$ of X such that for each U in $\boldsymbol{U}$ there is a V in $\mathcal{B}$ with $U \subset V$ and $\mathrm{B}_X(U) \subset \mathrm{B}_X(V)$. The set $M = \bigcup\{\, \mathrm{B}_X(U) : U \in \boldsymbol{U}\}$ need not be a member of the universe. But nonetheless it is covered by the σ-locally finite closed collection $\{\, \mathrm{B}_X(U) : U \in \boldsymbol{U}\}$ whose members satisfy $\mathcal{P}\text{-ind}\, \mathrm{B}_X(U) = \mathcal{P}\text{-Ind}\, \mathrm{B}_X(U) < n$.

Let $\boldsymbol{U} = \bigcup\{\, \boldsymbol{U}_k : k \geq 1\,\}$ where each $\boldsymbol{U}_k$ is locally finite. For each k we define the open sets

$$W_{0k} = \bigcup\{\, U \in \boldsymbol{U}_k : U \cap F_0 \neq \emptyset\,\},$$
$$W_{1k} = \bigcup\{\, U \in \boldsymbol{U}_k : U \cap F_0 = \emptyset\,\},$$

and then form the open sets

$$\widetilde{W}_{0k} = W_{0k} \setminus \bigcup\{\, \mathrm{cl}_X(W_{1m}) : m < k\,\},$$
$$\widetilde{W}_{1k} = W_{1k} \setminus \bigcup\{\, \mathrm{cl}_X(W_{0m}) : m \leq k\,\}.$$

Clearly, $\bigcup\{\, W_{0k} \cup W_{1k} : k \geq 1\,\} = X$. A simple calculation will show that $\widetilde{W}_{0i} \cap \widetilde{W}_{1j} = \emptyset$ for all i and j. So the open sets

$$\widetilde{W}_0 = \bigcup\{\, \widetilde{W}_{0k} : k \geq 1\,\} \quad \text{and} \quad \widetilde{W}_1 = \bigcup\{\, \widetilde{W}_{1k} : k \geq 1\,\}$$

are disjoint and the inclusions $F \subset \widetilde{W}_0$ and $G \subset \widetilde{W}_1$ hold. Denote the set $X \setminus (\widetilde{W}_0 \cup \widetilde{W}_1)$ by S. Let us show that S is contained in M. For x in S let i be the first integer such that $x \in W_{0i} \cup W_{1i}$.

Since $x \notin \widetilde{W}_{0i} \cup \widetilde{W}_{1i}$, we have that x is in $\bigcup\{\mathrm{B}_X(U) : U \in \boldsymbol{U}_j\}$ for some j not greater than i. So x is in M. Since S is a closed set of X, we have that S is in $\mathcal{U}$. From $S = \bigcup\{S \cap \mathrm{B}_X(U) : U \in \boldsymbol{U}\}$, we have that $\mathcal{P}$-Ind $S < n$ by Theorem 3.23. Thus we have shown that $\mathcal{P}$-Ind $X \leq n$.

The definition of σ-totally paracompactness appears to be rather technical. It can be shown that strongly paracompact spaces are σ-totally paracompact. A Hausdorff space is *strongly paracompact* if each of its open cover has a star-finite open refinement.

4. Embedding theorems

Many axiom schemes for the dimension function have utilized universal n-dimensional spaces in the separable metrizable space setting and universal 0-dimensional spaces in the general metrizable space setting. To prepare for the discussion in the next section on axioms for dimension, we shall give a brief account of the relevant embeddings.

The first embedding theorem will use the function space $(\mathbb{R}^{2n+1})^X$ and the notion of ε-mapping.

4.1. Definition. Let ε be a positive number and let $f\colon X \to Y$ be a continuous map of a metric space X to a topological space Y. Then f is said to be an *ε-mapping* if $\operatorname{diam}\left(f^{-1}(y)\right) < \varepsilon$ for every y in Y.

4.2. Theorem. *A compact metrizable space X with $\dim X \leq n$ can be embedded into the subspace N_n^{2n+1} of $\mathbb{R}^{2n+1}$ consisting of all points which have at most n rational coordinates.*

Proof. The complement $\mathbb{R}^{2n+1} \setminus N_n^{2n+1} = L_{n+1}^{2n+1}$ is the union of a countable family of linear n-varieties in $\mathbb{R}^{2n+1}$ defined by the equations $x_{i_1} = r_1$, $x_{i_2} = r_2$, ..., $x_{i_{n+1}} = r_{n+1}$, where $r_1, r_2, \ldots, r_{n+1}$ are rational numbers and $1 \leq i_1 < i_2 < \cdots < i_{n+1} \leq 2n+1$. Arrange this family of n-varieties into a sequence $H_1, H_2, \ldots$, and denote by Φ_i the subset of the function space $(\mathbb{R}^{2n+1})^X$ consisting of all $(1/i)$-mappings whose values miss H_i, $i = 1, 2, \ldots$.

Let us show that Φ_i is a dense open subset of $(\mathbb{R}^{2n+1})^X$. To prove that Φ_i is open, consider an $(1/i)$-mapping $f\colon X \to \mathbb{R}^{2n+1}$ such that $f[X] \cap H_i = \emptyset$. The closed subspace

$$A = \{(x, x') \in X \times X : d(x, x') \geq 1/i\}$$

is compact. Since $\|f(x) - f(x')\| > 0$ holds for each pair (x, x') in A, there exists a positive number δ such that

$$\|f(x) - f(x')\| \geq \delta, \qquad (x, x') \in A. \tag{1}$$

We may assume

$$\delta < \operatorname{dist}(f[X], H_i). \tag{2}$$

To see that f is an interior point of Φ_i we observe that every continuous map $g\colon X \to \mathbb{R}^{2n+1}$ with $\|f - g\| < \delta/2$ is a $(1/i)$-mapping with $g[X] \cap H_i = \emptyset$, where $\|f - g\|$ is the distance between f and g in the function space $(\mathbb{R}^{2n+1})^X$. Indeed, because $\|f(x) - f(x')\| < \delta$ follows from $x \in g^{-1}[y]$ and $x' \in g^{-1}[y]$, we have from (1) that every pair (x, x') of points of the set $g^{-1}[y]$ is not in A for any y. Thus, $\operatorname{diam} g^{-1}[y] < 1/i$ for every y in $\mathbb{R}^{2n+1}$, i.e., g is an $(1/i)$-mapping. The equality $g[X] \cap H_i = \emptyset$ is easily proved by way of contradiction by virtue of (2). So, Φ_i is open.

It now remains to be shown that Φ_i is dense in the function space $(\mathbb{R}^{2n+1})^X$. Consider a continuous map $f\colon X \to \mathbb{R}^{2n+1}$ and a positive number ε. The uniform continuity of f yields a positive number γ smaller than $1/i$ such that

$$\|f(x) - f(x')\| < \varepsilon/2 \quad \text{whenever} \quad d(x, x') < \gamma.$$

As $\dim X \leq n$, there is an open cover $\boldsymbol{U} = \{ U_j : j = 1, \ldots, r \}$ of X such that

$$\operatorname{ord} \boldsymbol{U} \leq n + 1, \tag{3}$$

$$\operatorname{diam}(U_j) < \gamma, \qquad j = 1, \ldots, r. \tag{4}$$

Consequently,

$$\operatorname{diam}\big(f[U_j]\big) < \varepsilon/2, \qquad j = 1, \ldots, r. \tag{5}$$

Select vertices $p_1, \ldots, p_r$ in $\mathbb{R}^{2n+1}$ in general position (i.e., no $m + 2$ of the vertices p_j lie in an m-dimensional linear variety of $\mathbb{R}^{2n+1}$, $m = 0, 1, \ldots, 2n$) such that every n-dimensional simplex in $\mathbb{R}^{2n+1}$

formed by these vertices is disjoint from the n-dimensional variety H_i and

$$\text{dist}\,(p_j, f[U_j]) < \varepsilon/2, \qquad j = 1, \dots, r. \tag{6}$$

For each point x of X define

$$w_j(x) = d(x, X \setminus U_j),\ j = 1, \dots, r, \quad \text{and} \quad w(x) = \textstyle\sum_{j=1}^{r} w_j(x)$$

and form the continuous map $g\colon X \to \mathbb{R}^{2n+1}$ by the formula

$$g(x) = \textstyle\sum_{j=1}^{r} \frac{w_j(x)}{w(x)} p_j, \qquad x \in X.$$

Using (3)–(6), we will get after a straightforward computation that g is an $(1/i)$-mapping with $\|f - g\| < \varepsilon$ and $g[X] \cap H_i = \emptyset$.

We have now proved that Φ_i is a dense, open subset of the function space $(\mathbb{R}^{2n+1})^X$. Since this function space is complete, the Baire category theorem yields that

$$\Phi = \bigcap \{\, \Phi_i : i = 1, 2, \dots \} \neq \emptyset.$$

As every map in Φ is an embedding of X into N_n^{2n+1}, the theorem is proved.

It is to be observed that the above argument can also be carried out in the function space $(\mathbb{I}^{2n+1})^X$ due to the convexity of $\mathbb{I}^{2n+1}$ in $\mathbb{R}^{2n+1}$. Consequently the set N_n^{2n+1} of the theorem can be replaced by the set $N_n^{2n+1} \cap \mathbb{I}^{2n+1}$.

It will be proved in Chapter VI that every finite dimensional separable metrizable space has a metrizable compactification of the same dimension (see Theorem VI.2.9). Thus we have the following theorem.

4.3. Theorem. *A separable metrizable space has finite dimension if and only if it can be embedded into a finite dimensional Euclidean space.*

We shall next describe a construction due to Hayashi [1990] which will be used to construct a universal n-dimensional space. The construction is similar to that of the Menger sponge M_n^m (see Engelking [1978], page 121). For each m-cell $\mathbb{I}^m$ there are two construction schemes. We shall call one the exclusion scheme and the other the inclusion scheme. The constructions are made on a base 6 procedure rather than the base 3 one used by Menger [1926]. To help the reader understand how the two schemes come into play, the construction will be illustrated in the 1-cell $\mathbb{I}^1$. In both constructions a Cantor set is built. As this special case is somewhat easier than the general one, the constructions can be made on a base 3 procedure. After completing the constructions for $\mathbb{I}^1$, we shall show how the base 6 procedure comes into the picture in $\mathbb{I}^2$.

For the exclusion scheme in the 1-cell $\mathbb{I}^1$ let us consider a subset X of $\mathbb{I}^1$ that is the union of a countable family of compact subsets X_i of the set $\mathbb{P}$ of irrational numbers. We want to construct a Cantor set in $\mathbb{I}^1$ that excludes the set X. We begin the construction by excluding X_1. Since $\{-1, 1\} \cap X_1 = \emptyset$, there are rational numbers a_1 and a_2 such that

$$-1 < a_1 < a_2 < 1 \quad \text{and} \quad X_1 \subset \{ x : a_1 < x < a_2 \}.$$

For each of the 1-cells $[-1, a_1]$ and $[a_2, 1]$ we perform a similar construction to exclude X_2. This results in four new 1-cells. Clearly the construction, when continued indefinitely, will eventually exclude each point of the set X from the Cantor set which consists of all points that are in some descending sequence of 1-cells, one from each stage of the construction.

To illustrate the inclusion scheme in the 1-cell $\mathbb{I}^1$, let X be a compact subset of $\mathbb{P} \cap \mathbb{I}^1$. We want to construct a Cantor set in $\mathbb{I}^1$ that includes the set X. Clearly the midpoint 0 of the 1-cell $\mathbb{I}^1$ is not in X. There are rational numbers l_1 and r_1 such that

$$l_1 < 0 < r_1 \quad \text{and} \quad X \subset [-1, l_1] \cup [r_1, 1].$$

This is the first step in the construction of the Cantor set. On each of the 1-cells $[-1, l_1]$ and $[r_1, 1]$ we perform a similar construction and remove a suitable middle open 1-cell. Obviously the space X is included in the resulting Cantor set.

Now let us indicate why the base 6 procedure is needed. The next space in the hierarchy of universal spaces is the 1-dimensional one contained in the 2-cell $\mathbb{I}^2$. The exclusion scheme of the construction will require a base 6 procedure. We consider a subset X of $\mathbb{I}^2$ that is the union of a countable family of compact subsets X_i of the set $\mathbb{P} \times \mathbb{P}$.

We want to construct a Sierpinski curve in $\mathbb{I}^2$ that excludes the set X. Since the combinatorial boundary of $\mathbb{I}^2$ is disjoint with $\mathbb{P} \times \mathbb{P}$, there are rational numbers a_1, a_2, b_1 and b_2 such that

$$-1 < a_1 < a_2 < 1, \qquad -1 < b_1 < b_2 < 1$$

and

$$X_1 \subset \{ (x, y) : a_1 < x < a_2, b_1 < y < b_1 \}.$$

For the first step of the the construction of the universal curve we remove the open 2-cell $(a_1, a_2) \times (b_1, b_2)$. In this way, the set X_1 is excluded from the the union of the remaining eight 2-cells. It is to be observed that there is no guarantee that $|a_1 - a_2|$ and $|b_1 - b_2|$ are less than 1. In the next stage we want to perform a similar construction for each of the remaining eight 2-cells in order to exclude X_2. The universal 1-dimensional space we wish to construct will consist of all points that are in the the intersection of some descending chain of 2-cells, one from each stage. To guarantee the topological uniqueness of the resulting 1-dimensional space, we must force the diameter of the chains of 2-cells to converge to 0. Using the rational midpoints c_1, c_2, c_3 and d_1, d_2, d_3 of the 1-cells contained in $\mathbb{I}^1$ that are formed by the points a_1, a_2 and b_1, b_2 respectively, we get the arrangements

$$-1 < c_1 < a_1 < c_2 < a_2 < c_3 < 1,$$
$$-1 < d_1 < b_1 < d_2 < b_2 < d_3 < 1,$$

where all of the 1-cells have length less than 1 and all of the 2-cells have diameter less that $\sqrt{2}$. (In this way, a base 6 procedure has emerged.) Now X_2 is excluded by removing simultaneously the middle open 2-cells from each of the 32 2-cells.

Now we shall describe the complete procedure more systematically. Both the exclusion and the inclusion schemes will use a base 6 construction. Let us describe the base 6 construction. For each positive integer i let $Q_i = \{ q_{i,k} : k = 0, 1, \ldots, 6^i \}$ be a finite subset of the set $\mathbb{Q}$ of rational numbers such that

(h1) $-1 = q_{i,0} < q_{i,1} < \cdots < q_{i,6^i} = 1$,
(h2) $Q_i \subset Q_{i+1}$,
(h3) $|q_{i,k} - q_{i,k-1}| < \frac{1}{2^{i-1}}$, $\qquad k = 1, \ldots, 6^i$,
(h4) each 1-cell $[q_{i,k}, q_{i,k+1}]$ formed by Q_i is subdivided into 6 parts by Q_{i+1}.

For each i we set $D_i = \{ q_{i,k} : 2 \text{ divides } k \}$. Then the 1-cells determined by D_i are subdivided into two parts by Q_i.

We can now describe Hayashi's modification of the Menger sponge construction. Let m and n be natural numbers with $n < m$. The m-cell $\mathbb{I}^m$ is divided into 6^{im} m-cells by the set Q_i and into 3^{im} m-cells by the set D_i. We shall denote the family of m-cells formed from Q_i by $\boldsymbol{L}_i$ and the family formed from D_i by $\boldsymbol{K}_i$. In the following way we inductively define a sequence of pairs of collections $\boldsymbol{F}_i$ and $\boldsymbol{G}_i$, $i = 0, 1, \ldots$, and a sequence F_i, $i = 0, 1, \ldots$, of closed subsets of $\mathbb{I}^m$ such that $\boldsymbol{F}_i \subset \boldsymbol{K}_i$, $\boldsymbol{G}_i \subset \boldsymbol{L}_i$ and $F_i \supset F_{i+1}$ hold for every i:

Let $\boldsymbol{F}_0 = \boldsymbol{G}_0 = \{ \mathbb{I}^m \}$ and $F_0 = \mathbb{I}^m$. Suppose that $\boldsymbol{F}_i$, $\boldsymbol{G}_i$ and F_i have been defined. With $S_i^{(n)}$ denoting the union of the faces S of the members of $\boldsymbol{G}_i$ with $\dim S \leq n$, let

$$\boldsymbol{F}_{i+1} = \{ K : K \in \boldsymbol{K}_{i+1},\ K \cap S_i^{(n)} \neq \emptyset \}, \qquad F_{i+1} = \bigcup \boldsymbol{F}_{i+1},$$
$$\boldsymbol{G}_{i+1} = \{ L : L \in \boldsymbol{L}_{i+1},\ L \subset F_{i+1} \}.$$

A *Hayashi sponge* H_n^m is defined to be the set $\bigcap \{ F_i : i \in \mathbb{N} \}$. Clearly a Hayashi sponge depends on the particular sequence Q_i, $i = 1, 2, \ldots$, possessing the conditions (h1)–(h4). But equally clearly from condition (h3) is the fact that any two Hayashi sponges are homeomorphic for a given pair m and n.

Let us construct an exclusion Hayashi sponge in the m-cell $\mathbb{I}^m$. Let $\mathbb{I}_n^m$ be the set of all points of $\mathbb{I}^m$ at most n of whose coordinates are irrational. Suppose X is the union of a countable family of compact subset X_i of $\mathbb{I}^m \setminus \mathbb{I}_n^m$. The set $S_0^{(n)}$ is the union of all n-dimensional faces of $\mathbb{I}^m$. Since $S_0^{(n)} \subset \mathbb{I}_n^m$, there is a positive rational number δ_0 such that $\operatorname{dist}(X_1, S_0^{(n)}) > m\delta_0$. Define

$$q_{1,1} = -1 + \tfrac{\delta_0}{2},\ q_{1,2} = -1 + \delta_0,\ q_{1,3} = 0,\ q_{1,4} = 1 - \delta_0,\ q_{1,5} = 1 - \tfrac{\delta_0}{2}.$$

Then the set F_1 of the Hayashi construction and X_1 are disjoint. Applying the next stage of the Hayashi construction with the compact set $X_1 \cup X_2$, one gets a set F_2 such that F_2 and $X_1 \cup X_2$ are disjoint. Proceeding inductively, one can construct a Hayashi sponge H_n^m that excludes X.

Let us next construct an inclusion Hayashi sponge. Here we make the assumption that $m > n > 0$. On inspecting the construction of the closed set F_1, one finds that $\mathbb{I}^m \setminus F_1$ is contained in the set A_1

which is the union of the family of all m-cells H formed by n factors of $\mathbb{I}^1$ and $m-n$ factors of the middle 1-cells formed by the set D_1. Let X be a compact subset of $\mathbb{I}^m \setminus \mathbb{I}^m_n$. Our aim is to find a set Q_1 such that $A_1 \cap X = \emptyset$. Consider the set

$$P_1 = \{ x \in \mathbb{I}^m : x_k = 0,\, k = 1, 2, \ldots, m-n \}.$$

(There are $\binom{m}{m-n}$ such sets formed by the combinations of $m-n$ coordinates.) Since $P_1 \subset \mathbb{I}^m_n$, there is a positive rational number δ less than 1 such that the m-cell

$$H_1 = \{ x \in \mathbb{I}^m : |x_k| \leq \delta,\, k = 1, 2, \ldots, m-n \}$$

does not meet X. Indeed, a common δ can be used for all $\binom{n}{m-n}$ sets H_j which form the set A_1. Consequently the set Q_1 given by $q_{1,1} = -\frac{1+\delta}{2}$, $q_{1,2} = -\delta$, $q_{1,3} = 0$, $q_{1,4} = \delta$ and $q_{1,5} = \frac{1+\delta}{2}$ will result in a set F_1 such that $X \subset F_1$. In a similar way, one can construct the set Q_2 so that $X \subset F_2$. Proceeding inductively, one can assure that $X \subset F_i$, $i \in \mathbb{N}$. The resulting Hayashi sponge H^m_n will contain the compact set X.

We are now ready to prove that the Hayashi sponge H^{2n+1}_n is a universal n-dimensional space.

4.4. Theorem. *For* $n \geq 1$, *every space* X *in* $\mathcal{M}_0$ *with* $\dim X \leq n$ *can be embedded in each Hayashi sponge* H^{2n+1}_n.

Proof. Since each space X with $\dim X \leq n$ can be embedded in a compact space of the same dimension, we may assume that X is compact. By Theorem 4.2, the space N^{2n+1}_n contains a copy of X. In the inclusion construction we have shown that there is a Hayashi sponge H^{2n+1}_n which contains this copy. Also we have shown that two Hayashi sponges are homeomorphic. So the theorem follows.

The following is an interesting application of the Hayashi sponge to the space $\mathbb{I}^{2n+1}_n$ of all points of $\mathbb{I}^{2n+1}$ at most n of whose coordinates are irrational. (Recall that $\mathcal{C}$ is the class of complete metrizable spaces.)

4.5. Theorem. *Every* $\mathcal{C}$*-hull of* $\mathbb{I}^{2n+1}_n$ *contains a copy of* H^{2n+1}_n.

Proof. Let Y be a $\mathcal{C}$-hull of $\mathbb{I}^{2n+1}_n$. By Lavrentieff's theorem (Theorem I.7.3), there is no loss in assuming that Y is a subset of $\mathbb{I}^{2n+1}$.

Let $X = \mathbb{I}^{2n+1} \setminus Y$. In the preceding discussion of the exclusion construction of the Hayashi sponges, we have shown that there is an H_n^{2n+1} that excludes X and thus is contained in Y.

The next proposition is a technical one whose roots are found in the first paper on axiom schemes by Menger [1929].

4.6. Proposition. *If X is a finite dimensional, metrizable continuum with more than one point, then there is a separable metrizable space Y such that Y is the union of a countable family of copies of X and such that every $\mathcal{C}$-hull Z of Y contains an arc.*

Proof. The construction of Y is a typical example of a construction of a self-similar set or fractal. By Theorem 4.3, the space X can be embedded in a unit cube $\mathbb{I}^m$. Indeed, there is no difficulty in arranging the embedding to contain a vertex of the cube and the center of the cube. Using this embedding, we can find 2^m such embeddings, one for each vertex of $\mathbb{I}^m$. The union Y_0 of all of these copies is connected and contains the center and all of the vertices of the cube. On dividing the cube into 2^m subcubes, we can use 2^m copies of Y_0, one for each of the subcubes, and take their union to form a connected set Y_1. Continuing inductively, we form the connected sets Y_n with the property that the 2^{nm} subcubes of $\mathbb{I}^m$ intersect the set Y_n in a geometrically similar way as Y_0. Clearly the set $Y = \bigcup\{Y_n : n \in \mathbb{N}\}$ is connected and each of the 2^{nm} subcubes of $\mathbb{I}^m$ contains a subset of Y which is geometrically similar to Y and dense in the subcube. Consequently the set Y is uniformly locally connected as well.

Let Z be any $\mathcal{C}$-hull of Y. By Lavrentieff's theorem, we may assume that Z is a subset of $\mathbb{I}^m$. So, $\mathbb{I}^m \supset Z \supset Y$. Clearly Z is connected and, from the construction of Y, uniformly locally connected. As Z is topologically complete, in a standard way, it follows that Z is arcwise connected.

Let us turn now to zero dimensional, metrizable spaces. It is not difficult to show that every separable such space can be embedded into the space $\mathbb{P}$ of irrational numbers. (Moreover, the space $\mathbb{P}$ can be embedded into the Cantor set.) For the nonseparable case, Morita [1954] has shown that they can be embedded into the generalized Baire space $N(D)$ of the same weight. For the purposes of the axiomatization of dimension we will need a lemma which is a sharpened version of Morita's theorem. In preparation for this lemma, some preliminary notation will be given.

Let D denote a discrete space. The Baire space $N(D)$ is the countable product

$$N(D) = \prod\{\, D_i : D_i = D,\ i = 1, 2, \ldots\},$$

where the metric ρ is defined by the formula

$$\rho(\langle a_i\rangle, \langle b_i\rangle) = \begin{cases} 1/i, & \text{when } a_k = b_k \text{ for } k < i \text{ and } a_i \neq b_i \\ 0, & \text{when } a_i = b_i \text{ for } i \geq 1. \end{cases}$$

Clearly the Baire space is complete. The r-neighborhood $\mathrm{S}_r(\langle a_i\rangle)$ of $\langle a_i\rangle$ is the set $\{a_1\} \times \cdots \times \{a_k\} \times \prod\{\, D_i : i > k\,\}$ whenever $\frac{1}{k+1} < r \leq \frac{1}{k}$. Consequently, if $\mathrm{S}_r(\langle a_i\rangle) \cap \mathrm{S}_r(\langle b_i\rangle)$ is not empty, then $\mathrm{S}_r(\langle a_i\rangle) = \mathrm{S}_r(\langle b_i\rangle)$. So the collection

$$\boldsymbol{U}_k = \{\, U : U = \mathrm{S}_{\frac{1}{k}}(\langle a_i\rangle) \text{ for some } \langle a_i\rangle\,\}$$

is an open cover of $N(D)$ by pairwise disjoint sets such that $\boldsymbol{U}_{k+1}$ refines $\boldsymbol{U}_k$.

Let p be a fixed point in D. Then, for each i, the finite product

$$N_i(D) = \prod\{\, D_j : D_j = D,\ 1 \leq j \leq i\,\}$$

is embedded into $N(D)$ by the map e_i given by the formula

$$e_i(\langle\, a_1, \ldots a_i\,\rangle) = \langle\, a_1, \ldots, a_i, p, p, \ldots\rangle.$$

Denote the natural projection of $N(D)$ onto $N_i(D)$ by φ_i and the natural projection of $N(D)$ onto D_i by π_i. Finally denote by G_i and G the subsets of $N(D)$ given by

$$G_i = e_i[\, N_i(D)\,] \quad \text{and} \quad G = \bigcup\{\, G_i : i = 1, 2, \ldots\}.$$

Clearly we have that

$$G = \{\, x \in N(D) : \pi(x) = p \text{ except for at most finitely many } i\,\}$$

and G is the union of a σ-discrete collection of singletons in $N(D)$.

4.7. Lemma. *Let D be an infinite discrete space with cardinality α and let p be a fixed point of D. With the notation given above, let H be a G_δ-set of $N(D)$ that contains G. Then $N(D)$ can be embedded into H.*

Proof. Let $E = D \setminus \{p\}$. Then $N(E)$ is a closed set of $N(D)$ that is disjoint from G. We may assume without loss of generality that $N(E) \cap H = \emptyset$. The set $N(D) \setminus H$ is the union of a collection $\{F_i : i = 1, 2, \dots\}$ of closed sets of $N(D)$ with $F_i \subset F_{i+1}$ and $F_1 = N(E)$.

To avoid confusion, let X denote the space $N(D)$ which is to be embedded into H. From the above discussion, for each i there exists a cover $\boldsymbol{U}_i$ of X such that

(a1) each member of $\boldsymbol{U}_i$ is open and closed,
(a2) any two distinct members of $\boldsymbol{U}_i$ are disjoint,
(a3) mesh $\boldsymbol{U}_i \leq 1/i$,
(a4) $\boldsymbol{U}_{i+1}$ is a refinement of $\boldsymbol{U}_i$.

Because the weight of X does not exceed α, the cardinality of $\boldsymbol{U}_i$ also does not exceed α.

We shall construct an embedding f of X into H by inductively defining continuous maps $f_n = \pi_n f$. To this end, for each n let $\widetilde{g}_n$ be a one-to-one correspondence between $\boldsymbol{U}_n$ and a subset of E. Define the continuous map $g_n\colon X \to E$ by $g_n(x) = \widetilde{g}_n(U)$, $x \in U \in \boldsymbol{U}_n$. These maps g_n will be used to define the maps f_n on each U in $\boldsymbol{U}_n$ and to define natural numbers $k(n, U)$ and $l(n, U)$ for each U in $\boldsymbol{U}_n$.

Let us start by observing that for each k and each $\langle a_1, \dots, a_n\rangle$ there exists an m with $\varphi_m^{-1}[\varphi_m e_n(\langle a_1, \dots, a_n\rangle)] \cap F_k = \emptyset$ because $e_n(\langle a_1, \dots, a_n\rangle)$ is not a member of F_k.

For $n = 1$ let $f_1 = g_1$. Obviously f_1 is constant on each U in $\boldsymbol{U}_1$. For each U in $\boldsymbol{U}_1$ let $k(1, U) = 1$ and let $l(1, U)$ be the minimum natural number $l \geq 1$ such that the set $\varphi_l^{-1}[\varphi_l e_1 f_1(x)] \cap F_{k(1,U)+1}$ is empty, where $x \in U$.

Let us go to $n = 2$. For U in $\boldsymbol{U}_2$ let U' be the unique member of $\boldsymbol{U}_1$ with $U \subset U'$. If $2 > l(1, U')$, we let $f_2(x) = g_2(x)$ for x in U and let $k(2, U) = k(1, U') + 1$. If $2 \leq l(1, U')$, we let $f_2(x) = p$ for x in U and let $k(2, U) = k(1, U')$. Let $l(2, U)$ be the minimum natural number $l \geq 2$ such that

$$\phi_l^{-1}[\phi_l e_2(f_1 \times f_2)(x)] \cap F_{k(2,U)+1} = \emptyset.$$

Note that $l(2,U) = l(1,U')$ and $k(2,U) = k(1,U')$ when $2 \leq l(1,U')$.

Continuing inductively, we have for each n, $n > 0$, a continuous map f_n, and natural numbers $k(n,U_n)$ and $l(n,U_n)$ that satisfy the following conditions:

(b) $k(1,U) = 1$ for each U in $\boldsymbol{U}_1$.

(c) f_n is constant on each U in $\boldsymbol{U}_n$.

(d) For each U in $\boldsymbol{U}_n$ the number $l(n,U)$ is the minimum natural number $l \geq n$ such that

$$\phi_l^{-1}[\phi_l e_n(f_1 \times \cdots \times f_n)(x)] \cap F_{k(n,U)+1} = \emptyset,$$

where x is any point in U.

(e) For each U in $\boldsymbol{U}_{n+1}$ let U' be the unique member of $\boldsymbol{U}_n$ such that $U \subset U'$. The following hold when x is in U.

(e1) If $n+1 > l(n,U')$, then $f_{n+1}(x) = g_{n+1}(x)$
and $k(n+1,U) = k(n,U') + 1$.

(e2) If $n+1 \leq l(n,U')$, then $f_{n+1}(x) = p$
and $k(n+1,U) = k(n,U')$.

Observe that when $U' \in \boldsymbol{U}_n$, $U \in \boldsymbol{U}_{n+1}$ and $U \subset U'$ hold we have $l(n,U') = l(n+1,U)$ under condition (e2).

In advance of proving f is an embedding of X let us prove that the set $\{ n : n+1 > l(n,U_n), x \in U_n \in \boldsymbol{U}_n \}$ is unbounded. Suppose that there is an m larger than any member of this set. Then $n \geq m$ implies that $n+1 \leq l(n,U_n)$ holds for the unique U_n in $\boldsymbol{U}_n$ such that $x \in U_n$. So we have $l(n,U_n) = l(n+1,U_{n+1})$ for $n \geq m$ from the earlier observation. That is, $l(n,U_n) = l(m,U_m)$ when $n \geq m$. But this contradicts the inequality $l(n,U_n) \geq n$ of condition (d).

To prove that f is an embedding of X, it is sufficient to show the maps f_n, $n > 0$, separate points and closed sets. (See Kelley [1955], page 116.) Let x be a point and L be a closed set with $x \notin L$. There exists an m such that $U_m \cap L = \emptyset$ when $x \in U_m$. Let n be such that $n > m$ and $n > l(n-1,U)$ when $x \in U \in \boldsymbol{U}_{n-1}$. Note that $U \subset U_m$. Then $f_n(x) = g_n(x)$ by (e1). As $\widetilde{g}_n$ is one-to-one on $\boldsymbol{U}_n$ and $g_n(x) \neq p$, it follows that $f_n[X \setminus U_m] \subset D \setminus \{ f_n(x) \}$. Because D is a discrete space, f_n separates x and L; thereby f is an embedding of X into $N(D)$.

Finally let us show that $f[X] \subset H$. For x in X and k in $\mathbb{N} \setminus \{0\}$ choose an n such that $k(n,U) = k$ and $x \in U \in \boldsymbol{U}_n$. (Such an n exists by condition (e1).) Then with $m = l(n,U)$ we have from (d) that $\varphi_m^{-1}[\varphi_m e_m(f_1 \times \cdots \times f_m)(x)] \cap F_{k+1} = \emptyset$ holds. So $f(x) \notin F_k$ follows since $F_k \subset F_{k+1}$.

The above proof will apply when X is a metrizable space with weight not exceeding α and $\operatorname{Ind} X = 0$. For, in this situation, the paracompactness of X will permit the construction of the covers $\boldsymbol{U}_i$ satisfying the conditions (a1)–(a4). Thus we have a sharpening of Morita's embedding theorem.

4.8. Theorem. *Let D be an infinite discrete space with cardinality α and let p be a fixed point of D. With the notation given above, let H be a G_δ-set of $N(D)$ that contains G. Then every metric space X with weight not exceeding α and* $\operatorname{Ind} X = 0$ *can be embedded into H.*

5. Axioms for the dimension function

An early problem in dimension theory was that of finding suitably simple properties of the dimension functions that can serve as axioms for dimension theory. One property is indisputable. Any such axiom scheme must have as an axiom the topological invariance of the function. That is, if d is a dimension function defined on a universe $\mathcal{U}$, then $\mathrm{d}(X)$ must be equal to $\mathrm{d}(Y)$ when X is in $\mathcal{U}$ and Y is homeomorphic to X. The other properties that will be emphasized in this section are those related to normal families. This section will be devoted to a discussion of axiom schemes that have been formulated around those of Menger.

In Menger [1929] a characterization of the dimension function for the universe of all subsets of the Euclidean plane was given and a conjecture was made for the universe of all subsets of Euclidean n-dimensional space. These axioms by Menger will be called the M axioms.

5.1. The M axioms. Let $\mathcal{U}_n$ be the universe of spaces X that are embeddable in the n-dimensional Euclidean space $\mathbb{R}^n$ and let d be an extended-integer valued function on $\mathcal{U}_n$ that satisfies the axioms:

(M1) $\mathrm{d}(\emptyset) = -1$, $\mathrm{d}(\{\emptyset\}) = 0$ and $\mathrm{d}(\mathbb{R}^m) = m$ for $m = 1, 2, \ldots, n$.
(M2) If Y is a subspace of a space X in $\mathcal{U}_n$, then $\mathrm{d}(Y) \leq \mathrm{d}(X)$.

(M3) If a space X in $\mathcal{U}_n$ is a union of a sequence $X_1, X_2, \ldots$ of closed subspaces, then $\mathrm{d}(X) \leq \sup\{\mathrm{d}(X_i) : i = 1, 2, \ldots\}$.

(M4) Every space X in $\mathcal{U}_n$ has a compactification $\widetilde{X}$ in $\mathcal{U}_n$ satisfying $\mathrm{d}(\widetilde{X}) = \mathrm{d}(X)$.

The axioms (M2) and (M3) and the first two conditions of axiom (M1) are related to Theorem 3.24 on normal families. The remaining axioms are much deeper properties from dimension theory. In [1929] Menger conjectured that ind is the only function d which satisfies the axioms (M1)–(M4) for each positive natural number n, and he showed that the conjecture was correct when $n \leq 2$. The conjecture is still unresolved for the cases of n greater than 2. Indeed, it is still not known whether d = ind satisfies axiom (M4) when $n > 3$.

Consider next the universe $\mathcal{U}_\omega$ of all spaces X that can be embedded in some n-dimensional Euclidean space and a function d that satisfies the M axioms (M1)–(M4). Then ind does satisfy all four of the M axioms. However, ind is not the only function to do so; each cohomological dimension $\dim_G$ with respect to a finitely generated Abelian group G also satisfy the M axioms in the universe $\mathcal{U}_\omega$. (See Example H5 below.)

The next set of axioms first appeared in Nishiura [1966]. The axioms were for functions d defined on the universe $\mathcal{M}_0$ of separable metrizable spaces. Clearly $\mathcal{U}_\omega$ is a proper subclass of $\mathcal{M}_0$. These axioms will be called the N axioms.

5.2. The N axioms. Let d be an extended-real valued function on the universe $\mathcal{M}_0$ that satisfies the axioms:

(N1) $\mathrm{d}(\{\emptyset\}) = 0$.

(N2) If Y is a subspace of a space X in $\mathcal{M}_0$, then $\mathrm{d}(Y) \leq \mathrm{d}(X)$.

(N3) If a space X in $\mathcal{M}_0$ is a union of a sequence $X_1, X_2, \ldots$ of closed subspaces, then $\mathrm{d}(X) \leq \sup\{\mathrm{d}(X_i) : i = 1, 2, \ldots\}$.

(N4) If a space X in $\mathcal{M}_0$ is a union of two subspaces X_1 and X_2, then $\mathrm{d}(X) \leq \mathrm{d}(X_1) + \mathrm{d}(X_2) + 1$.

(N5) Every space X in $\mathcal{M}_0$ has a compactification $\widetilde{X}$ in $\mathcal{M}_0$ satisfying $\mathrm{d}(\widetilde{X}) = \mathrm{d}(X)$.

(N6) If x is a point of a space X in $\mathcal{M}_0$, then every neighborhood U of x has a neighborhood V with $\mathrm{d}(\mathrm{B}(V)) \leq \mathrm{d}(X) - 1$ and $V \subset U$. (As usual, $+\infty - 1 = +\infty$.)

Clearly all of the N axioms except for (N5) are related to Theorem 3.24 for normal families. Indeed, from Theorems 3.25 and I.4.4 we see that ind satisfies these axioms. Only the axiom (N5) is a deeper property of the function ind. The next theorem shows that these axioms are independent and characterize the function ind.

5.3. Theorem. *For the universe $\mathcal{M}_0$, the* N *axioms are independent and the function* ind *is the only function* d *which satisfies the* N *axioms.*

Proof. Let us first prove that ind is a function d which satisfies the N axioms. From the coincidence theorem (Theorem I.4.4) and Theorem 3.25 we have that ind satisfies all of the N axioms except (N5). The coincidence theorem and Theorem VI.2.9 show that ind also satisfies (N5). So let us show that ind is the only such function. This will be done in four parts.

Part I. $\mathrm{d}(X) = -1$ *if and only if* $X = \emptyset$.

Proof. From axioms (N6) and (N1) we have $\mathrm{d}(\emptyset) \leq -1$. And from axioms (N4) and (N1) we have

$$0 = \mathrm{d}(\{\emptyset\}) = \mathrm{d}(\{\emptyset\} \cup \emptyset) \leq \mathrm{d}(\{\emptyset\}) + \mathrm{d}(\emptyset) + 1.$$

Hence, $\mathrm{d}(\emptyset) = -1$. Now suppose that $X \neq \emptyset$. Then by (N2) and (N1), we have $\mathrm{d}(X) \geq \mathrm{d}(\{\emptyset\}) = 0 > -1$. Thereby part I is proved.

Part II. $\operatorname{ind} X = 0$ *implies* $\mathrm{d}(X) = 0$.

Proof. By (N3) and (N1) we have that the set of rational numbers $\mathbb{Q}$ has $\mathrm{d}(\mathbb{Q}) = 0$. Axioms (N5) and (N2) then imply that there is a nonempty, compact dense-in-itself space X' with $\mathrm{d}(X') = 0$. Let X be a zero-dimensional space. Then X can be embedded in X'. From (N2) and part I we have $0 = \mathrm{d}(X') \geq \mathrm{d}(X) \geq 0$. Thus part II is proved.

Part III. *For each extended integer n with $n \geq -1$ we have*

$$\operatorname{ind} X \leq n \quad \textit{implies} \quad \mathrm{d}(X) \leq n.$$

Proof. Suppose that $\operatorname{ind} X \leq n < +\infty$. Then by the decomposition theorem (Theorem I.3.8) we have $X = \bigcup\{ X_i : i = 0, 1, \ldots, n \}$, where $\operatorname{ind} X_i \leq 0$ for each i. From axiom (N4) and part II we have $\mathrm{d}(X) \leq \sum_{i=0}^{n} \mathrm{d}(X_i) \leq n$. Part III is now proved.

Part IV. *For each extended integer n with $n \geq -1$ we have*

$$\mathrm{d}(X) \leq n \quad \textit{implies} \quad \operatorname{ind} X \leq n.$$

Proof. The statement is true for $n = -1$ by part I. Suppose that the statement is true for an integer n and let $\mathrm{d}(X) \leq n+1$. By (N6), each point of X has arbitrarily small open neighborhoods U whose boundaries $\mathrm{B}(U)$ have $\mathrm{d}(\mathrm{B}(U)) \leq \mathrm{d}(X) - 1 \leq n$. So $\operatorname{ind} \mathrm{B}(U) \leq n$. Thus we have shown that $\operatorname{ind} X \leq n+1$. The induction is completed and part IV now follows.

Parts III and IV give that ind is the only function d that satisfy the N axioms. The following six examples show the independence of the N axioms.

Example N1. Let $d = \operatorname{ind} + 1$.

Example N2. Let $\mathcal{P}_2$ be the class of zero-dimensional σ-compact spaces and the space $\emptyset$. Clearly $\mathcal{P}_2$ is a semi-normal family in the universe $\mathcal{M}_0$. From Theorem 3.29, we have that $\mathcal{P}_2\text{-}\operatorname{ind} X = \mathcal{P}_2\text{-}\operatorname{Ind} X$ for X in $\mathcal{M}_0$. So, by Theorem 3.23, $\mathcal{P}_2$-ind satisfies the axioms (N3) and (N4). Clearly, (N2) fails.

To see that axiom (N5) is satisfied, we observe for X in $\mathcal{M}_0$ that the inequality $\operatorname{ind} X - 1 \leq \mathcal{P}_2\text{-}\operatorname{ind} X$ can be shown to hold by a straightforward induction and that the inequality $\mathcal{P}_2\text{-}\operatorname{ind} X \leq \operatorname{ind} X$ holds by Proposition II.2.3. And obviously, if X is nonempty and compact, then $\operatorname{ind} X - 1 = \mathcal{P}_2\text{-}\operatorname{ind} X$. So, when X is not compact and $\operatorname{ind} X - 1 = \mathcal{P}_2\text{-}\operatorname{ind} X$ we let $\widetilde{X}$ be a dimension preserving compactification of X. Let us consider the remaining case where X is not compact and $\operatorname{ind} X = \mathcal{P}_2\text{-}\operatorname{ind} X$ holds. Let X_0 be a dimension preserving compactification of X and let x_0 be a point in the set $X_0 \setminus X$. Let $S = \{ x_n : n \in \mathrm{N} \}$ be a sequence in X converging to x_0 and let $D = \{ z_n : n \in \mathrm{N} \}$ be a dense subset of an absolute retract Z. Observe that if F is a closed subset of X_0 that contains at least m points of S then there is a continuous map of F into Z such that its image contains the first m points of D. Using the metrizability of X_0 and this observation, one can easily construct an embedding of $X_0 \setminus \{ x_0 \}$ into $X_0 \times Z$ such that its image Y has a closure $\widetilde{X}$ with $\widetilde{X} \setminus Y = \{ x_0 \} \times Z$. Let Z be the space $\mathbb{I}^{n+1}$ where $n = \operatorname{ind} X$. Then $\widetilde{X}$ is a compactification of X such that $\mathcal{P}_2\text{-}\operatorname{ind} X = \mathcal{P}_2\text{-}\operatorname{ind} \widetilde{X}$.

For the axioms (N1) and (N6), define d as follows: $\mathrm{d}(\emptyset) = -1$; $\mathrm{d}(X) = \mathcal{P}_2\text{-ind}\, X + 1$ if and only if $X \neq \emptyset$. This function d satisfies all the axioms except for (N2).

Example N3. Let d be defined as follows: $\mathrm{d}(X) = \mathrm{ind}\, X$ if and only if X is finite; $\mathrm{d}(X) = \mathrm{ind}\, X + 1$ if and only if X is infinite.

Example N4. Let d be defined as follows: $\mathrm{d}(X) = \mathrm{ind}\, X$ if and only if $\mathrm{ind}\, X \leq 0$; $\mathrm{d}(X) = \mathrm{ind}\, X + 1$ if and only if $\mathrm{ind}\, X > 0$.

Example N5. Let $\mathcal{P}_5$ be the class of countable spaces. Then $\mathcal{P}_5$ is a normal family in the universe $\mathcal{M}_0$. So by Theorem 3.24, the axioms (N2)–(N4) are satisfied by $\mathcal{P}_5$-ind and axiom (N5) fails. For the axioms (N1) and (N6), consider the function d defined as follows: $\mathrm{d}(\emptyset) = -1$; $\mathrm{d}(X) = \mathcal{P}_5\text{-ind}\, X + 1$ if and only if $X \neq \emptyset$.

Example N6. Let d be defined as follows: $\mathrm{d}(\emptyset) = -1$; $\mathrm{d}(X) = \mathrm{ind}\, X/(\mathrm{ind}\, X + 1)$ if and only if $-1 < \mathrm{ind}\, X < +\infty$; $\mathrm{d}(X) = 1$ if and only if $\mathrm{ind}\, X = +\infty$.

The proof of the theorem is now completed.

Obviously the N axioms restricted to the universe $\mathcal{U}_\omega$ of spaces embeddable in some Euclidean space also characterize ind and the axioms are independent. This fact relies on the equivalence of the finite-dimensionality of a space and its embeddability in some Euclidean space (Theorem 4.3). The axiom (N6) is closely tied to the small inductive dimension, that is, the **H** operation. The question of replacing this axiom with one more suitable to the covering dimension dim has been answered by Hayashi [1990] for the universe $\mathcal{U}_\omega$.

5.4. The H axioms. Let d be a real valued function on the universe $\mathcal{U}_\omega$ that satisfies the axioms:

(H1) $\mathrm{d}(\emptyset) = -1$, $\mathrm{d}(\{\emptyset\}) = 0$ and $\mathrm{d}(\mathbb{I}^m) = m$ for $m = 1, 2, \ldots$.

(H2) If Y is a subspace of a space X in $\mathcal{U}_\omega$, then $\mathrm{d}(Y) \leq \mathrm{d}(X)$.

(H3) If a space X in $\mathcal{U}_\omega$ is a union of a sequence $X_1, X_2, \ldots$ of closed subspaces, then $\mathrm{d}(X) \leq \sup\{\mathrm{d}(X_i) : i = 1, 2, \ldots\}$.

(H4) Every space X in $\mathcal{U}_\omega$ has a compactification $\widetilde{X}$ in $\mathcal{U}_\omega$ satisfying $\mathrm{d}(\widetilde{X}) = \mathrm{d}(X)$.

(H5) If X is a space in $\mathcal{U}_\omega$ with $\mathrm{d}(X) < n$ (where n is a positive integer), then there are n sets X_i, $i = 1, \ldots, n$, with $\mathrm{d}(X_i) \leq 0$ for every i and $X = \bigcup\{X_i : i = 1, \ldots, n\}$.

5.5. Theorem. *For the universe* $\mathcal{U}_\omega$, *the* H *axioms are independent and the function* dim *is the only function* d *which satisfies the* H *axioms.*

The proof of this theorem will use the facts that the class $\mathcal{U}_\omega$ is precisely the class of finite dimensional spaces in $\mathcal{M}_0$ and that Proposition 4.6 holds. These facts will also be used in the discussion of the following axiom scheme which is an extension of the H axioms. A proof of Theorem 5.5 will result from this axiom scheme.

5.6. The extended H axioms. Let d be an extended-real valued function on the universe $\mathcal{M}_0$ that satisfies the axioms:

(eH1) $\mathrm{d}(\emptyset) = -1$, $\mathrm{d}(\{\emptyset\}) = 0$ and $\mathrm{d}(\mathbb{I}^m) = m$ for $m = 1, 2, \ldots$.
(eH2) If Y is a subspace of a space X in $\mathcal{M}_0$, then $\mathrm{d}(Y) \leq \mathrm{d}(X)$.
(eH3) If a space X in $\mathcal{M}_0$ is a union of a sequence $X_1, X_2, \ldots$ of closed subspaces, then $\mathrm{d}(X) \leq \sup\{\mathrm{d}(X_i) : i = 1, 2, \ldots\}$.
(eH4) Every space X in $\mathcal{M}_0$ has a compactification $\widetilde{X}$ in $\mathcal{M}_0$ satisfying $\mathrm{d}(\widetilde{X}) = \mathrm{d}(X)$.
(eH5) If X is a space in $\mathcal{M}_0$ with $\mathrm{d}(X) < n$ (where n is a positive integer), then there are n sets X_i, $i = 1, \ldots, n$, with $\mathrm{d}(X_i) \leq 0$ for every i and $X = \bigcup\{X_i : i = 1, \ldots, n\}$.
(eH6) If X is a space in $\mathcal{M}_0$ with $\mathrm{d}(X) \leq 0$ then X is embeddable in some Euclidean space $\mathbb{R}^m$.

5.7. Proposition. *For the universe* $\mathcal{M}_0$, *the extended* H *axioms* (eH1) *through* (eH5) *are independent and* dim *is the only function* d *which satisfies the extended* H *axioms. Consequently, the restriction of the extended* H *axioms to the smaller universe* $\mathcal{U}_\omega$ *is equivalent to the* H *axioms.*

Proof. We already know that dim satisfies the extended H axioms. So let us show that a function d that satisfies the extended H axioms must be dim.

Let us first prove that the equivalence $\mathrm{d}(X) = 0$ if and only if X is totally disconnected holds for every nonempty compact space. Suppose that some nonempty compact space with a nondegenerate component C has $\mathrm{d}(X) = 0$. From (eH2), (eH3), (eH6) and Proposition 4.6 there exists a space Y with $\mathrm{d}(Y) = 0$ such that every compactification of Y contains an arc. From (eH4) we have $\mathrm{d}(\widetilde{Y}) = 0$ for some compactification $\widetilde{Y}$ of Y. We infer from (eH2) that $\mathrm{d}(\mathbb{I}) = 0$.

But this contradicts $\mathrm{d}(\mathbb{I}) = 1$ from (eH1). So, $\mathrm{d}(X) = 0$ implies X is totally disconnected when X is compact. For the converse implication let us show that $\mathrm{d}(X_C) = 0$ where X_C is the Cantor set. By (eH1) and (eH3) we have $\mathrm{d}(\mathbb{Q}) = 0$, where $\mathbb{Q}$ is the set of rational numbers. Axiom (eH4) gives a compactification $\widetilde{X}$ of $\mathbb{Q}$ such that $\mathrm{d}(\widetilde{X}) = 0$. As the space $\widetilde{X}$ is a nonempty, dense-in-itself and compact, it contains a copy of the Cantor set X_C. By (eH2) we have $\mathrm{d}(X_C) = 0$. As every compact totally disconnected space can be embedded in the Cantor set, we have by (eH2) the converse implication.

The equivalence $\dim X = 0$ if and only if $\mathrm{d}(X) = 0$ follows immediately from the one in the last paragraph and from (eH2) and (eH4).

We have $\dim X \leq \mathrm{d}(X)$ from the decomposition theorem and axiom (eH5). It remains to be shown that $\mathrm{d}(X) \leq \dim X$. Suppose that $0 < n = \dim X < \infty$. Recall that $\mathbb{I}_n^{2n+1}$ is the set of all points of $\mathbb{I}^{2n+1}$ at most n of whose coordinates are irrational. Axioms (eH1) and (eH3) give $\mathrm{d}(\mathbb{I}_n^{2n+1}) = n$. Let Y be a compactification of $\mathbb{I}_n^{2n+1}$ given by (eH4). By Theorem 4.5 the space Y contains a copy of the Hayashi sponge H_n^{2n+1} which, by Theorem 4.4, contains a copy of X. By axiom (eH2) we have $\mathrm{d}(X) \leq \mathrm{d}(Y) = n = \dim X$. Thus $\mathrm{d} = \dim$ has been established.

Obviously, axiom (eH6) is redundant in the universe $\mathcal{U}_\omega$. To establish the independence statement in the theorem it will be sufficient to prove the independence statement of Theorem 5.5 since $\mathrm{d}(X) = \infty$ if and only if $\dim X = \infty$.

Example H1. Let $\mathrm{d}(\emptyset) = -1$ and $\mathrm{d}(X) = 0$ for every nonempty space X in $\mathcal{U}_\omega$.

Example H2. Let d be the one given in Example N2.

Example H3. If X is embeddable in the interval $\mathbb{I}$, then define $\mathrm{d}(X) = \dim X$. If $\dim X = 1$ and X is not embeddable in $\mathbb{I}$, then define $\mathrm{d}(X) = 2$. For all other spaces X, define $\mathrm{d}(X) = \dim X$.

Example H4. Let d be the one given in Example N5.

Example H5. Let $Z(p)$ be the cyclic group of order p where p is a prime number. If $\mathrm{d}(X) = \text{c-dim}_{Z(p)} X$, where $\text{c-dim}_{Z(p)} X$ is the cohomological dimension of X for X in $\mathcal{U}_\omega$, then d satisfies the axioms (H1)–(H4) by a result of Shvedov (see Kuz′minov [1968]). Now, there is a space X in $\mathcal{U}_\omega$ such that $\text{c-dim}_{Z(p)} X \neq \dim X$. Consequently, d does not satisfy axiom (H5).

Both Theorem 5.5 and Proposition 5.7 are now proved.

The N axioms have been generalized to the universe $\mathcal{M}$ of metrizable spaces for Ind as follows.

5.8. The A axioms. Let d be an extended-real valued function on the universe $\mathcal{M}$ that satisfies the following axioms:

(A1) $\mathrm{d}(\{\emptyset\}) = 0$.

(A2) If Y is a subspace of a space X in $\mathcal{M}$, then $\mathrm{d}(Y) \leq \mathrm{d}(X)$.

(A3) If a space X in $\mathcal{M}$ is a union of a σ-locally finite collection $\boldsymbol{F}$ of closed subspaces, then $\mathrm{d}(X) \leq \sup\{\mathrm{d}(F) : F \in \boldsymbol{F}\}$.

(A4) If a space X in $\mathcal{M}$ is a union of two subpaces X_1 and X_2, then $\mathrm{d}(X) \leq \mathrm{d}(X_1) + \mathrm{d}(X_2) + 1$.

(A5) Every space X in $\mathcal{M}$ has a completion $\widetilde{X}$ in $\mathcal{M}$ satisfying $\mathrm{d}(\widetilde{X}) = \mathrm{d}(X)$.

(A6) If F is a nonempty closed subset of a space X in $\mathcal{M}$, then every neighborhood U of F has a neighborhood V satisfying $V \subset U$ and $\mathrm{d}\big(\mathrm{B}(V)\big) \leq \mathrm{d}(X) - 1$.

The A axioms are the natural generalizations of the N axioms. The following characterization theorem holds.

5.9. Theorem. *For the universe $\mathcal{M}$, the* A *axioms are independent and the function* Ind *is the only function* d *which satisfies the* A *axioms.*

The next lemma will be used in the proof of Theorem 5.9 and in the proof of the characterization by means of the S axioms which will be discussed after the proof of Theorem 5.9.

5.10. Lemma. *Let* d *be an extended-real valued function on the universe $\mathcal{M}$ that satisfies the axioms* (A1), (A2), (A3) *and* (A5). *Then $\mathrm{d}(N(D)) = 0$ for every discrete space D.*

Proof. Because of (A2), we may assume that D is infinite. The axioms (A1) and (A3) yield that the set G as defined in the discussion preceding Lemma 4.7 has $\mathrm{d}(G) = 0$. By (A5), there is a completion $\widetilde{G}$ of G such that $\mathrm{d}(\widetilde{G}) = 0$. By Lavrentieff's theorem, the identity map of G onto itself can be extended to a homeomorphism of a G_δ-set of $\widetilde{G}$ containing G onto a G_δ-set H of $N(D)$ containing G. So, $\mathrm{d}(H) = 0$. By Lemma 4.7 we have that $N(D)$ can be embedded into H. Axiom (A2) completes the proof.

Proof of Theorem 5.9. That Ind satisfies the A axioms follows from Theorem 3.25 and Theorem V.2.12.

The needed modifications of the proof of Theorem 5.3 occur in parts II, III and IV. Since the modifications for parts II and III are obvious, only the statement that $\mathrm{d}(X) = 0$ if and only if $\operatorname{Ind} X = 0$ requires a proof. To this end, suppose $\mathrm{d}(X) = 0$. Then, by (A6) and the analogue of part I, we have $\operatorname{Ind} X = 0$. For the converse we apply Lemma 5.10 and Theorem 4.8.

As for examples to show independence, only the axioms (A2) and (A5) require serious attention. For (A2), consider the absolute Borel class $\mathsf{A}(2)$ (that is, the $G_{\delta\sigma}$-class). As in Example N2, one can construct the required example by using $\mathsf{A}(2)$-Ind. For (A5), one uses the class of metric spaces that are the union of a σ-locally finite collection of singletons in place of the class $\mathcal{P}_5$ of countable spaces used in Example N5. (Observe that this class is $\mathbf{N}[\mathcal{P}_5 : \mathcal{M}]$, the normal family extension of the class $\mathcal{P}_5$ in the universe $\mathcal{M}$.)

According to Lemma 5.10, the A axioms imply the following S axioms.

5.11. The S axioms. Let d be an extended-real valued function on the universe $\mathcal{M}$ that satisfies the axioms:

(S1) For each discrete space D, $\mathrm{d}(N(D)) = 0$.
(S2) If Y is a subspace of a space X in $\mathcal{M}$, then $\mathrm{d}(Y) \leq \mathrm{d}(X)$.
(S3) If a space X in $\mathcal{M}$ is a union of two subpaces X_1 and X_2, then $\mathrm{d}(X) \leq \mathrm{d}(X_1) + \mathrm{d}(X_2) + 1$.
(S4) If F is a nonempty closed subset of a space X in $\mathcal{M}$, then every neighborhood U of F has a neighborhood V satisfying $V \subset U$ and $\mathrm{d}\big(\mathrm{B}(V)\big) \leq \mathrm{d}(X) - 1$.

5.12. Theorem. *For the universe $\mathcal{M}$, the* S *axioms are independent and the function* Ind *is the only function* d *which satisfies the* S *axioms.*

Proof. Axiom (S1) implies axiom (A1), and axiom (S4) is axiom (A6). Lemma 5.10 yields $\mathrm{d}(X) = 0$ for every metric space X with $\operatorname{Ind} X = 0$. Axiom (S4) yields $\mathrm{d}(X) = -1$ if and only if $X = \emptyset$, and $-1 < \mathrm{d}(X) \leq 0$ if and only if $\operatorname{Ind} X = 0$. The remainder of the proof is the same as that of Theorem 5.9. For the independence of the S axioms, only (S1) needs attention. In this case let $\mathrm{d}(X) = \operatorname{Ind} X + 1$.

The above axioms do not employ anything more than subsets and unions. And, the only maps involved are homeomorphisms. Another very useful tool in dimension theory is that of approximation of spaces by polyhedra. This approximation uses the concept of ε-mappings from a metric space to a topological space. The development of functions of dimensional types along the lines of approximations is yet to be made. The discussion on axioms will end with a short discussion of the axioms called the Al axioms and the Sh axioms. Since proofs of the theorems associated with these axioms will lead us afield, only the theorems will be stated.

5.13. The Al axioms. Let $\mathcal{K}_0$ be the universe of compact subspaces of Euclidean spaces and d be an integer valued function defined on $\mathcal{K}_0$ that satisfies the axioms:

(Al1) $\mathrm{d}(\emptyset) = -1$, $\mathrm{d}(\{\emptyset\}) = 0$ and $\mathrm{d}(\mathbb{I}^n) = n$ for $n = 1, 2, \ldots$.

(Al2) If a space X in $\mathcal{K}_0$ is the union of two closed subsets X_1 and X_2, then $\mathrm{d}(X) = \max\{\mathrm{d}(X_1), \mathrm{d}(X_2)\}$.

(Al3) For every space X in $\mathcal{K}_0$ there exists a positive number ε such that if $f\colon X \to Y$ is an ε-mapping of X onto a space Y in $\mathcal{K}_0$, then $\mathrm{d}(X) \leq \mathrm{d}(Y)$.

(Al4) If X is a space in $\mathcal{K}_0$ of cardinality larger than one, then there exists a closed subset L of X separating X with $\mathrm{d}(L) < \mathrm{d}(X)$.

5.14. Theorem. *For the universe* $\mathcal{K}_0$, *the* Al *axioms are independent and the function* dim *is the only function* d *which satisfies the* Al *axioms.*

5.15. The Sh axioms. Let $\mathcal{U}_\omega$ be the universe of subspaces of Euclidean spaces and d be an integer valued function defined on $\mathcal{U}_\omega$ that satisfies the axioms:

(Sh1) $\mathrm{d}(\emptyset) = -1$, $\mathrm{d}(\{\emptyset\}) = 0$ and $\mathrm{d}(\mathbb{I}^n) = n$ for $n = 1, 2, \ldots$.

(Sh2) If a space X in $\mathcal{U}_\omega$ is the union of a sequence $X_1, X_2, \ldots$ of closed subsets, then $\mathrm{d}(X) \leq \sup\{\mathrm{d}(X_i) : i = 1, 2, \ldots\}$.

(Sh3) For every space X in $\mathcal{U}_\omega$ there exists a positive number ε such that if $f\colon X \to Y$ is an ε-mapping of X onto a space Y in $\mathcal{U}_\omega$, then $\mathrm{d}(X) \leq \mathrm{d}(Y)$.

(Sh4) If X is a space in $\mathcal{U}_\omega$ of cardinality larger than one, then there exists a closed subset L of X separating X with $\mathrm{d}(L) < \mathrm{d}(X)$.

5.16. Theorem. *For the universe $\mathcal{U}_\omega$, the* Sh *axioms are independent and the function* dim *is the only function* d *which satisfies the* Sh *axioms.*

Finally we mention that the product theorem has not been used in the axioms. Its role appears not to be natural in the development of normal families. The formulation of the product theorems for general functions of inductive type is somewhat uncertain at this time. Perhaps an axiom using the product theorem would be a suitable substitute for the axiom using the addition theorem. In this way, it may be possible to avoid the decomposition theorem which is heavily used in the characterizations of ind and Ind. The decomposition theorem uses strongly the metrizability of the spaces; the product theorem does not use the metrizability and hence may afford a characterization to a universe larger than the metrizable one. In this vein, there is an inductive dimension that uses a subbase for the open sets of a topology (de Groot [1969]). It has been observed that the product theorem is easily proved with a definition that uses subbases. (See also van Douwen [1973] for a brief discussion of this subbasic dimension.)

6. Historical comments and unsolved problems

The terminology of functions of dimensional type was first introduced by Baladze [1982]. Earlier in [1964], Lelek introduced the terminology inductive invariant for the functions of inductive dimensional types $\mathcal{P}$-ind and $\mathcal{P}$-Ind.

The Example 1.2 and the material concerning the interconnection between the addition, point addition and the sum theorems (that is, the Theorems A, P and S) in Section 1 are taken from van Douwen [1973]. Similar results, in particular an example showing the failure of Theorem S, can be found in Przymusiński [1974]. The Proposition 1.15 is due to Hart, van Mill and Vermeer [1982] who showed that the small and large inductive completeness degrees disagreed in the universe of metrizable spaces. Example 1.16 generalizes this last result to other absolute Borel classes. These results should be compared with the fact that a normal space X has $\operatorname{Ind} X = 0$ if and only if every normal space Y obtained from X by the adjunction of a single point satisfies the condition $\operatorname{ind} Y = 0$, Isbell [1964].

The study of normal families began with Hurewicz [1927] in the universe $\mathcal{M}_0$ of separable metrizable spaces and by Morita [1954] in the universe $\mathcal{M}$ of metrizable spaces. The symbols for the operations **H** and **M** have been chosen to designate these origins. The various generalizations of normal families discussed in the chapter have appeared in Aarts [1972], Aarts and Nishiura [1973a]. The development found in Section 3 is essentially new. The influence of the work of Dowker [1953] is strongly felt in this development. The introduction of the operation **N**, the normal family extension, is the result of it.

The Dowker universe $\mathcal{D}$ is defined for the first time in Section 3. The definition is a culmination of the investigation in Section 3 of the connections between the operation **M**, **S** and **N**. There is a long history associated with this universe. It was known that the subspace theorem for Ind failed in the universe $\mathcal{N}$ of normal spaces. Even more, it fails for the universe $\mathcal{N}_H$ of hereditarily normal spaces. It was shown by Čech [1932] that the universe of perfectly normal spaces worked well for the subspace theorem. Subsequently, Dowker [1953] defined the universe $\mathcal{N}_T$ of totally normal spaces and proved the subspace theorem there. In Pasynkov [1967] and Lifanov and Pasynkov [1970], a space X is defined to be a *Dowker space* if it is a hereditarily normal space for which every open set is Dowker-open in the sense of Definition 3.4. (The class of Dowker spaces as defined by Lifanov and Pasynkov is not the Dowker universe $\mathcal{D}$ defined in this book.) In [1977] Nishiura defined a T_1-space X to be *super normal* if for every pair of separated set A and B there is a pair of disjoint open sets U and V such that $A \subset U$ and $B \subset V$ and U and V are the union of locally finite, in U and V respectively, families of cozero-sets of X. This definition was the first recognition of the connection between the operation **M** and the operation **S**. In Engelking [1978] the ideas in super normal and Dowker spaces were combined to yield the definition of strongly hereditarily normal spaces (see Definition 3.6). Clearly the Dowker universe contains the collection of strongly hereditarily normal spaces which, in turn, contains the class of super normal spaces. The Dowker universe contains all of the other classes properly. (See Engelking [1978] for a discussion of these inclusions.) Along these lines, Kotkin [1990] defined a new dimension function related to super normal spaces.

In Section 5 on axiomatics, the A axioms are due to Aarts [1971] and the S axioms are due to Sakai [1968]. Lemma 4.7 and Theorem 4.8 are essentially from Aarts [1971]. The Al axioms are due to Alexandroff [1932] and the modification of these axioms to the Sh axioms are due to Shchepin [1972]. The reader is referred to Alexandroff and Pasynkov [1973], Engelking [1978] and Arkhangel′skiĭ and Fedorchuk [1990] for further discussions on these axioms. The theorem that cohomological dimension with respect to a finitely generated Abelian group satisfies the M axioms in the universe $\mathcal{U}_\omega$ of subspaces of Euclidean spaces was proved by Shvedov; the proof was first published in Kuz′minov [1968]. The H axioms are due to Hayashi [1990]. The preliminary embedding theorems involving the Hayashi sponge H_n^m found in Section 4 are essentially due to Hayashi. The extended H axioms are a natural outgrowth of Hayashi's proof of Theorem 5.5.

As mentioned in the Section 5, the additivity axioms (N4), (A4) and (S3) are closely tied to the metrizability of the spaces X. A suitable substitute for these axioms that would yield characterizations in larger universes would be welcomed. In this regard, see Baladze [1982] for a characterization theorem along the lines of the S axioms for nonmetrizable spaces. (The theorem of the last reference suffers from the requirement that the spaces satisfy the Lindelöf condition as well as other conditions that imply the additivity theorem.)

Unsolved problems

1. The extended H axioms have not been shown to be independent. The axiom (eH6) has been used in the proof of Proposition 5.7. This was made necessary by the requirement that the space X in Proposition 4.6 be finite dimensional.

Can this requirement be deleted from the hypothesis of Proposition 4.6?

It is known that there are infinite dimensional continua that have no positive, finite dimensional, closed subsets. (See Henderson [1967].)

2. Are the extended H axioms independent?

3. Can one prove the equivalence (in the universe $\mathcal{U}_\omega$) of the three axiom schemes N, H and Sh without passing through the dimension function dim?

4. A set of axioms characterizing the covering dimension dim in the universe of all (not necessarily metrizable) compact spaces whose dimension dim is finite was given by Lokucievskiĭ [1973].

Find axioms characterizing Ind or dim for universes $\mathcal{U}$ larger than $\mathcal{M}$.

5. Are the Menger sponge M_n^m and the Hayashi sponge H_n^m homeomorphic?

6. The class $\mathcal{E}$ of strongly hereditarily normal spaces has been characterized as those hereditarily normal spaces X for which every regularly open set U is the union of a point-finite family of open F_σ-sets in X (see page 168 of Engelking [1978]).

Characterize those spaces X which form the Dowker universe $\mathcal{D}$.

CHAPTER IV

FUNCTIONS OF COVERING DIMENSIONAL TYPE

The chapter deals with covering dimension modulo a class $\mathcal{P}$. Parallel to Chapter III, the functions that are associated with the covering approach will be designated as *functions of covering dimensional type.* They include $\mathcal{P}$-dim, $\mathcal{P}$-Dim, $\mathcal{P}$-sur and $\mathcal{P}$-def. The investigation of functions of covering dimensional type that began in Chapter II will be continued in this chapter. Theorems on dimension of finite unions will be discussed in the first section. The other theorems of dimension theory such as the countable sum theorems and the subspace theorems are naturally studied, as they were for functions of inductive dimensional type, in the context of cosmic and normal families. These will be discussed in the second section. Also, the coincidence of $\mathcal{P}$-dim and $\mathcal{P}$-Dim will be established in Section 2. This investigation of cosmic and normal families will lead to the conclusion that the Dowker universe $\mathcal{D}$ is, in a sense, an optimal universe in which the theory of functions of covering dimensional type can exist without anomalies. Section 3 will be devoted to this discussion of the Dowker universe. The dimensional relationships of the domain and image of closed mappings (or open mappings) are usually stated in terms of the covering dimension. Section 4 has a short discussion on related mapping theorems for dimensions modulo classes $\mathcal{P}$.

Although some parts of the discussion do not require the full strength of the agreement for the chapter, the simplicity of the exposition gained from the agreement will compensate for the lack of a fuller discussion.

Agreement. *The universe is closed-monotone and open-monotone and is contained in $\mathcal{N}_H$. The classes contain the empty space $\emptyset$ as a member and are closed-monotone, open-monotone and strongly closed-additive in the universe. Unless indicated otherwise, X will always be a space in the universe.*

1. Finite unions

The initial part of this section will be devoted to $\mathcal{P}$-dim. Let us begin with the theorem that has the easiest proof.

1.1. Theorem. *If A and B are closed subsets of a space X such that $X = A \cup B$, then*

$$\mathcal{P}\text{-dim}\, X = \max\{\, \mathcal{P}\text{-dim}\, A,\ \mathcal{P}\text{-dim}\, B\,\}.$$

Proof. In view of Theorem II.5.6 it is sufficient to prove the maximum is not less than $\mathcal{P}$-dim X. Let $\max\{\, \mathcal{P}\text{-dim}\, A,\ \mathcal{P}\text{-dim}\, B\,\} \leq n$ with $n \geq 0$ and suppose that $\boldsymbol{U} = \{\, U_i : i = 0, 1, \ldots, k\,\}$ is a finite $\mathcal{P}$-border cover of X with enclosure F. The restricted collection $\{\, U_i \cap A : i = 0, 1, \ldots, k\,\}$ is a $\mathcal{P}$-border cover of A and hence it has a $\mathcal{P}$-border cover shrinking $\boldsymbol{V} = \{\, V_i : i = 0, 1, \ldots, k\,\}$ with enclosure F' such that $\operatorname{ord} \boldsymbol{V} \leq n + 1$. For each i let $U_i' = V_i \cup (U_i \setminus A)$ and denote the collection of such U_i' by $\boldsymbol{U}'$. Then $\boldsymbol{U}'$ is a $\mathcal{P}$-border cover of X with enclosure $F \cup F'$ that shrinks $\boldsymbol{U}$. We now perform a symmetric construction with $\boldsymbol{U}'$ and B to complete the proof.

The stronger form of the above finite sum theorem will require a technical lemma on $\mathcal{P}$-border covers. The Proposition II.4.4 has the following generalization which will hold for normal spaces as well.

1.2. Lemma. *In a space X let $\boldsymbol{F} = \{\, F_\alpha : \alpha \in A\,\}$ be a closed collection and $\boldsymbol{U} = \{\, U_\alpha : \alpha \in A\,\}$ be a locally finite open collection such that $F_\alpha \subset U_\alpha$ for each α. Then there exists an open collection $\boldsymbol{V} = \{\, V_\alpha : \alpha \in A\,\}$ such that $F_\alpha \subset V_\alpha \subset U_\alpha$ for each α and the collections $\boldsymbol{F}$ and $\{\, \operatorname{cl}(V_\alpha) : \alpha \in A\,\}$ are combinatorially equivalent, that is,*

$$F_{\alpha_1} \cap \cdots \cap F_{\alpha_m} \neq \emptyset$$
$$\textit{if and only if}$$
$$\operatorname{cl}(V_{\alpha_1}) \cap \cdots \cap \operatorname{cl}(V_{\alpha_m}) \neq \emptyset$$

for each finite collection of distinct indices $\alpha_1, \ldots, \alpha_m$.

Proof. One merely applies a transfinite induction on the indexing set A to attain the appropriate modification of the proof of Proposition II.4.4.

1.3. Lemma. *Let X be a space and Y be a subspace of X. Suppose that $\boldsymbol{U} = \{U_\alpha : \alpha \in A\}$ is an open collection of X such that $\boldsymbol{U}$ is locally finite in the space $\bigcup \boldsymbol{U}$ and such that the restricted collection $\boldsymbol{U}_Y = \{U_\alpha \cap Y : \alpha \in A\}$ has* ord $\boldsymbol{U}_Y \leq m$. *Then there exists an open collection $\boldsymbol{W} = \{W_\alpha : \alpha \in A\}$ that shrinks $\boldsymbol{U}$ such that $(\bigcup \boldsymbol{W}) \cap Y = (\bigcup \boldsymbol{U}) \cap Y$ and* ord $\boldsymbol{W} \leq m$ *hold.*

Proof. We may assume $X = \bigcup \boldsymbol{U}$ without loss of generality. Then $Y' = \{x : \operatorname{ord}_x \boldsymbol{U} \leq m\}$ is a closed subset of X that contains Y. In the closed subspace Y' let $\boldsymbol{F} = \{F_\alpha : \alpha \in A\}$ be a closed cover that shrinks $\boldsymbol{U}$. Then $\operatorname{ord}_x \boldsymbol{F} \leq m$ for each x in Y'. The required collection $\boldsymbol{W}$ is obtained by an application of Lemma 1.2.

1.4. Lemma. *Let Y be a subspace of a space X and suppose $\boldsymbol{U} = \{U_i : i = 0, 1, \ldots, k\}$ is a finite open collection of X such that its restriction $\boldsymbol{U}_Y$ to Y is a $\mathcal{P}$-border cover of Y with enclosure F. If $\mathcal{P}$-dim $Y \leq n$, then there is a set G and there is an open collection $\boldsymbol{W} = \{W_i : i = 0, 1, \ldots, k\}$ of X such that the following conditions hold:*

(a) $W_i \subset U_i$, $i = 0, 1, \ldots, k$.
(b) $G = Y \setminus \bigcup \boldsymbol{W} \in \mathcal{P}$.
(c) ord $\boldsymbol{W} \leq n + 1$.

Moreover, if Y is also closed in X, then the further condition

(d) $(\bigcup \boldsymbol{U}) \setminus G = \bigcup \boldsymbol{W}$

holds provided the condition (c) *is changed to the following:*

(c*) *There is an open set V of X with $\operatorname{ord}_x \boldsymbol{W} \leq \operatorname{ord}_x \boldsymbol{U}$ for each x in $X \setminus V$ such that the inclusions $Y \setminus G \subset V \subset \{x : \operatorname{ord}_x \boldsymbol{W} \leq n + 1\}$ hold modulo the set $H = X \setminus \bigcup \boldsymbol{U}$.*

Proof. As $\emptyset \in \mathcal{P} \cap \mathcal{U}$, the case of $\mathcal{P}$-dim $Y = -1$ is trivially true. So we shall assume $0 \leq \mathcal{P}$-dim $Y \leq n$. Without loss of generality, we may also assume $X = (\bigcup \boldsymbol{U}) \cup Y$.

The restricted $\mathcal{P}$-border cover $\boldsymbol{U}_Y = \{U_i \cap Y : i = 0, 1, \ldots, k\}$ has a $\mathcal{P}$-border cover shrinking $\boldsymbol{Z} = \{Z_i : i = 0, 1, \ldots, k\}$ with enclosure G such that ord $\boldsymbol{Z} \leq n + 1$. For each i let $U_i^!$ be an open set of X such that $U_i^! \cap Y = Z_i$ and $U_i^! \subset U_i$. For the open collection $\boldsymbol{U}' = \{U_i^! : i = 0, 1, \ldots, k\}$ we have $\operatorname{ord}_x \boldsymbol{U}' \leq n + 1$ for each x in $Y \setminus G$. By Lemma 1.3 there is an open collection $\boldsymbol{V} = \{V_i : i = 0, 1, \ldots, k\}$ such that $V_i \subset U_i^!$ for each i, ord $\boldsymbol{V} \leq n + 1$ and

$$(\bigcup \boldsymbol{V}) \cap Y = (\bigcup \boldsymbol{U}') \cap Y.$$

Since $\left(\bigcup \boldsymbol{U}'\right) \cap Y = \left(\bigcup \boldsymbol{Z}\right) \cap Y = Y \setminus G$, the conditions (a)–(c) are easily verified for the set G and the collection $\boldsymbol{W} = \mathbf{V}$.

Suppose that Y is also closed in X. Then we have that the set G is closed in X, the open set $X \setminus G$ is equal to the set $(X \setminus Y) \cup \left(\bigcup \mathbf{V}\right)$ and the set $Y \setminus G$ is closed in the subspace $X \setminus G$. So there is an open set V in X such that $Y \setminus G \subset V \subset X \setminus G$ and $\mathrm{cl}_{X \setminus G}(V) \subset \bigcup \mathbf{V}$. Define the two collections $\boldsymbol{W}_0$ and $\boldsymbol{W}_1$ to be

$$\boldsymbol{W}_0 = \{\, V_i \cup (U_i \setminus (\mathrm{cl}_X(V) \cup Y)) : U_i \cap V \neq \emptyset \,\},$$
$$\boldsymbol{W}_1 = \{\, U_i \setminus Y : U_i \cap V = \emptyset \,\}.$$

Then $\boldsymbol{W} = \boldsymbol{W}_0 \cup \boldsymbol{W}_1$ is an open collection such that $\boldsymbol{W}$ shrinks $\boldsymbol{U}$ and $\bigcup \boldsymbol{W} = X \setminus G = \left(\bigcup \boldsymbol{U}\right) \setminus G$. Hence the condition (d) holds. If $x \in V$, then $\mathrm{ord}_x \boldsymbol{W} = \mathrm{ord}_x \boldsymbol{W}_0 = \mathrm{ord}_x \mathbf{V} \le n+1$. If $x \in X \setminus V$, then $\mathrm{ord}_x \boldsymbol{W} \le \mathrm{ord}_x \boldsymbol{U}$ because $\boldsymbol{W}$ shrinks $\boldsymbol{U}$. As $X = \left(\bigcup \boldsymbol{U}\right) \cup Y$ has been assumed, we have $H = F \subset G$. Thereby the condition (c*) holds.

1.5. Theorem. *If $X = A \cup B$ where A is open, then*

$$\mathcal{P}\text{-}\dim X \le \max\{\, \mathcal{P}\text{-}\dim A,\ \mathcal{P}\text{-}\dim B \,\}.$$

Proof. From Theorem II.5.6 there is no loss in generality in assuming the further condition $A \cap B = \emptyset$.

Let $\max\{\, \mathcal{P}\text{-}\dim A, \mathcal{P}\text{-}\dim B \,\} \le n$ with $n \ge 0$ and suppose that $\boldsymbol{U} = \{\, U_i : i = 0, 1, \ldots, k \,\}$ is a finite $\mathcal{P}$-border cover of X with enclosure F. By the closed monotonicity conditions on $\mathcal{P}$ and $\mathcal{U}$, the trace $\boldsymbol{U}_B$ of $\boldsymbol{U}$ on B is a $\mathcal{P}$-border cover of B. From Lemma 1.4 we have an open collection $\boldsymbol{W} = \{\, W_i : i = 0, 1, \ldots, k \,\}$ of X that shrinks $\boldsymbol{U}$, a set G and an open set V such that

$$G = B \setminus \bigcup \boldsymbol{W} \in \mathcal{P} \cap \mathcal{U}, \qquad \left(\bigcup \boldsymbol{U}\right) \setminus G = \bigcup \boldsymbol{W},$$
$$B \setminus G \subset V \subset \{\, x : \mathrm{ord}_x \boldsymbol{W} \le n+1 \,\} \mod F,$$
$$\mathrm{ord}_x \boldsymbol{W} \le \mathrm{ord}_x \boldsymbol{U}, \qquad x \in X \setminus V,$$

Since $X \setminus \bigcup \boldsymbol{W} = F \cup G$ and since $\mathcal{P}$ is strongly closed-additive, we have that $\boldsymbol{W}$ is a $\mathcal{P}$-border cover of X with enclosure $F \cup G$. The collection $\boldsymbol{W}_A = \{\, W_i \cap A : i = 0, 1, \ldots, k \,\}$ is a $\mathcal{P}$-border cover of A with enclosure $F \cap A$. As $\mathcal{P}\text{-}\dim A \le n$, there is a collection

$\mathbf{W}' = \{ W_i' : i = 0, 1, \ldots, k \}$ such that $\mathbf{W}'$ is a $\mathcal{P}$-border cover of A with enclosure G' that shrinks $\mathbf{W}_A$ and $\operatorname{ord} \mathbf{W}' \leq n+1$. The sets $A \setminus (V \cup F)$ and $B \setminus (G \cup F)$ are separated. Let V' be an open set such that $A \setminus (V \cup F) \subset V'$ and $\big(B \setminus (G \cup F)\big) \cap \operatorname{cl}_X(V') = \emptyset$. For each i, define the open set $Z_i = (W_i \cap V) \cup (W_i' \cap V')$. Then the open collection $\mathbf{Z} = \{ Z_i : i = 0, 1, \ldots, k \}$ is a $\mathcal{P}$-border cover of X with enclosure $F \cup G \cup (G' \setminus V)$ and $\operatorname{ord} \mathbf{Z} \leq n+1$.

1.6. Corollary. *Let A and B be open subsets of a space X such that $X = A \cup B$. Then*

$$\mathcal{P}\text{-dim}\, X \leq \max\{ \mathcal{P}\text{-dim}\, A,\ \mathcal{P}\text{-dim}\, B \}.$$

Another consequence of Theorem 1.5 is the following. In a sense, the theorem justifies the "modulo a class $\mathcal{P}$" terminology.

1.7. Theorem. *For every closed $\mathcal{P}$-kernel F of a space X,*

$$\mathcal{P}\text{-dim}\, X = \mathcal{P}\text{-dim}\,(X \setminus F).$$

Proof. By Theorem 1.5, $\mathcal{P}\text{-dim}\, X \leq \mathcal{P}\text{-dim}\,(X \setminus F)$. To prove the opposite inequality, assume $\mathcal{P}\text{-dim}\, X \leq n$ and let $\mathbf{U}$ be a finite $\mathcal{P}$-border cover of $X \setminus F$ with enclosure G. As $\mathcal{P}$ is strongly closed-additive, $\mathbf{U}$ is a finite $\mathcal{P}$-border cover of X with enclosure $F \cup G$. Let $\mathbf{W}$ be a $\mathcal{P}$-border cover of X with enclosure H such that $\mathbf{W}$ refines $\mathbf{U}$ and has $\operatorname{ord} \mathbf{W} \leq n+1$. Open monotonicity implies that the restriction $\mathbf{W}'$ of $\mathbf{W}$ to $X \setminus F$ is a finite $\mathcal{P}$-border cover of $X \setminus F$ of order less than or equal to $n+1$. Therefore $\mathcal{P}\text{-dim}\,(X \setminus F) \leq n$.

In passing, we state the analogous theorem that provides the same justification for $\mathcal{P}$-Ind. Its proof is an immediate consequence of Theorems III.1.8 and III.3.23.

1.8. Theorem. *Suppose that the universe is contained in the Dowker universe $\mathcal{D}$ and that the class $\mathcal{P}$ is a semi-normal family. For every closed $\mathcal{P}$-kernel F of a space X,*

$$\mathcal{P}\text{-Ind}\, X = \mathcal{P}\text{-Ind}\,(X \setminus F).$$

1.9. Theorem (Point addition theorem). *If $X \notin \mathcal{P}$, then*

$$\mathcal{P}\text{-dim}\,(X \cup \{p\}) \le \mathcal{P}\text{-dim}\, X.$$

Proof. As the inequality is trivial when $p \in X$, we assume $p \notin X$. Theorem 1.5 will complete the proof.

Finally we have an addition theorem.

1.10. Theorem (Addition theorem). *Suppose that the universe is monotone and that $\mathcal{P}$ is monotone and additive. Let A and B be subsets of a space X such that $X = A \cup B$. Then*

$$\mathcal{P}\text{-dim}\, X \le \mathcal{P}\text{-dim}\, A + \mathcal{P}\text{-dim}\, B + 1.$$

Proof. We may assume that $m = \mathcal{P}\text{-dim}\, A$ and $n = \mathcal{P}\text{-dim}\, B$ are finite because the inequality is trivial in the contrary case.

Let $\boldsymbol{U}$ be a finite $\mathcal{P}$-border cover of X with enclosure F. Since $\mathcal{U}$ is monotone and the class $\mathcal{P}$ is monotone, the restriction $\boldsymbol{U}_A$ of $\boldsymbol{U}$ to A is a $\mathcal{P}$-border cover of A. By Lemma 1.4 there is a set G_A and an open collection $\boldsymbol{W}_A$ such that $\boldsymbol{W}_A$ refines $\boldsymbol{U}$ and satisfies $G_A = A \setminus \bigcup \boldsymbol{W}_A \in \mathcal{P} \cap \mathcal{U}$ and $\operatorname{ord} \boldsymbol{W}_A \le m + 1$. Similarly, there is a set G_B and an open collection $\boldsymbol{W}_B$ such that $\boldsymbol{W}_B$ refines $\boldsymbol{U}$ and satisfies $G_B = B \setminus \bigcup \boldsymbol{W}_B \in \mathcal{P} \cap \mathcal{U}$ and $\operatorname{ord} \boldsymbol{W}_B \le n + 1$. Let $\boldsymbol{W} = \boldsymbol{W}_A \cup \boldsymbol{W}_B$. Clearly, $\boldsymbol{W}$ refines $\boldsymbol{U}$ and $\operatorname{ord} \boldsymbol{W} \le m + n + 2$. Obviously, $X \setminus \bigcup \boldsymbol{W} = F \cup \left(G_A \setminus \bigcup \boldsymbol{W}_B\right) \cup \left(G_B \setminus \bigcup \boldsymbol{W}_A\right)$. As $\mathcal{P}$ is monotone and additive, we have that $\boldsymbol{W}$ is a finite $\mathcal{P}$-border cover of X. Therefore, $\mathcal{P}\text{-dim}\, X \le m + n + 1$.

Let us now go to $\mathcal{P}$-Dim. The closed monotonicity of the function $\mathcal{P}$-Dim is not immediately obvious without further conditions. So some of the corresponding statements to the above theorems cannot be proved by analogy. The following lemma and theorems do not rely on the closed monotonicity of $\mathcal{P}$-Dim and so their proofs are straightforward modifications of the corresponding ones for $\mathcal{P}$-dim.

1.11. Lemma. *Let Y be a subspace of a space X and suppose that $\boldsymbol{U} = \{U_\alpha : \alpha \in A\}$ is an open collection of X such that $\boldsymbol{U}$ is locally finite in the subspace $\bigcup \boldsymbol{U}$ and the restriction $\boldsymbol{U}_Y$ of $\boldsymbol{U}$ to Y is a $\mathcal{P}$-border cover of Y with enclosure F. If $\mathcal{P}\text{-Dim}\, Y \le n$, then* *there is a set* G *and there is an open collection* $\mathbf{W} = \{ W_\alpha : \alpha \in A \}$ *of* X *such that the following conditions hold:*

(a) $W_\alpha \subset U_\alpha$ for each α in A.
(b) $G = Y \setminus \bigcup \mathbf{W} \in \mathcal{P}$.
(c) $\operatorname{ord} \mathbf{W} \leq n + 1$.

Moreover, if Y *is also closed in* X*, then the further condition*

(d) $(\bigcup \mathbf{U}) \setminus G = \bigcup \mathbf{W}$

holds provided the condition (c) *is changed to the following:*

(c*) *There is an open set* V *of* X *with* $\operatorname{ord}_x \mathbf{W} \leq \operatorname{ord}_x \mathbf{U}$ *for each* x *in* $X \setminus V$ *such that the inclusions* $Y \setminus G \subset V \subset \{ x : \operatorname{ord}_x \mathbf{W} \leq n + 1 \}$ *hold modulo the set* $H = X \setminus \bigcup \mathbf{U}$.

1.12. Theorem. *If* A *and* B *are closed subsets of a space* X *such that* $X = A \cup B$*, then*

$$\mathcal{P}\text{-Dim}\, X \leq \max\{ \mathcal{P}\text{-Dim}\, A, \mathcal{P}\text{-Dim}\, B \}.$$

Corresponding to Theorem 1.5 is the following. Notice that there is the stronger hypothesis of $A \cap B = \emptyset$.

1.13. Theorem. *Let* A *and* B *be disjoint subsets of a space* X *such that* $X = A \cup B$ *and* A *is open. Then*

$$\mathcal{P}\text{-Dim}\, X \leq \max\{ \mathcal{P}\text{-Dim}\, A, \mathcal{P}\text{-Dim}\, B \}.$$

Proof. The proof of Theorem 1.5 assumes $A \cap B = \emptyset$ because this condition is a consequence of Theorem II.5.6. Since the analogous theorem for $\mathcal{P}$-Dim is not known, we must assume this added condition. Otherwise, the proof is the same when the obvious changes have been made.

1.14. Theorem (Addition theorem). *Suppose that the universe is monotone and that* $\mathcal{P}$ *is monotone and additive. Let* A *and* B *be subsets of a space* X *such that* $X = A \cup B$*. Then*

$$\mathcal{P}\text{-Dim}\, X \leq \mathcal{P}\text{-Dim}\, A + \mathcal{P}\text{-Dim}\, B + 1.$$

2. Normal families

The functions of covering dimensional type will be studied in the context of normal families. The main concerns are the monotonicity theorems and the various closed additivity theorems for $\mathcal{P}$-dim and $\mathcal{P}$-Dim. It will be shown by cosmic family arguments that the equality $\mathcal{P}\text{-dim} = \mathcal{P}\text{-Dim}$ holds and thereby the closed monotonicity will be establish for $\mathcal{P}$-Dim in the Dowker universe $\mathcal{D}$. The following theorem concerns the monotonicity of functions of covering dimensional type.

2.1. Theorem. *Let the universe be monotone and $\mathcal{P}$ be monotone. For each space X, $\mathcal{P}\text{-dim}\, Y \leq \mathcal{P}\text{-dim}\, X$ holds for every subspace Y of X if and only if $\mathcal{P}\text{-dim}\, U \leq \mathcal{P}\text{-dim}\, X$ holds for every open subspace U of X.*

Proof. As the proof of one of the implications is obvious, we shall give only the other one. Suppose that $\mathcal{P}\text{-dim}\, U \leq n$ for each open subset U of X and let Y be a subset of X. We may assume $0 \leq n = \mathcal{P}\text{-dim}\, X$ because the contrary case follows from definitions. Let $\mathbf{W}$ be a finite $\mathcal{P}$-border cover of Y with enclosure F. For each W in $\mathbf{W}$ let W' be an open set of X such that $W = W' \cap Y$. Denote the collection of these open sets W' by $\mathbf{W}'$. Then we have $Y \setminus F \subset U = \bigcup \mathbf{W}'$. Recalling that $\emptyset \in \mathcal{P} \cap \mathcal{U}$, we have that $\mathbf{W}'$ is a finite $\mathcal{P}$-border cover of U. From $\mathcal{P}\text{-dim}\, U \leq n$, there is a $\mathcal{P}$-border cover $\mathbf{V}'$ of the open subspace U with enclosure G' such that $\operatorname{ord} \mathbf{V}' \leq n+1$ and $\mathbf{V}'$ refines $\mathbf{W}'$. Denote $\{ V' \cap Y : V' \in \mathbf{V}' \}$ by $\mathbf{V}$ and let $G = G' \cap Y$. Since the universe $\mathcal{U}$ is monotone and $\mathcal{P}$ is monotone and strongly closed-additive, we have that $G \cup F$ is in $\mathcal{P}$. It follows that $\mathbf{V}$ is a $\mathcal{P}$-border cover of Y with $\operatorname{ord} \mathbf{V} \leq n+1$.

Let us precede our investigation of normal families by studying regular and cosmic families separately. We extend Lemma 1.4.

2.2. Lemma. *For a space X let $\mathbf{U} = \{ U_j : j = 0, 1, \ldots, k \}$ be a finite $\mathcal{P}$-border cover with enclosure F and let A be an open set contained in $\{ x : 0 < \operatorname{ord}_x \mathbf{U} \leq n+1 \}$. Suppose that Y is a closed set of X with $\mathcal{P}\text{-dim}\, Y \leq n$. Then there is a $\mathcal{P}$-border cover $\mathbf{W} = \{ W_j : j = 0, 1, \ldots, k \}$ with enclosure G and there is an open set B that satisfy the conditions*

(a) *$W_j \subset U_j$ for $j = 0, 1, \ldots, k$,*

(b) $A \subset B \subset \bigcup \boldsymbol{W} = (\bigcup \boldsymbol{U}) \setminus G$,
(c) $G = F \cup (Y \setminus B)$ *and* $Y \setminus G = Y \cap B$,
(d) $W_j \cap A = U_j \cap A$ *for all* j,
(e) $\operatorname{ord}\{ W_j \cap B : j = 0, 1, \dots, k \} \leq n + 1$.

Proof. As A is open, we have $\mathcal{P}$-dim $(Y \setminus A) \leq n$. By Lemma 1.4 there is an open collection $\boldsymbol{W}' = \{ W_j' : j = 0, 1, \dots, k \}$ of X with $G' = Y \setminus (A \cup \bigcup \boldsymbol{W}') \in \mathcal{P}$ and there is an open set V such that $\boldsymbol{W}'$ shrinks $\boldsymbol{U}$ modulo G', $V \subset \bigcup \boldsymbol{W}'$ and

$$(1) \qquad (\bigcup \boldsymbol{U}) \setminus G' = \bigcup \boldsymbol{W}', \quad G' \subset Y \setminus A,$$

$$(2) \qquad \operatorname{ord}_x \boldsymbol{W}' \leq \operatorname{ord}_x \boldsymbol{U}, \qquad x \in X \setminus V,$$

$$(3) \qquad Y \setminus (A \cup G') \subset V \subset \{ x : \operatorname{ord}_x \boldsymbol{W}' \leq n + 1 \} \mod F.$$

For each j let $W_j = (U_j \cap A) \cup W_j'$ and denote the collection of open sets so constructed by $\boldsymbol{W}$. Also let $B = A \cup (V \setminus G')$. As $F \cap A = \emptyset$, we have from (1) that $A \cap G = \emptyset$ and $X \setminus \bigcup \boldsymbol{W} = F \cup G'$. Since G' is in $\mathcal{P} \cap \mathcal{U}$, the collection $\boldsymbol{W}$ is a $\mathcal{P}$-border cover of X with enclosure $G = F \cup G'$ that shrinks $\boldsymbol{U}$. So conditions (a) and (b) hold. We infer from (3) that $Y \setminus (A \cup G) \subset V$ holds. So it follows that $Y = (Y \cap (A \cup G)) \cup (Y \setminus (A \cup G)) \subset (A \cup G) \cup V = B \cup G$. Consequently $Y \setminus G = (Y \cap B) \setminus G = Y \setminus G$, where the last equality follows from (b). As $B \cap G$ is empty, we have $Y \setminus G = Y \cap B$ and thus $Y \setminus B \subset G$. Moreover, as $G \setminus F \subset Y$, we have $G \setminus F \subset Y \setminus B$. So we have arrived at $G \subset F \cup (G \setminus F) \subset F \cup (Y \setminus B) \subset G$ and thereby (c) is verified. Condition (d) follows from the definition of W_j; and condition (e) is a consequence of (d) and (3).

2.3. Theorem. *Suppose that* $\mathcal{P}$ *is a regular family and let* $\{ X_i : i = 1, 2, \dots \}$ *be a countable closed cover of a space* X. *Then*

$$\mathcal{P}\text{-dim}\, X = \sup \{ \mathcal{P}\text{-dim}\, X_i : i = 1, 2, \dots \}.$$

Proof. We must show that the supremum is no less than $\mathcal{P}$-dim X. By Theorem 1.1 we may assume $X_i \subset X_{i+1}$ for each i. Let $\boldsymbol{U}_0 = \{ U_{0,j} : j = 0, 1, \dots, k \}$ be a finite $\mathcal{P}$-border cover of X with enclosure F_0. Define $A_0 = \emptyset$. With the aid of Lemma 2.2 we can construct for each i a $\mathcal{P}$-border cover $\boldsymbol{U}_i = \{ U_{i,j} : j = 0, 1, \dots, k \}$ with enclosure F_i and an open set A_i that satisfy the following conditions: For $i \geq 0$

(a-i) $\boldsymbol{U}_{i+1}$ shrinks $\boldsymbol{U}_i \mod F_{i+1}$;

and for $i \geq 1$

(b-i) $A_{i-1} \subset A_i \subset \bigcup \boldsymbol{U}_i = (\bigcup \boldsymbol{U}_{i-1}) \setminus F_i$,
(c-i) $X_i \setminus F_i = X_i \cap A_i$,
(d-i) $A_{i-1} \cap U_{i,j} = A_{i-1} \cap U_{i-1,j}$,
(e-i) $\operatorname{ord}\{ U_{i,j} \cap A_i : j = 0, 1, \ldots, k \} \leq n + 1$.

For each j let $U_j = \bigcup\{ U_{i,j} \cap A_i : i = 1, 2, \ldots \}$ and let $\boldsymbol{U}$ denote the collection of open sets so constructed. By (a-i), (c-i) and (d-i) we have $U_j = \bigcap\{ U_{i,j} : i = 1, 2, \ldots \}$ and hence it follows that $X \setminus \bigcup \boldsymbol{U} = \bigcup\{ F_i : i = 0, 1, 2, \ldots \}$ by condition (b-i). Since $\mathcal{P}$ is a regular family and $U_i \subset U_{0,i}$, the collection $\boldsymbol{U}$ is a $\mathcal{P}$-border cover of X that shrinks $\boldsymbol{U}_0$ mod $\bigcup\{ F_i : i = 0, 1, 2, \ldots \}$. Let x be a point of X that is not in the enclosure of $\boldsymbol{U}$. Then x is in $X_i \cap A_i$ for some i by condition (c-i). Consequently $\operatorname{ord}_x \{ U_{i,j} \cap A_i : j = 0, 1, \ldots, k \} \leq n + 1$ for that i by condition (e-i). So $\operatorname{ord} \boldsymbol{U} \leq n + 1$ by condition (d-i). Thereby we have shown $\mathcal{P}\text{-}\dim X \leq n$.

Cosmic families will be considered next. The following lemma will prove useful in the next section as well as in the present one.

2.4. Lemma. *Suppose that $\mathcal{P}$ is a cosmic family and that n is in $\mathbb{N}$. Let X be a space and $\boldsymbol{F}$ be a locally finite closed cover of X such that $\mathcal{P}\text{-}\dim F \leq n$ for each F in $\boldsymbol{F}$. If $\boldsymbol{U} = \{ U_s : s \in S \}$ is a $\mathcal{P}$-border cover of X with enclosure G such that each member of $\boldsymbol{F}$ meets only finitely many sets U_s, then there is a $\mathcal{P}$-border cover with enclosure H that shrinks $\boldsymbol{U}$ modulo H and has order not exceeding $n + 1$.*

Proof. For the purposes of a transfinite inductive construction, adjoin the set $F_0 = \emptyset$ to the collection $\boldsymbol{F}$ and arrange the members of this collection into a transfinite sequence F_α, $\alpha \leq \xi$, of type $\xi + 1$. Let us inductively define for each α a closed $\mathcal{P}$-kernel G_α of F_α and a $\mathcal{P}$-border cover $\boldsymbol{U}_\alpha = \{ U_{\alpha,s} : s \in S \}$ of X such that G_α, $\boldsymbol{U}_\alpha$ and the closed set $H_\alpha = \bigcup\{ G_\gamma : \gamma \leq \alpha \}$ satisfy the condition

$$X \setminus \bigcup \boldsymbol{U}_\alpha = G \cup H_\alpha, \tag{4}$$

and satisfy the next conditions modulo the set H_α

$$U_{\alpha,s} \subset U_{\beta,s} \quad \text{if } \alpha > \beta \quad \text{and} \quad U_{0,s} \subset U_s, \qquad s \in S, \tag{5}$$

$$\operatorname{ord}\{ F_\alpha \cap U_{\alpha,s} : s \in S \} \leq n, \tag{6}$$

$$U_{\beta,s} \setminus U_{\alpha,s} \subset \bigcup\{ F_\gamma : \beta \leq \gamma \leq \alpha \}, \qquad \beta < \alpha \text{ and } s \in S. \tag{7}$$

Every condition will be satisfied for $\alpha = 0$ if we define $U_{0,s} = U_s$ for all s in S and define $G_0 = \emptyset$. Assume that the $\mathcal{P}$-border cover $\boldsymbol{U}_\alpha$ and closed $\mathcal{P}$-kernels G_α of F_α satisfying (4)–(7) are defined for all α less than α_0, where $\alpha_0 \geq 1$. Let us first consider the collection $\boldsymbol{U}'_{\alpha_0} = \{ U'_{\alpha_0,s} : s \in S \}$ and the set H'_{α_0} defined by

$$H'_{\alpha_0} = \bigcup\{ G_\alpha : \alpha < \alpha_0 \},$$
$$U'_{\alpha_0,s} = \bigcap\{ U_{\alpha,s} : \alpha < \alpha_0 \} \setminus H'_{\alpha_0}$$

for each s in S. We shall show that this collection is open in X. This is clear when $\alpha_0 = \eta + 1$, because we will then have that $H'_{\alpha_0} = H_\eta$ is a closed set and $U'_{\alpha_0,s} = U_{\eta,s} \setminus H_\eta$. So we shall assume that α_0 is a limit ordinal number.

Consider a point x that is not in the closed set $G \cup H'_{\alpha_0}$. As $\mathcal{P}$ is a cosmic family and $\boldsymbol{F}$ is a locally finite collection in X, we have that $G \cup H'_{\alpha_0}$ is in $\mathcal{P} \cap \mathcal{U}$. Also there exists a neighborhood U in X of x and an ordinal number β less than α_0 such that $U \cap F_\gamma = \emptyset$ whenever $\beta \leq \gamma < \alpha_0$. As $\boldsymbol{U}_\beta$ is a $\mathcal{P}$-border cover of X with enclosure $G \cup H_\beta$, there exists an s in S such that x is in $U_{\beta,s}$. It follows from (7) that $x \in U_{\alpha,s}$ whenever $\beta < \alpha < \alpha_0$, so $x \in U'_{\alpha_0,s}$. Hence $\boldsymbol{U}'_{\alpha_0}$ will be a $\mathcal{P}$-border cover of X with enclosure $G \cup H'_{\alpha_0}$ once it is shown that each $U'_{\alpha_0,s}$ containing the point x is a neighborhood of x. To this end, consider the open set U and the ordinal number β determined above for the point x. From (5) we have $x \in (U \cap U_{\beta,s}) \setminus H_\beta$. Since x is not in the closed set H'_{α_0}, we have that x is in the open set $(U \cap U_{\beta,s}) \setminus H'_{\alpha_0}$ and this open set is contained in $U'_{\alpha_0,s}$ by (7). Thereby, it has been shown that $\boldsymbol{U}'_{\alpha_0}$ is a $\mathcal{P}$-border cover of X.

The restricted collection $\{ F_{\alpha_0} \cap U'_{\alpha_0,s} : s \in S \}$ of $\boldsymbol{U}'_{\alpha_0}$ to F_{α_0} is a finite $\mathcal{P}$-border cover of F_{α_0} because each member of the family $\boldsymbol{F}$ meets at most finitely many members of the collection $\boldsymbol{U}$ and condition (5) holds. Consequently, from $\mathcal{P}\text{-dim}\, F_{\alpha_0} \leq n$, this $\mathcal{P}$-border cover of F_{α_0} has a $\mathcal{P}$-border cover shrinking $\{ V_s : s \in S \}$ with enclosure G_{α_0} of order not exceeding $n + 1$. Define $U_{\alpha_0,s}$ to be the open set $(U'_{\alpha_0,s} \setminus F_{\alpha_0}) \cup V_s$ for each s in S. Then $\boldsymbol{U}_{\alpha_0} = \{ U'_{\alpha_0,s} : s \in S \}$ is readily seen to be a $\mathcal{P}$-border cover of X with enclosure $H'_{\alpha_0} \cup G_{\alpha_0}$ that satisfies the conditions (4)–(7) for $\alpha = \alpha_0$. Hence the construction of the $\mathcal{P}$-border covers $\boldsymbol{U}_\alpha$ of X satisfying the conditions (4)–(7) for $\alpha \leq \xi$ is completed.

Now it follows from (6) that ord $\boldsymbol{U}_\xi \leq n$. By virtue of (5), $\boldsymbol{U}_\xi$ is a shrinking of $\boldsymbol{U}$ modulo H and the lemma is established.

The next locally finite sum theorem for $\mathcal{P}$-dim is an immediate consequence of Lemma 2.4.

2.5. Theorem. *Suppose that $\mathcal{P}$ is a cosmic family and let X be a space and $\boldsymbol{F}$ be a locally finite closed cover of X. Then*

$$\mathcal{P}\text{-dim}\, X = \sup\{\,\mathcal{P}\text{-dim}\, F : F \in \boldsymbol{F}\,\}.$$

Combining Theorems 2.3 and 2.5, we have the following theorem on semi-normal families.

2.6. Theorem. *Suppose that $\mathcal{P}$ is a semi-normal family. Then the class $\{\, X : \mathcal{P}\text{-dim}\, X \leq n \,\}$ is a semi-normal family for each n.*

Finally we come to the analogue of Dowker's theorem that characterizes dim by means of locally finite open covers.

2.7. Theorem. *Suppose that $\mathcal{P}$ is a cosmic family. Then*

$$\mathcal{P}\text{-dim}\, X = \mathcal{P}\text{-Dim}\, X.$$

for each space X.

Proof. Obviously $\mathcal{P}\text{-dim}\, X \leq \mathcal{P}\text{-Dim}\, X$. So let n be in $\mathbb{N}$ and assume $\mathcal{P}\text{-dim}\, X \leq n$. Let $\boldsymbol{U}$ be a $\mathcal{P}$-border cover of X with enclosure G such that $\boldsymbol{U}$ is locally finite in $X \setminus G$. By Theorem 1.7 we have $\mathcal{P}\text{-dim}\,(X \setminus G) \leq n$. The collection $\boldsymbol{U} = \{\, U_s : s \in S \,\}$ is a locally finite open cover of $X \setminus G$. Denote by $\boldsymbol{T}$ the family of nonempty finite subsets of S and for every T in $\boldsymbol{T}$ define

$$F_T = \left(\bigcap\{\,\mathrm{cl}_{X\setminus G}(U_s) : s \in T\,\}\right) \cap \left(\bigcap\{\,(X \setminus G) \setminus U_s : s \notin T\,\}\right).$$

We have $\mathcal{P}\text{-dim}\, F_T \leq n$ because F_T is closed in $X \setminus G$. The collection $\boldsymbol{F}$ is a locally finite closed cover of $X \setminus G$. By Lemma 2.4 there is a $\mathcal{P}$-border cover $\boldsymbol{V}$ of $X \setminus G$ with enclosure H that shrinks $\boldsymbol{U}$ modulo H and has ord $\boldsymbol{V} \leq n + 1$. As $X \setminus \bigcup \boldsymbol{V} = G \cup H$ and $G \cap H = \emptyset$ hold and $\mathcal{P}$ is strongly closed-additive, we have that $\boldsymbol{V}$ is a $\mathcal{P}$-border cover of X. Therefore, $\mathcal{P}\text{-Dim}\, X \leq n$.

Remark. As we have found in the remark at the end of Section II.5, the spaces under consideration are required to be hereditarily normal due to the possibility that $\mathcal{P} \neq \{\,\emptyset\,\}$ is true. It was found there that many of the arguments found in that section held true for all normal spaces when $\mathcal{P} = \{\emptyset\}$. This observation is also true for the present section. Indeed, this is especially true for the last three theorems. We shall state these facts next.

2.8. Theorem. *Let* $\boldsymbol{F}$ *be a countable, closed cover of a normal space* X. *Then*

$$\dim X = \sup\{\dim F : F \in \boldsymbol{F}\}.$$

2.9. Theorem. *Let* $\boldsymbol{F}$ *be a locally finite, closed cover of a normal space* X. *Then*

$$\dim X = \sup\{\dim F : F \in \boldsymbol{F}\}.$$

2.10. Theorem. *For every normal space* X,

$$\dim X = \operatorname{Dim} X.$$

3. The Dowker universe $\mathcal{D}$

It was shown in the last section that the function of covering dimensional type $\mathcal{P}$-dim behaves very nicely for semi-normal families $\mathcal{P}$. Indeed, the family $\{X : \mathcal{P}\text{-dim}\, X \leq n\}$ is guaranteed to be semi-normal with fewer conditions than the corresponding family for the function of inductive dimensional type $\mathcal{P}$-Ind. But, as it is for $\mathcal{P}$-Ind, the situation for normal families will require a stronger condition on the universe. The difficulty is due to the lack of open monotonicity of these function in the case of the general universe. Thus the Dowker universe $\mathcal{D}$ will come to the fore in that the universe will be required to be contained in $\mathcal{D}$.

Agreement. *In addition to the agreement made in the introduction of the chapter, the universe will be contained in the Dowker universe.*

3.1. Lemma. *Suppose that* $\mathcal{P}$ *is a cosmic family. If* Y *is a cozero-set in a space* X, *then*

$$\mathcal{P}\text{-dim}\, Y \leq \mathcal{P}\text{-dim}\, X.$$

Proof. Select a continuous function $f\colon X \to [0,1]$ such that Y is $f^{-1}[(0,1]]$. As the sets $f^{-1}[[2^{-i+1}, 2^{-i}]]$, $i = 0, 1, \dots$, are closed in X and form a locally finite cover of the subspace Y, we have $\mathcal{P}\text{-dim}\, Y \le \mathcal{P}\text{-dim}\, X$ by Theorems II.5.6 and 2.5.

The next lemma deals with Dowker-open sets.

3.2. Lemma. *Suppose that $\mathcal{P}$ is a semi-normal family. If Y is a Dowker-open subset of a space X, then*

$$\mathcal{P}\text{-dim}\, Y \le \mathcal{P}\text{-dim}\, X.$$

Proof. Assume $0 \le \mathcal{P}\text{-dim}\, X = n$. By definition, the set Y is the union of a point-finite collection $\{\, U_s : s \in S \,\}$ of cozero-sets U_s of X. For each positive integer i let $\boldsymbol{T}_i$ be the family of all subsets of the index set S that have exactly i elements. By Lemma III.3.3, the collection of subsets K_i of Y consisting of the points of X that are in U_s for exactly i members s in S, where i is a positive integer, has the following properties:

(1) $Y = \bigcup\{\, K_i : i \ge 1 \,\}$.
(2) $K_i \cap K_j = \emptyset$ for $i \ne j$.
(3) $F_i = \bigcup\{\, K_j : j \le i \,\}$ is closed in Y for $i \ge 1$.
(4) $K_i = \bigcup\{\, K_T : T \in \boldsymbol{T}_i \,\}$, where the sets K_T, defined by letting $K_T = \bigcap\{\, K_i \cap U_s : s \in T \,\}$ for $T \in \boldsymbol{T}_i$, are open in K_i and pairwise disjoint.

As each U_s is a cozero-set of X, the set K_T is a cozero-set of K_i by (4). From (2) and (4) and from the identity

$$K_T = \big(F_i \cup (X \setminus Y)\big) \cap \big(\bigcap\{\, U_s : s \in T \,\}\big),$$

we have that K_T is a countable union of closed sets of X for T in $\boldsymbol{T}_i$. By Theorem 2.6 we have $\mathcal{P}\text{-dim}\, K_T \le n$. So $\mathcal{P}\text{-dim}\, K_i \le n$ by the same theorem because of (4). Theorem 1.5 gives $\mathcal{P}\text{-dim}\, F_i \le n$. Finally $\mathcal{P}\text{-dim}\, Y \le n$ is established by Theorem 2.6.

In passing, we comment that the first two lemmas do not use the condition $\mathcal{U} \subset \mathcal{D}$. The next lemma deals with the class $\mathcal{E}$ of strongly hereditarily normal spaces.

3.3. Lemma. *Suppose that the universe is contained in the class $\mathcal{E}$ and that $\mathcal{P}$ is a semi-normal family. If Y is an open subspace of a space X, then*

$$\mathcal{P}\text{-dim}\, Y \leq \mathcal{P}\text{-dim}\, X.$$

Proof. Assume $0 \leq \mathcal{P}\text{-dim}\, X = n$. Let $\boldsymbol{U} = \{ U_i : i = 0, 1, \ldots, k \}$ be a finite $\mathcal{P}$-border cover of Y with enclosure F. The sets U_i and $X \setminus \mathrm{cl}_X(U_i)$ are separated for each i. Since X is in $\mathcal{E}$, there is a Dowker-open set V_i of X such that $U_i = V_i \cap Y$. Let $\widetilde{Y}$ be the open set $\bigcup\{ V_i : i = 0, 1, \ldots, k \}$. Then $\boldsymbol{V} = \{ V_i : i = 0, 1, \ldots, k \}$ is a finite $\mathcal{P}$-border cover of $\widetilde{Y}$. By Lemma 3.2 and Corollary 1.6, we have $\mathcal{P}\text{-dim}\, \widetilde{Y} \leq n$. Let $\boldsymbol{V}' = \{ V_i' : i = 0, 1, \ldots, k \}$ be a $\mathcal{P}$-border cover of $\widetilde{Y}$ with enclosure F' such that $\boldsymbol{V}'$ shrinks $\boldsymbol{V}$ modulo F' and $\mathrm{ord}\, \boldsymbol{V}' \leq n + 1$. The collection $\boldsymbol{U}' = \{ V_i' \cap Y : i = 0, 1, \ldots, k \}$ clearly satisfies $\mathrm{ord}\, \boldsymbol{U}' \leq n + 1$ and shrinks $\boldsymbol{U}$ modulo F'. Since $\mathcal{P}$ is open-monotone, we have that $G = F' \cap Y$ is in $\mathcal{P}$ and hence $\boldsymbol{U}'$ is a $\mathcal{P}$-border cover of Y with enclosure $F \cup G$. So $\mathcal{P}\text{-dim}\, Y \leq n$.

3.4. Theorem. *Suppose that $\mathcal{P}$ is a semi-normal family. Then $\{ X : \mathcal{P}\text{-dim}\, X \leq n \}$ is an open-monotone, strongly closed-additive semi-normal family for each n in $\mathbb{N}$.*

Proof. By definition, the Dowker universe is the normal family extension of the class $\mathcal{E}$ in the universe $\mathcal{N}_H$. (See Definition III.3.1.) So X is the union of a σ-locally finite collection of closed sets of X that are in the class $\mathcal{E}$. The theorem follows easily from Lemma 3.3 and Theorems 1.5 and 2.6.

3.5. Theorem. *Suppose that the universe is monotone and $\mathcal{P}$ is a normal family. Then $\{ X : \mathcal{P}\text{-dim}\, X \leq n \}$ is a strongly closed-additive normal family for each n in $\mathbb{N}$.*

Proof. The theorem follows easily from Theorems 2.1 and 3.4.

The aim for the rest of the section is to prove a characterization of $\mathcal{P}$-dim by means of partitions. We shall use the following definition.

3.6. Definition. Let $\boldsymbol{G}$ be a $\mathcal{P}$-border cover of a space X with enclosure G and let Y be a subspace of X. Then the *relative dimension of Y with respect to $\boldsymbol{G}$*, denoted $\mathcal{P}\text{-dim}_{\boldsymbol{G}}\, Y$, is at most n if the restricted collection $\mathbf{G}|Y$ admits a shrinking by a $\mathcal{P}$-border cover $\boldsymbol{H}$ of $Y \setminus G$ with order at most $n+1$.

The following general covering lemma will prove useful.

3.7. Lemma. *Let X be a normal space and F_1 and F_2 be a pair of disjoint closed sets of X. If $\mathbf{G}$ is a locally finite open cover of X refining the binary cover $\{X \setminus F_1, X \setminus F_2\}$, then there exists a closed set S of X and a locally finite open cover $\boldsymbol{H}$ of X satisfying the conditions:*

(a) *S is a partition between F_1 and F_2.*
(b) *$\boldsymbol{H}$ shrinks $\mathbf{G}$.*
(c) $\operatorname{ord}_x \boldsymbol{H} < \operatorname{ord}_x \mathbf{G}$ *for any point x in S.*

Proof. Since $\mathbf{G} = \{G_\alpha : \alpha \in A\}$ is a locally finite open covering of the normal space X, there is an open shrinking $\boldsymbol{U} = \{U_\alpha : \alpha \in A\}$ of $\mathbf{G}$ such that $\operatorname{cl}(U_\alpha) \subset G_\alpha$ for every α. Let $\boldsymbol{U}_1 = \{U_\alpha : \alpha \in A_1\}$ be the family of all members of $\boldsymbol{U}$ that meet F_1. Define

$$S = \mathrm{B}(\bigcup \boldsymbol{U}_1).$$

Then S is a partition between F_1 and F_2. Let x be any point of S. Obviously x is not in U_α for each α in A_1. But there is an α in A_1 with $x \in \operatorname{cl}(U_\alpha)$ since $\mathbf{G}$, whence $\boldsymbol{U}$, is locally finite. Since $\boldsymbol{U}$ covers X there exists a β that is not in A_1 with $x \in U_\beta$. From this observation it follows easily that

$$\boldsymbol{H} = \{G_\alpha : \alpha \in A_1\} \cup \{G_\alpha \setminus S : \alpha \notin A_1\}$$

satisfies conditions (b) and (c).

3.8. Lemma. *In a space X let $\mathbf{G} = \{G_\alpha : \alpha \in A\}$ be a $\mathcal{P}$-border cover with enclosure G and let B be a closed set such that $\mathbf{G}$ is locally finite in the space $X \setminus G$ and $\mathcal{P}\text{-}\dim_{\mathbf{G}} B \leq n$. Then there exists a $\mathcal{P}$-border cover $\boldsymbol{H} = \{H_\alpha : \alpha \in A\}$ of X with enclosure H and there exists an open set V such that $\boldsymbol{H}$ shrinks $\mathbf{G}$ $\bmod H$, $B \subset V$ $\bmod H$ and $\operatorname{ord} \boldsymbol{H}|V \leq n+1$.*

Proof. As $\mathcal{P}\text{-}\dim_{\mathbf{G}} B \leq n$, there is a $\mathcal{P}$-border cover $\mathbf{G}_0$ of B with enclosure G_0 such that $\operatorname{ord} \mathbf{G}_0 \leq n+1$ holds and $\mathbf{G}_0$ is a shrinking of $\mathbf{G}|B$. We write $\mathbf{G}_0 = \{G_\alpha : a \in A'\}$ with $A' \subset A$. Because B is closed in X and $\mathcal{P}$ is strongly closed-additive, we have that $G \cup G_0$

is a closed $\mathcal{P}$-kernel of X. We shall perform the remainder of the construction in the subspace $X \setminus (G \cup G_0)$. There will be no loss of generality if we let $G \cup G_0$ be the empty set. Select a closed cover $\boldsymbol{K} = \{ K_\alpha : \alpha \in A' \}$ of B that shrinks $\mathbf{G}_0$ with $\operatorname{ord} \boldsymbol{K} \leq n+1$. By Lemma 1.2 there is for each α in A' an open set H'_α of X with $K_\alpha \subset H'_\alpha \subset G_\alpha$ such that $\operatorname{ord}\{ H'_\alpha : \alpha \in A' \} \leq n+1$. Let V be an open set of X with $B \subset V \subset \operatorname{cl}(V) \subset \bigcup \{ H'_\alpha : \alpha \in A' \}$ and let $H_\alpha = H'_\alpha \cup (G_\alpha \setminus \operatorname{cl}(V))$ for $\alpha \in A'$. Then V and the collection $\boldsymbol{H} = \{ H_\alpha : \alpha \in A \}$ satisfy the desired conditions.

The next theorem provides a characterization of $\mathcal{P}$-dim by means of partitions.

3.9. Theorem. *For every space X, $\mathcal{P}\text{-}\dim X \leq n$ if and only if for each pair of closed sets F_1 and F_2 of X and for each finite $\mathcal{P}$-border cover $\mathbf{G}$ with enclosure G such that $F_1 \cap F_2 = \emptyset \mod G$ there exists a closed set S such that S is a partition between F_1 and F_2 modulo G and $\mathcal{P}\text{-}\dim_{\mathbf{G}} S \leq n-1$.*

Proof. Suppose $\mathcal{P}\text{-}\dim X \leq n$. Since $F_1 \cap F_2 = \emptyset \mod G$, there are closed sets F'_1 and F'_2 with $F_1 \subset F'_1$ and $F_2 \subset F'_2$ such that $F'_1 \setminus G$ and $F'_2 \setminus G$ are neighborhoods of $F_1 \setminus G$ and $F_2 \setminus G$ respectively in the subspace $X \setminus G$ and

$$F'_1 \cap F'_2 \subset G. \tag{1}$$

Let

$$\mathbf{G}_1 = \mathbf{G} \wedge \{ X \setminus (F'_1 \cup G),\ X \setminus (F'_2 \cup G) \}.$$

Since $\mathbf{G}_1$ is a finite $\mathcal{P}$-border cover of X with enclosure G, there is a finite $\mathcal{P}$-border cover $\mathbf{G}_2$ with enclosure G_2 that refines $\mathbf{G}_1$ such that $\operatorname{ord} \mathbf{G}_2 \leq n+1$. In the subspace $X \setminus G_2$ there exist by Lemma 3.7 a closed subset B that is a partition between $F'_1 \setminus G_2$ and $F'_2 \setminus G_2$ and a finite open cover $\boldsymbol{H}$ of $X \setminus G_2$ that shrinks $\mathbf{G}_2$ with $\operatorname{ord} \boldsymbol{H}|B \leq n$. From (1) we infer the existence of a closed set S that separates F_1 and F_2 modulo G and such that $S \cap (X \setminus G_2) \subset B$. Clearly $\boldsymbol{H}$ is a $\mathcal{P}$-border cover of X that refines $\mathbf{G}$. Since S is closed, we have that $\boldsymbol{H}|S$ is a $\mathcal{P}$-border cover of S with enclosure $G_2 \cap S$. Therefore $\boldsymbol{H}|S$ is a $\mathcal{P}$-border cover of S that refines $\mathbf{G}|S$ with $\operatorname{ord} \boldsymbol{H}|S \leq n$. That is, $\mathcal{P}\text{-}\dim_{\mathbf{G}} S \leq n-1$.

Conversely, suppose that the condition is satisfied and consider a finite $\mathcal{P}$-border cover $\boldsymbol{G} = \{ G_i : i = 1, \ldots, k \}$ of X with enclosure G. Since X is hereditarily normal, there is a closed collection $\{ F_i : i = 1, \ldots, k \}$ that covers X modulo G with $F_i \subset G_i$ modulo G for each i. For the purposes of the construction let $\boldsymbol{H}_0 = \boldsymbol{G}$ and let the enclosure of $\boldsymbol{H}_0$ be denoted by H_0. Select an open set U_1 of X that satisfies

$$F_1 \subset U_1 \subset \mathrm{cl}(U_1) \subset G_1 \mod H_0,$$
$$\mathcal{P}\text{-}\dim_{\boldsymbol{H}_0} \mathrm{B}(U_1) \leq n - 1.$$

By Lemma 3.8 there is a $\mathcal{P}$-border cover $\boldsymbol{H}_1 = \{ H_{1,i} : i = 1, \ldots, k \}$ of X with enclosure H_1 and an open set V_1 of X that satisfies

$$\boldsymbol{H}_1 \text{ shrinks } \boldsymbol{H}_0 \mod H_1,$$
$$\mathrm{B}(U_1) \subset V_1 \mod H_1,$$
$$\mathrm{ord}\, \boldsymbol{H}_1 | V_1 \leq n.$$

Analogously, by repeated application of the condition of the theorem together with the repeated application of Lemma 3.8, we can select for $i = 2, \ldots, k$ an open set U_i of X and we can get a $\mathcal{P}$-border cover $\boldsymbol{H}_i = \{ H_{i,j} : j = 1, \ldots, k \}$ of X with enclosure H_i and an open set V_i of X that satisfies

$$F_i \subset U_i \subset \mathrm{cl}(U_i) \subset G_i \mod H_i,$$
$$\mathcal{P}\text{-}\dim_{\boldsymbol{H}_{i-1}} \mathrm{B}(U_i) \leq n - 1,$$
$$\boldsymbol{H}_i \text{ shrinks } \boldsymbol{H}_{i-1} \mod H_i,$$
$$\mathrm{B}(U_i) \subset V_i \mod H_i,$$
$$\mathrm{ord}\, \boldsymbol{H}_i | V_i \leq n.$$

Clearly we have

$$\mathrm{ord}\, \boldsymbol{H}_k | \big(\bigcup \{ V_i : i = 1, \ldots, k \} \big) \leq n.$$

With

$$\boldsymbol{L}_i = \{ U_i,\ X \setminus \mathrm{cl}(U_i) \}, \qquad i = 1, \ldots, k,$$

set

$$\boldsymbol{L} = \bigwedge\{\, \boldsymbol{L}_i : i = 1, \ldots, k \,\}.$$

Then $\boldsymbol{L}$ covers $X \setminus \bigcup\{\, \mathrm{B}(U_i) : i = 1, \ldots, k \,\}$ and $\operatorname{ord} \boldsymbol{L} \leq 1$ holds. Moreover, $\boldsymbol{L}$ refines $\boldsymbol{G}$ modulo H_k since $\bigcap\{\, X \setminus \operatorname{cl}(U_i) : i = 1, \ldots, k \,\}$ is contained in H_k. Define $\boldsymbol{W}$ to be the collection

$$\{\, L \setminus H_k : L \in \boldsymbol{L} \,\} \cup \{\, H_{k,j} \cap \left(\bigcup\{\, V_i : i = 1, \ldots, k \,\}\right) : j = 1, \ldots, k \,\}.$$

Then $\boldsymbol{W}$ is a $\mathcal{P}$-border cover of X with enclosure H_k that refines $\boldsymbol{G}$ for which $\operatorname{ord} \boldsymbol{W} \leq n + 1$ holds. Therefore, $\mathcal{P}\text{-}\dim X \leq n$.

We are now able to prove the following useful lemma.

3.10. Lemma. *For a space X let F be a closed set and suppose that $\boldsymbol{G}$ is a $\mathcal{P}$-border cover of X with enclosure G such that*

(a) *the collection $\boldsymbol{G}$ is locally finite in $X \setminus G$,*
(b) *$\mathcal{P}\text{-}\dim_{\boldsymbol{G}} F \leq n$,*
(c) *$\sup\{\, \mathcal{P}\text{-}\dim Z : Z$ is closed and $Z \cap F \in \mathcal{P} \,\} \leq n$.*

Then

$$\mathcal{P}\text{-}\dim_{\boldsymbol{G}} X \leq n.$$

Proof. Denote by $\{\, G_\alpha : \alpha \in A \,\}$ the given $\mathcal{P}$-border cover $\boldsymbol{G}$ of X. Since $\mathcal{P}\text{-}\dim_{\boldsymbol{G}} F \leq n$, by Lemma 3.8 there is a $\mathcal{P}$-border cover $\boldsymbol{H}_1 = \{\, H_{1\alpha} : \alpha \in A \,\}$ with enclosure H_1 and there is an open set V_1 such that $\boldsymbol{H}_1$ shrinks $\boldsymbol{G}$ modulo H_1, $\operatorname{ord} \boldsymbol{H}_1 | V_1 \leq n + 1$ and $F \subset V_1$ modulo H_1. The sets $F \cap V_1$ and $X \setminus V_1$ are separated modulo H_1. There is an open set U of X such that $F \cap V_1 \subset U \subset \operatorname{cl}(U) \subset V_1$ modulo H_1. As $F \setminus U \in \mathcal{P}$, from (c) we have $\mathcal{P}\text{-}\dim (X \setminus U) \leq n$. Again by Lemma 3.8 there is a $\mathcal{P}$-border cover $\boldsymbol{H}_2 = \{\, H_{2\alpha} : \alpha \in A \,\}$ with enclosure H_2 and there is an open set V_2 such that $\boldsymbol{H}_2$ shrinks $\boldsymbol{H}_1$ modulo H_2, $\operatorname{ord} \boldsymbol{H}_2 | V_2 \leq n + 1$ and $X \setminus U \subset V_2$ modulo H_2. Let $\boldsymbol{W}$ be the collection of open sets

$$W_\alpha = \left((H_{1\alpha} \cap V_1) \cup (H_{2\alpha} \cap V_2)\right) \setminus H_2, \qquad \alpha \in A.$$

Clearly $\boldsymbol{W}$ shrinks $\boldsymbol{G}$ and $\boldsymbol{W}$ is a $\mathcal{P}$-border cover of X with enclosure H_2. Let us show that $\operatorname{ord} \boldsymbol{W} \leq n + 1$. To this end let x be in $V_1 \setminus H_2$ and let $i_1, \ldots, i_m$ be distinct indices such that x is in W_{i_j} for $j = 1, \ldots, m$. Since $\boldsymbol{H}_2$ shrinks $\boldsymbol{H}_1$, we have

$$W_{i_j} \subset (H_{1i_j} \cap V_1) \cup (H_{2i_j} \cap V_2) = H_{1i_j}, \qquad j = 1, \ldots, m.$$

So we have $m \leq n+1$, that is, $\operatorname{ord}_x \boldsymbol{W} \leq n+1$. Finally suppose that x is in $(X \setminus V_1) \setminus H_2$ and let $i_1 \ldots, i_m$ be distinct indices such that x is in W_{i_j} for $j = 1, \ldots, m$. From $x \in (W_{i_j} \setminus V_1) \subset W_{i_j} \cap V_2$ we have that x is in $H_{2i_j} \cap V_2$ for $j = 1, \ldots, m$. So we have $m \leq n+1$, that is, $\operatorname{ord}_x \boldsymbol{W} \leq n$. This completes the proof of $\operatorname{ord} \boldsymbol{W} \leq n+1$. Thereby we have shown $\mathcal{P}\text{-}\dim_{\boldsymbol{G}} X \leq n$.

Note that the discussion of $\mathcal{P}\text{-}\dim_{\boldsymbol{G}} X \leq n$ did not use the condition $\mathcal{U} \subset \mathcal{D}$. Also, Lemma 3.8 can be proved in the universe $\mathcal{N}[\mathcal{P}]$, the normal spaces modulo $\mathcal{P}$ defined in Section II.5.

The next theorem, due to Morita [1953], concerning ordinary covering dimension will be used in the following section. As its proof is quite long, we shall not present it here; in addition to the original proof in Morita [1953] there is another one on pages 127–129 of Nagami [1970]. We shall use $\dim_{\boldsymbol{G}}$ for $\mathcal{P}\text{-}\dim_{\boldsymbol{G}}$ when $\mathcal{P} = \{\emptyset\}$.

3.11. Theorem. *Let X be a space, $\boldsymbol{G}$ be a locally finite open covering of X, $\boldsymbol{F} = \{F_\alpha : \alpha \in A\}$ be a closed covering of X and $\boldsymbol{U} = \{U_\alpha : \alpha \in A\}$ be a locally finite open collection satisfying the following conditions:*

(a) *$F_\alpha \subset U_\alpha$ for each α in A.*
(b) *$\dim_{\boldsymbol{G}} F_\alpha \leq n$ for each α in A.*
(c) *$\dim(F_\alpha \cap F_\beta) \leq n-1$ whenever $\alpha \neq \beta$.*

Then

$$\dim_{\boldsymbol{G}} X \leq n.$$

4. Dimension and mappings

The agreement of the last section will be continued.

Agreement. *In addition to the agreement made in the introduction of the chapter, the universe will be contained in the Dowker universe.*

Mappings were already encountered in Chapter II. Theorem II.6.6 characterized $\mathcal{P}$-dim by means of mappings into spheres for classes $\mathcal{P}$ that satisfied rather mild conditions. Also theorems concerning mappings into spheres and $\mathcal{P}$-Ind were proved in Chapter II. These theorems can be used to compare the functions $\mathcal{P}$-dim and $\mathcal{P}$-Ind.

4.1. Theorem. *Suppose that $\mathcal{P}$ is a semi-normal family. For every space X,*

$$\mathcal{P}\text{-}\dim X \leq \mathcal{P}\text{-}\operatorname{Ind} X.$$

4.2. Theorem. *Suppose that the universe is contained in the class $\mathcal{N}_P$ of perfectly normal spaces and that $\mathcal{P}$ is a cosmic family. For every space X,*

$$\mathcal{P}\text{-}\dim X \leq \mathcal{P}\text{-}\operatorname{Ind} X.$$

Proofs. Theorem 4.1 is an immediate consequence of Proposition II.7.1 and Theorem III.3.23.

For Theorem 4.2 we observe that a semi-normal family is cosmic but not conversely. Consequently, more restrictions have been placed on the universe. When $\mathcal{U}$ is contained in the class $\mathcal{N}_P$ of perfectly normal spaces, a cosmic family $\mathcal{P}$ is automatically open-monotone and strongly closed-additive. Thus $\mathcal{P}\text{-}\operatorname{Ind} Y \leq \mathcal{P}\text{-}\operatorname{Ind} X$ follows by Theorem III.3.26 whenever Y is an open subspace of X. Proposition II.7.1 will complete the proof.

The equivalence of the inequalities $\operatorname{Ind} X \leq 0$ and $\dim X \leq 0$ for normal spaces X is given in Corollary II.5.8. We also have the following theorem from Theorem II.5.7.

4.3. Theorem. *Under the conditions of Theorems 4.1 or 4.2, for each space X,*

$$\mathcal{P}\text{-}\operatorname{Ind} X \leq 0 \quad \textit{if and only if} \quad \mathcal{P}\text{-}\dim X \leq 0.$$

Proof. Only $0 \geq \mathcal{P}\text{-}\dim X \geq \mathcal{P}\text{-}\operatorname{Ind} X$ requires a proof. Suppose $\mathcal{P}\text{-}\dim X = 0$ and let F_0 and F_1 be disjoint closed subsets of X. Let U_0 and U_1 be open sets with $F_1 \cap \operatorname{cl}(U_0) = \emptyset$ and $F_0 \cap \operatorname{cl}(U_1) = \emptyset$ such that $X = U_0 \cup U_1$. The $\mathcal{P}$-border cover $\boldsymbol{U} = \{U_0, U_1\}$ has by Theorem II.5.7 a $\mathcal{P}$-border cover shrinking $\boldsymbol{V} = \{V_0, V_1\}$ with enclosure G such that $\operatorname{ord} \boldsymbol{V} \leq 1$. Define S to be $X \setminus (W_0 \cup W_1)$ where $W_0 = V_0 \cup \big(X \setminus \operatorname{cl}(U_1)\big)$ and $W_1 = V_1 \cup (X \setminus \operatorname{cl}(U_0))$. As S is a closed subset of G, we have that S is in $\mathcal{P}$. Also $F_0 \subset W_0$ and $F_1 \subset W_1$. So S will be a partition between F_0 and F_1 in X if $W_0 \cap W_1 = \emptyset$. This last condition obviously holds. Thereby we have shown $\mathcal{P}\text{-}\operatorname{Ind} X \leq 0$.

In passing, let us remark that the equality $\operatorname{Ind} X = \dim X$ for every metrizable space X will be established in Chapter V by means of basic dimension theory, that is, a dimension theory that relies on the existence of certain bases for the open sets of a space X.

Let us turn to other mapping theorems. In dimension theory, there are theorems on dimension raising maps and dimension lowering maps. These theorems usually deal with mappings that satisfy additional conditions such as being closed maps. Some analogues of these theorems also hold for other functions of dimensional type. In another line of development, there are theorems that require no additional conditions on the mapping. These theorems establish relationships between dimensions and other functions of dimensional type. The remainder of the section will be a short exposition on these mapping theorems.

4.4. Definition. Let $f: X \to Y$ be a continuous function. The *order of* f, denoted d_f, is the least cardinal number α such that the cardinality of $f^{-1}[y]$ is less than or equal to α for every y in Y. The *dimension of* f, denoted $\dim f$, is the extended integer $\sup\{\dim f^{-1}[y] : y \in Y\}$.

The following dimension-lowering closed mapping theorem is due to Morita [1956].

4.5. Theorem. *Let f be a closed, continuous mapping of a normal space X onto a nonempty paracompact space Y. Then*

$$\dim X \leq \dim f + \operatorname{Ind} Y.$$

Proof. Obviously we may assume $Y = f[X]$. Recall that the equivalence $\operatorname{Ind} X \leq 0$ if and only if $\dim X \leq 0$ holds for the universe $\mathcal{N}$ of normal spaces (Corollary II.5.8). So when $\operatorname{Ind} Y \leq 0$ the inequality that is to be proved becomes $\dim X \leq \dim f + \dim Y$. We shall give here a straightforward proof of this inequality. Clearly we may assume that $\dim f$ is finite. Let $\mathbf{G}$ be a finite open cover of X. For each point y of Y we have $\dim f^{-1}[y] \leq \dim f$. We infer from Lemma 3.8 the existence of an open subset H_y of X such that $f^{-1}[y] \subset H_y$ and $\dim_{\mathbf{G}} H_y \leq \dim f$. With $V_y = Y \setminus f[X \setminus H_y]$ for each y in Y, we have that $\mathbf{V} = \{V_y : y \in Y\}$ is an open cover of Y. As Y is a paracompact space, it has is a locally finite open cover $\{ U_\alpha : \alpha \in A \}$ that refines $\boldsymbol{V}$. We have $\operatorname{Dim} Y = \dim Y$ by Theorem 2.10. Consequently there is an open cover $\boldsymbol{W} = \{ W_\alpha : \alpha \in A \}$ of Y such that $\operatorname{ord} \boldsymbol{W} \leq 1$ and $W_\alpha \subset U_\alpha$ for each α in A. Now for each α with $W_\alpha \neq \emptyset$ we select a member y_α of Y such that $W_\alpha \subset V_{y_\alpha}$. Then $f^{-1}[W_\alpha] \subset H_{y_\alpha}$. So $\{ f^{-1}[W_\alpha] : \alpha \in A \}$ is an open cover of X by the mutually disjoint sets $f^{-1}[W_\alpha]$ with $\dim_{\boldsymbol{G}} f^{-1}[W_\alpha] \leq \dim f$. We infer from this that the finite open cover $\boldsymbol{G}$ of X can be refined by an open cover whose order does not exceed $\dim f + 1$. Thereby we have shown that $\dim X \leq \dim f + \operatorname{Ind} Y$ when $\operatorname{Ind} Y \leq 0$.

We shall now complete the proof by establishing the inductive step of the induction on $\operatorname{Ind} Y$. Assume that $\dim f$ is finite and let $\operatorname{Ind} Y = m \geq 1$ and $\boldsymbol{G}$ be a finite open cover of X. The proof will begin in the same way as in the preceding paragraph. For each y in Y let H_y and V_y be as above and let $\{ U_\alpha : \alpha \in A \}$ be a locally finite open cover of Y that refines $\boldsymbol{V} = \{ V_y : y \in Y \}$. Assume that the indexing set A is well-ordered. Let $\boldsymbol{F} = \{ F_\alpha : \alpha \in A \}$ be a closed cover of Y such that $F_\alpha \subset U_\alpha$ for each α. Then choose for each α an open set W_α of Y such that

$$F_\alpha \subset W_\alpha \subset \operatorname{cl}_Y(W_\alpha) \subset U_\alpha, \qquad \operatorname{Ind} \mathrm{B}_Y(W_\alpha) \leq m - 1.$$

Set

$$\begin{aligned} K_0 &= W_0, \\ K_\alpha &= W_\alpha \setminus \bigcup \{ \operatorname{cl}_Y(W_\beta) : \beta < \alpha \}, \qquad \alpha > 0. \end{aligned}$$

It follows that $\operatorname{Ind} \big(\operatorname{cl}_Y(K_\alpha) \cap \operatorname{cl}_Y(K_\beta)\big) \leq m - 1$ whenever $\alpha \neq \beta$ and that $Y = \bigcup \{ \operatorname{cl}_Y(K_\alpha) : \alpha \in A \}$. By the induction hypothesis we have

$$\dim \big(f^{-1}[\operatorname{cl}_Y(K_\alpha)] \cap f^{-1}[\operatorname{cl}_Y(K_\beta)]\big) \leq \dim f + m - 1, \qquad \alpha \neq \beta.$$

Since $\dim_{\boldsymbol{G}} f^{-1}[\operatorname{cl}_Y(K_\alpha)] \leq \dim f$, we have by Theorem 3.11 that

$$\dim_{\boldsymbol{G}} X \leq \dim f + m.$$

Hence $\dim X \leq \dim f + m$ and the induction is completed.

In general the classes $\mathcal{P} \cap \mathcal{U}$ will contain nonempty spaces. Consequently, extensions of the last theorem to general classes $\mathcal{P}$ become more complicated. To get an easy extension, we make the following definition.

4.6. Definition. For a class $\mathcal{P}$ the *dimension offset* in the universe is the least nonnegative extended-integer $\Phi(\mathcal{P})$ such that

$$\dim X \leq \dim f + \Phi(\mathcal{P})$$

holds for every closed map $f\colon X \to Y$ with $X \in \mathcal{U}$ and $Y \in \mathcal{P} \cap \mathcal{U}$.

Obviously the only interesting values of the offset are the finite ones. In this case, we have that $\dim X \leq \Phi(\mathcal{P})$ for each X in $\mathcal{P} \cap \mathcal{U}$.

4.7. Theorem. *Suppose that $\mathcal{P}$ is a semi-normal family. Let $f\colon X \to Y$ be a closed continuous map, where Y is a paracompact space. Then*

$$\dim X \leq \dim f + \mathcal{P}\text{-Ind}\, f[X] + 1 + \Phi(\mathcal{P}).$$

Proof. The proof will be by induction on $\mathcal{P}$-Ind $f[X]$. The case where $\mathcal{P}\text{-Ind}\, f[X] = -1$ follows from the definition of the $\mathcal{P}$-offset. Suppose $\mathcal{P}\text{-Ind}\, f[X] = m \geq 0$ and let $n = \dim f + m + 1 + \Phi(\mathcal{P})$. The inductive step of the proof will follow those of Theorem 4.5. Let $\mathbf{G}$ be a finite open cover of X. In exactly the same manner as in Theorem 4.5 we can find a locally finite collection $\{ K_\alpha : \alpha \in A \}$ of open sets of Y such that

$$\begin{aligned} Y &= \bigcup \{ \operatorname{cl}_Y(K_\alpha) : \alpha \in A \}, \\ \mathcal{P}\text{-Ind}\big(\operatorname{cl}_Y(K_\alpha) \cap \operatorname{cl}_Y(K_\beta)\big) &\leq m - 1, \qquad \alpha \neq \beta, \\ \dim_{\mathbf{G}} f^{-1}[\operatorname{cl}_Y(K_\alpha)] \leq \dim f &\leq n, \qquad \alpha \in A. \end{aligned}$$

By the induction hypothesis we have

$$\dim\big(f^{-1}[\operatorname{cl}_Y(K_\alpha)] \cap f^{-1}[\operatorname{cl}_Y(K_\beta)]\big) \leq n - 1 \qquad \alpha \neq \beta.$$

Now we have by Theorem 3.11 that $\dim_{\mathbf{G}} X \leq n$ holds. As $\mathbf{G}$ is an arbitrary finite open cover of X, we have $\dim X \leq n$.

4.8. Examples.

a. Let the universe be the class $\mathcal{M}_0$ of separable metrizable spaces and $\mathcal{P}$ be the class of countable spaces. Then $\Phi(\mathcal{P}) = 0$. So we have $\dim X \leq \dim f + \mathcal{P}\text{-Ind}\, Y + 1$ for every closed continuous map $f\colon X \to Y$.

b. Let the universe be $\mathcal{M}_0$ and $\mathcal{Q}$ be a semi-normal family. Define $\mathcal{P}$ to be the class $\{ X \in \mathcal{Q} : \operatorname{Ind} X \leq m \}$. Then $\mathcal{P}$ is a semi-normal family and $\Phi(\mathcal{P}) = m$. Therefore, for a closed continuous map $f : X \to Y$ we have $\dim X \leq \dim f + \mathcal{P}\text{-}\operatorname{Ind} f[X] + m + 1$.

In particular, suppose that $\mathcal{Q} = \mathsf{A}(1)$, the class of absolute additive Borel class 1, and suppose that Y is a space with the property that each of its compact subsets K has $\dim K \leq m$. Then we have $\dim X \leq \dim f + \mathsf{A}(1)\text{-}\operatorname{Ind} Y + m + 1$ for closed continuous maps $f : X \to Y$.

4.9. Definition. For a class $\mathcal{P}$ the *order offset* in the universe is the least nonnegative extended-integer $\Omega(\mathcal{P})$ such that

$$\dim Y \leq d_f + \Omega(\mathcal{P}) - 2$$

holds whenever f is a closed map of a nonempty space X in $\mathcal{P} \cap \mathcal{U}$ onto a space Y in $\mathcal{U}$. The order offset $\Omega(\{ \emptyset \})$ is defined to be 0.

Using a singleton space for Y, we observe that $\Omega(\mathcal{P}) \geq 1$ if there exists a nonempty space X in $\mathcal{P} \cap \mathcal{U}$. Also, when $\Omega(\mathcal{P})$ is finite we have $\dim Y \leq \Omega(\mathcal{P}) - 1$ for each Y in $\mathcal{P} \cap \mathcal{U}$.

4.10. Theorem. *Suppose that $\mathcal{P}$ is a semi-normal family. Let $f : X \to Y$ be a closed, continuous, onto map of a nonempty space X. Then*

$$\dim Y \leq \mathcal{P}\text{-}\operatorname{Ind} X + d_f + \Omega(\mathcal{P}) - 1.$$

Proof. The proof will be by a double induction on $n = \mathcal{P}\text{-}\operatorname{Ind} X$ and $k = d_f - 1$. If $\mathcal{P} \cap \mathcal{U}$ contains a nonempty space, then the case $n = -1$ and $k \geq 0$ follows from the definition of $\Omega(\mathcal{P})$. Suppose $k = 0$ and $n \geq 0$. We shall prove $\dim Y \leq n + \Omega(\mathcal{P})$ by applying Theorem 3.9. Let $\mathbf{G}$ be an open cover of Y and let F_0 and F_1 be disjoint closed sets of Y. Note that f is a homeomorphism. There is a partition S between $f^{-1}[F_0]$ and $f^{-1}[F_1]$ in X such that $\mathcal{P}\text{-}\operatorname{Ind} S \leq n - 1$. So $f[S]$ is a partition between F_0 and F_1 in Y. As $n + \Omega(\mathcal{P}) \geq 0$, it is sufficient to show $\dim f[S] \leq n + \Omega(\mathcal{P}) - 1$ when $S \neq \emptyset$. This follows by induction on n for $k = 0$. We have $\dim_{\mathbf{G}} f[S] \leq n + \Omega(\mathcal{P}) - 1$. Theorem 3.9 yields $\dim Y \leq n + \Omega(\mathcal{P})$ when $k = 0$.

Finally suppose $n \geq 0$ and $k \geq 1$. Let $\mathbf{G}$ be an open cover of Y and let F_0 and F_1 be disjoint closed sets of Y. There is a partition S between $f^{-1}[F_0]$ and $f^{-1}[F_1]$ in X such that $\mathcal{P}\text{-}\operatorname{Ind} S \leq n - 1$.

Then, as $n+k+\Omega(\mathcal{P}) \geq 1$, we have $\dim f[S] \leq n+k+\Omega(\mathcal{P})-1$ by the induction hypothesis. Let U_0 and U_1 be disjoint open sets of X with $f^{-1}[F_0] \subset U_0$ and $f^{-1}[F_1] \subset U_1$ such that $X \setminus S = U_0 \cup U_1$. Because f is a closed map, we have that $V_0 = X \setminus f[S \cup U_1]$ and $V_1 = X \setminus f[S \cup U_0]$ are disjoint and open in Y with $F_0 \subset V_0$ and $F_1 \subset V_1$. Denote $X \setminus (V_0 \cup V_1)$ by K. Then $K = (f[U_0] \cap f[U_1]) \cup f[S]$. We shall use Lemma 3.10 to show $\dim K \leq n+k+\Omega(\mathcal{P})-1$. Let M be a nonempty closed subset of K (and hence of Y) that is contained in $K \setminus f[S]$. The set $U_0 \cap f^{-1}[y]$ consists of at most k points for each point y in M because

$$M \subset K \setminus f[S] \subset f[U_0] \cap f[U_1]$$

and $U_0 \cap U_1 = \emptyset$. The restriction of f to $\widetilde{X} = \operatorname{cl}_X(U_0) \cap f^{-1}[M]$ and the induction hypothesis give

$$\dim M \leq n+k+\Omega(\mathcal{P})-1$$

because $\mathcal{P}\text{-Ind}\,\widetilde{X} \leq \mathcal{P}\text{-Ind}\,X \leq n$. From Lemma 3.10 we have

$$\dim_{\mathbf{G}} K \leq n+k+\Omega(\mathcal{P})-1.$$

Theorem 3.9 yields $\dim Y \leq n+k+\Omega(\mathcal{P})$ and the induction is completed.

The appearance of the offset functions in the last two theorems permits inductive proofs. The finiteness of these offset functions forces an upper bound on the dimensions of the spaces in $\mathcal{P} \cap \mathcal{U}$. The next theorem replaces the dimension offset function with functions of dimensional type and permits the use of continuous maps that are not necessarily closed. We shall use the class $\mathcal{L}$ of locally compact spaces to define the function of dimensional type that will replace the dimension offset.

4.11. Notation. For each space X let

$$\operatorname{loccom} X = \mathcal{L}\text{-ind}\, X.$$

Also, for a continuous map $f\colon X \to Y$ let

$$\operatorname{loccom} f = \sup\{\operatorname{loccom} f^{-1}[y] : y \in Y\}$$

and

$$\operatorname{loccom}(\partial f) = \sup\{\operatorname{loccom} \mathrm{B}_X(f^{-1}[y]) : y \in Y\}.$$

Agreement. *The universe is the class $\mathcal{M}_0$ for the remainder of the section.*

4.12. Theorem. *Let $f\colon X \to Y$ be a continuous map. Then*

$$\dim X \leq \dim Y + \max\{\dim f, \operatorname{def} X\} + \operatorname{loccom} f + 1.$$

The proof of this theorem will rely on the existence of a special closed extension of the mapping (Theorem 4.15 below). Consequently its proof will be delayed.

Clearly the class $\mathcal{L}$ is closed-monotone and open-monotone. But it fails to be strongly closed-additive. The next lemma is easily proved. Its proof will be left to the reader.

4.13. Lemma. *Let $f\colon X \to Y$ be a closed continuous map. Then*

(a) $\mathrm{B}_X(f^{-1}[y])$ *is compact for each y in Y,*
(b) $\operatorname{loccom}(\partial f) = -1$,
(c) *if $\bigcup\{\mathrm{B}_X(f^{-1}[y]) : y \in Y\} \subset \widetilde{X} \subset X$, then $f|\widetilde{X}$ is a closed map of $\widetilde{X}$ onto $f[\widetilde{X}]$.*

The proof of our closed extension theorem will follow from the next lemma.

4.14. Lemma. *Suppose $X \subset Y$. If C is a closed subset of Y and $Z = C \cap X$, then*

$$\dim C \leq \max\{\dim Z, \dim(Y \setminus X)\} + \operatorname{loccom} Z + 1.$$

Proof. Select a metric for Y. The lemma is inductively proved by claims involving the following two statements:

Statement Δ_n. Let X, Y, C and Z be as in the hypothesis of the lemma. If $\operatorname{loccom} Z \leq n$, then for each z in Z and for $\varepsilon > 0$ there is a subset U of C such that $z \in U$, U is open in C, $\operatorname{diam} U < \varepsilon$ and

$$\dim \mathrm{B}_C(U) \leq \max\{\dim Z, \dim(Y \setminus X)\} + n.$$

Statement Γ_n. Let X, Y, C and Z be as in the hypothesis of the lemma. If $\operatorname{loccom} Z \leq n$, then

$$\dim C \leq \max\{\dim Z, \dim(Y \setminus X)\} + n + 1.$$

Claim. *Γ_{-1} is a true statement.*

Proof. If $\operatorname{loccom} Z = -1$, then Z is locally compact and hence is open in $\operatorname{cl}_C(Z)$. So each of the sets $C \setminus \operatorname{cl}_C(Z)$, $\operatorname{cl}_C(Z) \setminus Z$ and Z are F_σ-sets in C. Consequently, from the sum and subspace theorems for dim we have

$$\dim C \leq \max\{\dim Z, \dim(Y \setminus X)\}.$$

Claim. *The validity of Γ_{n-1} implies the validity of Δ_n for $n \geq 0$.*

Proof. Suppose $\operatorname{loccom} Z \leq n$, $z \in Z$ and $\varepsilon > 0$. There exists by Proposition II.2.20 an open neighborhood U of z in the space C such that $\operatorname{diam} U < \varepsilon$ and $\operatorname{loccom}(\mathrm{B}_C(U) \cap Z) \leq n-1$. We have from statement Γ_{n-1} that for $C' = \mathrm{B}_C(U)$ and $Z' = \mathrm{B}_C(U) \cap Z$ the inequality $\dim C' \leq \max\{\dim Z', \dim(Y \setminus X)\} + n$ holds. The inequality $\dim \mathrm{B}_C(U) \leq \max\{\dim Z, \dim(Y \setminus X)\} + n$ also holds because $Z' \subset Z$. Thereby statement Δ_n is valid.

Claim. *The validity of Δ_n implies the validity of Γ_n for $n \geq 0$.*

Proof. Suppose that Δ_n is valid. Let $\operatorname{loccom} Z \leq n$. Then for each positive natural number m there is countable base $\mathcal{B}_m$ for the open sets of C with $\operatorname{diam} \mathrm{B}_C(U) \leq \frac{1}{m}$ such that $\dim \mathrm{B}_C(U) \leq \max\{\dim Z, \dim(Y \setminus X)\} + n$ for each U in $\mathcal{B}_m$. Define the sets G and H by

$$G = \bigcap\{\bigcup \mathcal{B}_m : m = 1, 2, \ldots\},$$
$$H = \bigcup\{\bigcup\{\mathrm{B}_C(U) : U \in \mathcal{B}_m\} : m = 1, 2, \ldots\}.$$

Then G is a G_δ-set of C containing Z and H is an F_σ-set of C. We have $\dim(C \setminus G) \leq \dim(C \setminus Z) \leq \dim(Y \setminus X)$ by the subspace theorem for dim. And $\dim H \leq \max\{\dim Z, \dim(Y \setminus X)\} + n$ by the sum theorem. As $C \setminus G$ is an F_σ-set of C, we have by the addition theorem that

$$\begin{aligned}\dim C &\leq \dim\big((C \setminus G) \cup H\big) + \dim(G \setminus H) + 1 \\ &\leq \max\{\dim Z, \dim(Y \setminus X)\} + n + 1 + \dim(G \setminus H).\end{aligned}$$

That $\dim(G \setminus H) \leq 0$ remains to be shown. This will follow from Proposition I.3.6 and Theorem I.8.10 with the aid of the fact that the restriction of $\mathcal{B}_m$ to the set $G \setminus H$ for each m will yield a base for $G \setminus H$.

The proof the lemma is completed by the observation that Γ_n is valid for $n \geq -1$.

We are now ready to prove the closed extension theorem. Of course, the existence of a closed extension of a continuous map is not surprising. It is the inequality in the next theorem that is of interest.

4.15. Theorem. *Let $f: X \to Y$ be a continuous, onto map. Then there is a closed continuous map $\widetilde{f}: \widetilde{X} \to Y$, where X is dense in $\widetilde{X}$, such that $f = \widetilde{f}|X$ and*

$$\dim f \leq \dim \widetilde{f} \leq \max\{\dim f, \operatorname{def} X\} + \operatorname{loccom}(\partial f) + 1.$$

Proof. Let K be a metrizable compactification of X such that $\operatorname{def} X = \dim(K \setminus X)$. The natural projection $p: K \times Y \to Y$ is a closed map. Let G be the graph of f and H be the closure of G in $K \times Y$. Define g to be $p|H$. Then g is a closed mapping of H onto Y. With $H_0 = \bigcup\{\mathrm{B}_H(g^{-1}[y]) : y \in Y\}$ define $\widetilde{X}$ to be the set $H_0 \cup G$ and $\widetilde{f}$ to be $p|\widetilde{X}$. By Lemma 4.13 one readily sees that $\widetilde{f}$ is a closed map of $\widetilde{X}$ onto Y that extends f and that X is dense in $\widetilde{X}$. We must prove that $\dim \widetilde{f}$ satisfies the required inequality. Let $y \in Y$. Then

$$\widetilde{f}^{-1}[y] = \mathrm{B}_H(g^{-1}[y]) \cup (f^{-1}[y] \times \{y\}).$$

It is clear from the sum theorem that

$$\dim \widetilde{f}^{-1}[y] \leq \max\{\dim \mathrm{B}_H(g^{-1}[y]), \dim f\}.$$

We shall compute an upper bound for $\dim \mathrm{B}_H(g^{-1}[y])$. Observe that

$$K \times \{y\} \supset g^{-1}[y],$$
$$X \times \{y\} \supset f^{-1}[y] \times \{y\},$$
$$g^{-1}[y] \cap (X \times \{y\}) = g^{-1}[y] \cap G = f^{-1}[y] \times \{y\}.$$

Also we have from the denseness of G in H that a point (x, y) in G is an interior point of $g^{-1}[y] \cap G$ in the space G if and only if (x, y) is an interior point of $g^{-1}[y]$ in the space H. So

$$\begin{aligned} \mathrm{B}_H(g^{-1}[y]) \cap (X \times \{y\}) &= \mathrm{B}_G(g^{-1}[y] \cap G) \\ &= \mathrm{B}_X(f^{-1}[y]) \times \{y\}. \end{aligned}$$

We can now apply Lemma 4.12 to the sets $X \times \{y\}$, $K \times \{y\}$, $\mathrm{B}_H(g^{-1}[y])$ and $\mathrm{B}_X(f^{-1}[y]) \times \{y\}$ to conclude that

$$\begin{aligned}\dim \mathrm{B}_H(g^{-1}[y]) &\\ &\le \max\{\dim \mathrm{B}_X(f^{-1}[y]), \dim(K \setminus X)\}\\ &\qquad + \operatorname{loccom} \mathrm{B}_X(f^{-1}[y]) + 1\\ &\qquad\le \max\{\dim f, \operatorname{def} X\} + \operatorname{loccom}(\partial f) + 1.\end{aligned}$$

Let us now prove Theorem 4.12.

Proof of Theorem 4.12. Let $\widetilde{f}\colon \widetilde{X} \to Y$ be the closed extension of f provided by Theorem 4.15. Theorem 4.5 applied to $\widetilde{f}$ will complete the proof because $\dim X \le \dim \widetilde{X}$ and $\operatorname{loccom}(\partial f) \le \operatorname{loccom} f$. The last inequality follows from the fact that $\operatorname{loccom} \mathrm{B}_X(f^{-1}[y]) \le \operatorname{loccom} f^{-1}[y]$ for every y in Y.

Let us now eliminate the invariant loccom from Theorems 4.12 and 4.15. We will need the following proposition.

4.16. Proposition. *Let Z be a space with $\operatorname{def} Z \ge 0$. Then $\operatorname{loccom} Z \le \operatorname{cmp} Z$.*

Proof. The proof is by induction on $\operatorname{cmp} Z$. Since $\operatorname{cmp} Z = 0$ if and only if $\operatorname{def} Z = 0$ by de Groot's theorem, the induction must start with $\operatorname{cmp} Z = 0$. Suppose that $\operatorname{cmp} Z = 0$. Then each point z of Z has arbitrarily small neighborhoods with compact boundaries. So $\operatorname{loccom} Z \le 0$. The inductive step is proved in the same manner.

4.17. Lemma. *Let $f\colon X \to Y$ be a continuous map. Then*

$$\operatorname{loccom}(\partial f) \le \min\{\dim f, \operatorname{def} X\}.$$

Proof. Let y be a point in Y. That $\operatorname{loccom} \mathrm{B}(f^{-1}[y]) \le \operatorname{def} X$ is obvious when $\operatorname{def} \mathrm{B}(f^{-1}[y]) = -1$. If $\operatorname{def} \mathrm{B}(f^{-1}[y]) \ge 0$, then the above proposition yields

$$\operatorname{loccom} \mathrm{B}(f^{-1}[y]) \le \operatorname{cmp} \mathrm{B}(f^{-1}[y]) \le \operatorname{def} \mathrm{B}(f^{-1}[y]) \le \operatorname{def} X,$$

where the last inequality holds because $\mathrm{B}(f^{-1}[y])$ is closed in X. So $\operatorname{loccom}(\partial f) \le \operatorname{def} X$ follows. Obviously, $\operatorname{loccom}(\partial f) \le \dim f$. The lemma is proved.

The next theorem is now easily established.

4.18. Theorem. *Let $f\colon X \to Y$ be a continuous, onto map. Then there is a closed continuous map $\widetilde{f}\colon \widetilde{X} \to Y$, where X is dense in $\widetilde{X}$, such that $f = \widetilde{f}|X$ and*

$$\dim f \leq \dim \widetilde{f} \leq \dim f + \operatorname{def} X + 1.$$

Consequently,

$$\dim X \leq \dim Y + \dim f + \operatorname{def} X + 1.$$

The following application of Theorem 4.18 will use the fact that for every space X there is a continuous map f of X into the Cantor set such that $f^{-1}[y]$ is a quasi-component of X for each y in $f[X]$. (See Section VI.3 where quasi-components are discussed in more detail.)

4.19. Theorem. *Let X be a space such that* $\dim Q \leq n$ *for each quasi-component Q of X. Then*

$$\dim X \leq \operatorname{def} X + n + 1.$$

Proof. Let $f\colon X \to f[X]$ be a continuous map into the Cantor set such that $f^{-1}[y]$ is a quasi-component of X for each y in $f[X]$. Then $\dim f \leq n$. Theorem 4.18 will complete the proof.

An immediate consequence of Theorem 4.12 follows.

4.20. Theorem. *Let X be a space such that*

(a) *each quasi-component of X is locally compact,*
(b) $\dim Q \leq n$ *for each quasi-component Q of X,*
(c) $\operatorname{def} X \leq n$.

Then $\dim X \leq n$.

Proof. As in the proof of the previous theorem, we have from Theorem 4.12 that $\dim X \leq n + \operatorname{loccom} f + 1$ because of conditions (b) and (c). Condition (a) gives $\operatorname{loccom} f = -1$.

5. Historical comments and unsolved problems

The development of $\mathcal{P}$-dim and $\mathcal{P}$-Dim found in the chapter is essentially new. The reader is referred to the earlier work of Baladze [1982] concerning $\mathcal{P}$-dim for general topological spaces and Aarts and Nishiura [1973a]. One can see that the theory of the covering dimension dim in the universe $\mathcal{N}$ of normal spaces has strongly influenced the development of the chapter. Of particular importance are the works of Dowker and of Morita. Theorems 4.7 and 4.10 concerning closed, continuous mappings are new in the sense that the offset functions are new. (Although one can find closed, continuous mapping theorems in Baladze [1982] which do not use these offset functions, the proofs found there seem incomplete.) There are many more mapping theorems in dimension theory, in particular, mappings into polyhedra. These have not been investigated in the context of $\mathcal{P}$-dim. The Hurewicz closed mapping theorems for continuous maps on compact metrizable spaces have been generalized to Theorems 4.12 and 4.15. They originally appeared in Nishiura [1972]. Theorems 4.19 and 4.20 are the main theorems of Nishiura [1964]. Results related to the last two references can be found in Lelek [1964] and Reichaw [1972].

Unsolved problems

1. The Morita-Hurewicz closed mapping theorems have been generalized as Theorems 4.7 and 4.10. The statements of these theorems have a mixture of dim and $\mathcal{P}$-dim.

Find suitable offset functions so that the function dim in these theorems can be replaced by $\mathcal{P}$-dim.

2. If possible, develop a theory of mappings into polyhedra for $\mathcal{P}$-dim analogous to that of dim.

3. Inverse sequences have been successfully exploited in the study of $\dim X$.

Is $\mathcal{P}\text{-}\dim X$ amenable to similar exploitations?

CHAPTER V

FUNCTIONS OF BASIC DIMENSIONAL TYPE

One of the subjects of this chapter is the relation between the dimension-like functions and the existence of extensions that satisfy some dimensional properties. Theorem I.7.11 is a typical example of the results that will be derived. Because of their different nature, the compactness dimension functions and compactifications will be discussed in the next chapter instead of here. Another subject is the theory of excision. The notion of excision is complementary to that of extension. Theorem I.10.5 may serve as an example of the results that are discussed in this context. The idea of complementarity has already been seen in Proposition II.10.2. What unifies excision and extension is the basic inductive dimension function modulo a class $\mathcal{P}$. This and other basic dimension functions, the definition of which involves the existence of special bases, will be exposed in this chapter. With the help of the basic dimension functions we are able to line up the dimension functions modulo $\mathcal{P}$ under mild conditions on $\mathcal{P}$. Metrizability will play a dominant role in the development of this chapter; it is in fact an implicit part of the definition of the basic dimension functions.

Agreement. *Every space is metrizable, that is,* $\mathcal{U} = \mathcal{M}$.

1. The basic inductive dimension

Let us begin with the basic inductive dimension as a motivation for the basic inductive dimension modulo a class $\mathcal{P}$. These basic dimensions are the key to the excision and extension theorems of the next section. At the end of this section we shall prove some fundamental theorems for Ind.

We have seen earlier in Proposition I.3.6 a characterization of the small inductive dimension ind that uses a base for the open sets.

A well-known characterization of the large inductive dimension Ind reads as follows:

Let n be a natural number. A space X has $\operatorname{Ind} X \leq n$ *if and only if there exists a σ-locally finite base $\mathcal{B}$ for the open sets such that* $\operatorname{Ind} \mathrm{B}(U) \leq n-1$ *for every U in $\mathcal{B}$.*

This characterization will be a natural by-product of our development (Theorem 1.13 below). The definition of the basic inductive dimension is a mixture of this characterization of Ind and the characterization of ind by means of bases.

1.1. Definition. Let $\mathcal{P}$ be a class of spaces and let X be a (metrizable) space. One assigns the *basic inductive dimension modulo $\mathcal{P}$*, denoted $\mathcal{P}$-Bind X, as follows.

(i) $\mathcal{P}\text{-Bind}\, X = -1$ if and only if $X \in \mathcal{P}$.

(ij) For each natural number n, $\mathcal{P}\text{-Bind}\, X \leq n$ if there exists a σ-locally finite base $\mathcal{B}$ for the open sets of X such that $\mathcal{P}\text{-Bind}\, \mathrm{B}(U) \leq n-1$ for every U in $\mathcal{B}$.

Since the class $\mathcal{P}$ is topologically invariant, the function $\mathcal{P}$-Bind is also topologically invariant. The function $\{\emptyset\}$-Bind will be denoted by Bind. In view of the Nagata-Smirnov-Bing Metrization Theorem, the metrizability of the space X is part of the definition of $\mathcal{P}$-Bind X.

The first proposition easily follows from the characterization of ind in Theorem II.2.10. The proof is by induction on $\mathcal{P}$-Bind and is left to the reader.

1.2. Proposition. *For every space X,*

$$\mathcal{P}\text{-ind}\, X \leq \mathcal{P}\text{-Bind}\, X.$$

For separable spaces we have the equality of $\mathcal{P}$-ind and $\mathcal{P}$-Bind.

1.3. Proposition. *For every separable space X,*

$$\mathcal{P}\text{-ind}\, X = \mathcal{P}\text{-Bind}\, X.$$

Proof. In view of the preceding proposition it will suffice to prove that $\mathcal{P}\text{-Bind}\, X \leq \mathcal{P}\text{-ind}\, X$ for every separable space X. For the inductive step of the proof, suppose that the theorem has been proved for all spaces X with $\mathcal{P}\text{-ind}\, X \leq n-1$. Assume $\mathcal{P}\text{-ind}\, X \leq n$. By

Theorem II.2.10 there is a base $\mathcal{B}$ for the open sets of X such that $\mathcal{P}$-indB $(U) \leq n-1$ for each U in $\mathcal{B}$. As X is second countable, we may assume that $\mathcal{B}$ is countable. By the induction hypothesis, $\mathcal{P}$-Bind B $(U) \leq n-1$ for all U in $\mathcal{B}$. It follows that $\mathcal{P}$-Bind $X \leq n$.

The easy inductive proof of the following proposition will be left to the reader.

1.4. Proposition. *If $\mathcal{P}$ and $\mathcal{Q}$ are classes with $\mathcal{P} \supset \mathcal{Q}$, then* $\mathcal{P}$-Bind $\leq \mathcal{Q}$-Bind. *In particular,* $\mathcal{P}$-Bind $\leq$ Bind.

The agreements made on the classes $\mathcal{P}$ allow easy inductive proofs for the following propositions.

1.5. Proposition. *For every closed subspace Y of a space X,*

$$\mathcal{P}\text{-Bind}\, Y \leq \mathcal{P}\text{-Bind}\, X.$$

1.6. Proposition. *The class $\mathcal{P}$ is monotone if and only if the dimension function* $\mathcal{P}$-Bind *is monotone, i.e., for every subspace Y of a space X,*

$$\mathcal{P}\text{-Bind}\, Y \leq \mathcal{P}\text{-Bind}\, X.$$

The corresponding statements hold for F_σ-, G_δ- and open-monotone classes.

The next open subspace theorem is similar to Theorem II.2.17.

1.7. Proposition. *Let X be such that* $0 \leq \mathcal{P}$-Bind X. *For every open subspace Y of X,* $\mathcal{P}$-Bind $Y \leq \mathcal{P}$-Bind X.

Now we shall compare $\mathcal{P}$-Bind and $\mathcal{P}$-Ind. To this end, we introduce the notion of a framework of a base.

1.8. Definition. A pair $\big(\{ U_\gamma : \gamma \in \Gamma \}, \{ F_\gamma : \gamma \in \Gamma \}\big)$ is called a *framework of a base of X* if

(i) $\{ U_\gamma : \gamma \in \Gamma \}$ $\big(\{ F_\gamma : \gamma \in \Gamma \}\big)$ is an open (closed) collection,
(ij) $\emptyset \neq F_\gamma \subset U_\gamma$ for every γ in Γ,
(iij) if $\{ V_\gamma : \gamma \in \Gamma \}$ is any collection of open sets of X such that $F_\gamma \subset V_\gamma \subset U_\gamma$ for each γ in Γ, then it is a σ-locally finite base for the open sets of X.

We have the following existence proposition.

1.9. Proposition. *For each space X there exists a framework of a base.*

Proof. For each i in $\mathbb{N}$ let $\boldsymbol{W}_i$ be the collection of all open balls with diameter less than $\frac{1}{i+1}$. Let $\{ U^i_\alpha : \alpha \in A_i \}$ be a locally finite cover of X that refines $\boldsymbol{W}_i$. Because this covering is locally finite, it has a closed shrinking $\{ F^i_\alpha : \alpha \in A_i \}$. We may assume that each F^i_α is nonempty. Then $\big(\{ U^i_\alpha : \alpha \in A_i,\, i \in \mathbb{N} \}, \{ F^i_\alpha : \alpha \in A_i,\, i \in \mathbb{N} \}\big)$ is a framework of a base of X.

The relation between $\mathcal{P}$-Bind and $\mathcal{P}$-Ind is as follows.

1.10. Theorem. *For every space X,*

$$\mathcal{P}\text{-Bind}\, X \leq \mathcal{P}\text{-Ind}\, X.$$

Proof. The proof is by induction on $\mathcal{P}$-Ind. Suppose that the theorem holds for spaces X with $\mathcal{P}\text{-Ind}\, X \leq n-1$. Let $\mathcal{P}\text{-Ind}\, X \leq n$ and let $\big(\{ U_\gamma : \gamma \in \Gamma \}, \{ F_\gamma : \gamma \in \Gamma \}\big)$ be a framework of a base. For each γ in Γ select an open set V_γ such that $F_\gamma \subset V_\gamma \subset U_\gamma$ and $\operatorname{Ind} \mathrm{B}\,(V_\gamma) \leq n-1$. We have $\mathcal{P}\text{-Bind}\, \mathrm{B}\,(V_\gamma) \leq n-1$ by the induction hypothesis for each γ in Γ and hence $\mathcal{P}\text{-Bind}\, X \leq n$ follows.

We shall discuss the special case $\mathcal{P} = \{ \emptyset \}$ to conclude this section. In the discussion we shall use only the sum theorem for Ind (see Theorem III.3.25 (b) and (c)). This is permissible because the semi-normal family arguments of Section III.3 do not require the subspace theorem nor the addition theorem. So in this way we can find a second proof for the subspace and addition theorems in the universe $\mathcal{M}$. We shall state here for convenience the sum theorem for Ind.

1.11. Theorem (Sum theorem). *If $X = \bigcup \boldsymbol{F}$ for a σ-locally finite closed collection $\boldsymbol{F}$ in X, then*

$$\operatorname{Ind} X \leq \sup \{ \operatorname{Ind} F : F \in \boldsymbol{F} \}.$$

Using the sum theorem, we now can prove the following important theorem.

1.12. Theorem. *For every space X,*

$$\operatorname{Bind} X = \operatorname{Ind} X.$$

Proof. Only $\operatorname{Ind} X \leq \operatorname{Bind} X$ requires proof. Assume $\operatorname{Bind} X \leq 0$. By definition, there is a σ-locally finite base $\mathcal{B}$ such that $\mathrm{B}(U) = \emptyset$ for every U in $\mathcal{B}$. We may write $\mathcal{B} = \{ U^i_\alpha : \alpha \in A_i, i \in \mathbb{N} \}$ where the collection $\{ U^i_\alpha : \alpha \in A_i \}$ is locally finite for each i in $\mathbb{N}$. Suppose that F and G are disjoint closed sets of X. For each j in $\mathbb{N}$ define

$$U_j = X \setminus \bigcup \{ U^i_\alpha : \alpha \in A_i,\ i \leq j,\ U^i_\alpha \cap F = \emptyset \}.$$

Observe that $\{ U^i_\alpha : \alpha \in A_i,\ i \leq j,\ U^i_\alpha \cap F = \emptyset \}$ is locally finite. It follows that U_j is an open-and-closed set. In this way we get a sequence $\{ U_i : i \in \mathbb{N} \}$ of open-and-closed sets such that

$$U_0 \supset U_1 \supset U_2 \supset \cdots \supset F \quad \text{and} \quad \bigcap \{ U_i : i \in \mathbb{N} \} = F.$$

In a similar way, we can construct a sequence $\{ V_i : i \in \mathbb{N} \}$ of open-and-closed sets such that

$$V_0 \supset V_1 \supset V_2 \supset \cdots \supset G \quad \text{and} \quad \bigcap \{ V_i : i \in \mathbb{N} \} = G.$$

A straightforward argument will show that $W = \bigcup \{ U_i \setminus V_i : i \in \mathbb{N} \}$ is an open-and-closed set such that $F \subset W \subset X \setminus G$. Hence $\emptyset$ is a partition between F and G.

Now suppose for $n > 0$ that the inequality has been proved for all spaces X with $\operatorname{Bind} X \leq n - 1$. Assume $\operatorname{Bind} X \leq n$. Choose a σ-locally finite base $\mathcal{B}$ for the open sets with $\operatorname{Bind} \mathrm{B}(U) \leq n - 1$ for each U in $\mathcal{B}$. By the induction hypothesis, $\operatorname{Ind} \mathrm{B}(U) \leq n - 1$ for each U in $\mathcal{B}$. We consider the set $D = \bigcup \{ \mathrm{B}(U) : U \in \mathcal{B} \}$. As the collection $\{ \mathrm{B}(U) : U \in \mathcal{B} \}$ is σ-locally finite, by the sum theorem we have $\operatorname{Ind} D \leq n - 1$. Obviously, $\operatorname{Bind}(X \setminus D) \leq 0$. By the induction hypothesis, $\operatorname{Ind}(X \setminus D) \leq 0$. Now suppose that F and G are disjoint closed subsets of X. By Proposition I.4.6 there is a partition S in X between F and G such that $\operatorname{Ind}\big(S \cap (X \setminus D)\big) = -1$, that is, $S \subset D$. It follows that $\operatorname{Ind} S \leq n - 1$, whence $\operatorname{Ind} X \leq n$.

There are several consequences of this theorem. An obvious one is the sum theorem for Bind. The second one is the characterization theorem.

1.13. Theorem. *For every space X and for every natural number n, $\operatorname{Ind} X \leq n$ if and only if there exists a σ-locally finite base $\mathcal{B}$ for the open sets such that $\operatorname{Ind} \mathrm{B}(U) \leq n-1$ for every U in $\mathcal{B}$.*

The subspace theorem for Ind follows from Proposition 1.6. A more general result has been obtained in Theorem III.3.25 (a).

1.14. Theorem (Subspace theorem). *For every subspace Y of a space X,*

$$\operatorname{Ind} Y \leq \operatorname{Ind} X.$$

The decomposition theorem can be established quite easily. The ideas of the proof will be generalized in the next section.

1.15. Theorem (Decomposition theorem). *For each natural number n, $\operatorname{Ind} X \leq n$ if and only if X can be partitioned into $n+1$ disjoint subsets X_i, $i = 0, 1, \ldots, n$, with $\operatorname{Ind} X_i \leq 0$ for every i.*

Proof. To prove necessity, using Theorem 1.12 and Definition 1.1, we select a σ-locally finite base $\mathcal{B}$ for the open sets of X such that $\operatorname{Bind} \mathrm{B}(U) \leq n-1$ for each U in $\mathcal{B}$. Write $X_1 = \bigcup\{ \mathrm{B}(U) : U \in \mathcal{B} \}$. The sum theorem for Bind gives $\operatorname{Bind} X_1 \leq n-1$. Clearly we also have $\operatorname{Bind}(X \setminus X_1) \leq 0$. And thus $\operatorname{Ind}(X \setminus X_1) \leq 0$ holds by Theorem 1.12. Repeating this process, we find that X can be partitioned into $n+1$ zero-dimensional subsets.

The proof of the sufficiency is by induction and makes use of Propositions I.4.3 and I.4.6. The reader is asked to provide the details.

Finally the addition theorem and the product theorem for general metrizable spaces can now be easily established. A proof of the product theorem can be based on Theorem 1.13. We leave the details to the reader.

1.16. Theorem (Addition theorem). *If $X = Y \cup Z$, then*

$$\operatorname{Ind} X \leq \operatorname{Ind} Y + \operatorname{Ind} Z + 1.$$

1.17. Theorem (Product theorem). *Let $X \times Y$ be the topological product of two spaces X and Y, at least one of which is not empty. Then*

$$\operatorname{Ind}(X \times Y) \leq \operatorname{Ind} X + \operatorname{Ind} Y.$$

Due to Theorem 1.12 we shall henceforth freely interchange Bind and Ind in the sum, subspace, decomposition, addition and product theorems.

2. Excision and extension

The relations between $\mathcal{P}$-Sur and $\mathcal{P}$-Bind and between $\mathcal{P}$-Def and $\mathcal{P}$-Bind will be discussed. The similarities as well as the differences of the notions of surplus and deficiency will be clearly exposed.

Let us discuss the surplus function first. The fundamental result is the excision theorem. To prove this theorem we need the next two lemmas.

2.1. Lemma. *Let $\{ B_\gamma : \gamma \in \Gamma \}$ be a σ-locally finite closed collection of subsets of a space X. For each γ in Γ let $\{ B_\gamma^\delta : \delta \in \Delta_\gamma \}$ be a σ-locally finite (closed) collection of subsets of the subspace B_γ. Then $\{ B_\gamma^\delta : \gamma \in \Gamma,\ \delta \in \Delta_\gamma \}$ is a σ-locally finite (closed) collection of subsets of X.*

Proof. It is sufficient to prove the lemma for locally finite covers $\{ B_\gamma : \gamma \in \Gamma \}$. We may assume that the indexing of the collection by Γ is one-to-one. As the collection $\{ B_\gamma^\delta : \delta \in \Delta_\gamma \}$ is a σ-locally finite collection of subsets of the subspace B_γ for each γ, we may write $\Delta_\gamma = \bigcup \{ \Delta_\gamma^i : i \in \mathbb{N} \}$ where $\{ B_\gamma^\delta : \delta \in \Delta_\gamma^i \}$ is locally finite for each i. We only need to show for any i that $\{ B_\gamma^\delta : \gamma \in \Gamma,\ \delta \in \Delta_\gamma^i \}$ is locally finite. Let x be in X. First select a neighborhood U of x such that $U \cap B_\gamma = \emptyset$ for each γ in $\Gamma \setminus \Gamma_x$, where Γ_x is some finite subset of Γ. (Here we have used the fact that the indexing is one-to-one.) Because B_γ is a closed set, for every γ in Γ_x we can select a neighborhood U_γ of x such that the set $U_\gamma \cap B_\gamma$ meets only finitely many B_γ^δ with δ in Δ_γ^i. It follows then that $U \cap \left(\bigcap \{ U_\gamma : \gamma \in \Gamma_x \} \right)$ is a neighborhood of x meeting at most finitely many members of $\{ B_\gamma^\delta : \gamma \in \Gamma,\ \delta \in \Delta_\gamma^i \}$.

2.2. Lemma. *In each space X with $\mathcal{P}$-Bind $X \le n$ there exists a σ-locally finite closed collection $\boldsymbol{F}$ of subsets of X such that*

(a) *every member of $\boldsymbol{F}$ belongs to $\mathcal{P}$,*
(b) Bind $(X \setminus \bigcup \boldsymbol{F}) \le n$.

Proof. The proof is by induction on $\mathcal{P}$-Bind. For the first step, suppose $\mathcal{P}$-Bind $X \le 0$. By definition there exists a σ-locally finite base $\mathcal{B}$ for the open sets of X such that $\mathrm{B}(V) \in \mathcal{P}$ for every V in $\mathcal{B}$. Clearly $\boldsymbol{F} = \{ \mathrm{B}(V) : V \in \mathcal{B} \}$ satisfies all of the required conditions. Now suppose that the lemma holds for all spaces X

with $\mathcal{P}$-Bind $X \leq n-1$. Assume $\mathcal{P}$-Bind $X \leq n$. Then there exists a σ-locally finite open base $\mathcal{B}$ of X with $\mathcal{P}$-Bind $\mathrm{B}(V) \leq n-1$ for every V in $\mathcal{B}$. By the induction hypothesis, for every V in $\mathcal{B}$ there exists a σ-locally finite closed collection $\boldsymbol{F}_V$ of subsets of the subspace $\mathrm{B}(V)$ such that

$$\boldsymbol{F}_V \subset \mathcal{P}, \qquad \operatorname{Bind}\big(\mathrm{B}(V) \setminus \bigcup \boldsymbol{F}_V\big) \leq n-1.$$

The collection $\boldsymbol{F} = \bigcup\{\boldsymbol{F}_V : V \in \mathcal{B}\}$ is σ-locally finite and closed by the previous lemma. Observe that $\mathrm{B}(V) \setminus \bigcup \boldsymbol{F}$ is a closed set in $X \setminus \bigcup \boldsymbol{F}$ for every V in $\mathcal{B}$. From the sum theorem for Bind we have $\operatorname{Bind}\big((\bigcup\{\mathrm{B}(V) : V \in \mathcal{B}\}) \setminus \bigcup \boldsymbol{F}\big) \leq n-1$. It is clear that $\operatorname{Bind}(X \setminus \bigcup \boldsymbol{F}) \leq 0$. So $\operatorname{Bind}\big((X \setminus \bigcup\{\mathrm{B}(V) : V \in \mathcal{B}\}) \setminus \bigcup \boldsymbol{F}\big) \leq 0$ holds by the subspace theorem for Bind. Finally $\operatorname{Bind}(X \setminus \bigcup \boldsymbol{F}) \leq n$ follows from the addition theorem for Bind.

The excision theorem is fundamentally related to $\mathcal{P}$-Sur. Its statement will use classes that are semi-normal families.

2.3. Theorem (Excision theorem). *Suppose that $\mathcal{P}$ is a semi-normal family. Then for every space X with $\mathcal{P}$-Bind $X \leq n$ there exists an F_σ $\mathcal{P}$-kernel Y with $\operatorname{Ind}(X \setminus Y) \leq n$. In particular,*

$$\mathcal{P}\text{-Sur}\, X \leq \mathcal{P}\text{-Bind}\, X.$$

Proof. By Lemma 2.2 there exists a σ-locally finite closed collection $\boldsymbol{F}$ of $\mathcal{P}$-kernels such that $\operatorname{Bind}(X \setminus \bigcup \boldsymbol{F}) \leq n$. As $\mathcal{P}$ is a σ-locally finitely closed-additive class of spaces, we have that $\bigcup \boldsymbol{F}$ is in $\mathcal{P}$. Theorem 1.12 will complete the proof.

The additive absolute Borel classes may serve as examples of semi-normal families (Corollary II.9.4). The excision theorem has many interesting consequences.

2.4. Theorem (Coincidence theorem). *Suppose that $\mathcal{P}$ is a semi-normal family. For every space X,*

$$\mathcal{P}\text{-Bind}\, X = \mathcal{P}\text{-Ind}\, X = \mathcal{P}\text{-Sur}\, X.$$

Proof. This follows from Theorems 1.10, II.3.6 and 2.3.

Having established the coincidence theorem, we easily obtain the sum and decomposition theorems. We state these theorems for $\mathcal{P}$-Ind only. Of course they are also valid for $\mathcal{P}$-Bind as well as for $\mathcal{P}$-Sur.

2.5. Theorem (Sum theorem). *Suppose that $\mathcal{P}$ is a semi-normal family. If $X = \bigcup \boldsymbol{F}$ where $\boldsymbol{F}$ is a σ-locally finite closed collection in X, then*

$$\mathcal{P}\text{-Ind}\, X = \sup \{ \mathcal{P}\text{-Ind}\, F : F \in \boldsymbol{F} \}.$$

Proof. By the excision theorem there exists for each F in $\boldsymbol{F}$ an F_σ $\mathcal{P}$-kernel Y_F such that $\text{Bind}\, Z_F \leq n$ where $Z_F = F \setminus Y_F$. The set $Y = \bigcup \{ Y_F : F \in \boldsymbol{F} \}$ is a member of $\mathcal{P}$ by Lemma 2.1. As $F \setminus Y$ coincides with $Z_F \setminus Y$ and is closed in $X \setminus Y$ for each F in $\boldsymbol{F}$, we have by the sum and subspace theorems for Bind that $\text{Bind}\,(X \setminus Y) \leq n$. It will follow from Theorem 1.12 that $\mathcal{P}\text{-Sur}\, X \leq n$. The theorem follows by the coincidence theorem.

2.6. Theorem (Decomposition theorem). *Suppose that $\mathcal{P}$ is a semi-normal family and that n is in $\mathbb{N}$. Then $\mathcal{P}\text{-Ind}\, X \leq n$ if and only if X can be partitioned into $n+1$ subsets X_i, $i = 0, 1, \ldots, n$, such that $\mathcal{P}\text{-Ind}\, X_0 \leq 0$ and $\text{Ind}\, X_i \leq 0$ for $i = 1, \ldots, n$.*

Proof. To prove the necessity, by the excision theorem we select an F_σ $\mathcal{P}$-kernel Y such that $\text{Ind}\,(X \setminus Y) \leq n$. Then by Theorem 1.15 the set $X \setminus Y$ can be partitioned into $n+1$ sets of dimension at most 0. One of these sets is then added to Y to obtain the desired decomposition.

The sufficiency will follow from the 0-dimensional case of Proposition I.4.6 and induction.

In connection with the last theorem there is an example which shows the failure of the addition theorem for the large inductive dimension modulo a class of spaces, even in the case where the class is a semi-normal family.

2.7. Example. Let $\mathcal{Z} = \{ X : \text{Ind}\, X \leq 0 \}$. From the sum theorem it follows that the family $\mathcal{Z}$ is semi-normal. An easy computation will show $\mathcal{Z}\text{-Ind}\, \mathbb{R}^3 = 2$. Because $\mathbb{R}^3$ is the union of two one-dimensional sets, it is the union of two sets with $\mathcal{Z}$-Ind equal to zero. This shows the failure of the addition theorem.

Another consequence of the coincidence theorem is the open subspace theorem for $\mathcal{P}$-Bind, $\mathcal{P}$-Ind and $\mathcal{P}$-Sur.

2.8. Theorem. *Suppose that $\mathcal{P}$ is a semi-normal family. For every open subspace Y of a space X,*

$$\mathcal{P}\text{-Ind}\, Y \leq \mathcal{P}\text{-Ind}\, X.$$

Proof. From the conditions on $\mathcal{P}$ it easily follows that $\mathcal{P}$ is open-monotone. The theorem now follows from Proposition 1.6 and the coincidence theorem.

We shall turn now to extension theorems. First we introduce the operation of expanding open sets of a space to open sets of an extension of the space. This operation will apply to any topological space.

2.9. Definition. Suppose that Y is an extension of a space X. For each open subset of the space X the *expansion of U in Y* is defined to be the open set $\text{ex}\,(U) = Y \setminus \text{cl}_Y(X \setminus U)$.

The set $\text{ex}\,(U)$ depends on Y. If confusion is likely to arise, a subscript will be used to indicate the space in which the expansion operator is applied. Some properties of the expansion operator will be collected in the next lemma.

2.10. Lemma. *Suppose that X is a dense subspace of a space Y. Let U and V be open subsets of the space X. The following properties hold.*

(a) *If W is an open set in Y with $W \cap X \subset U$, then $W \subset \text{ex}\,(U)$. Thus $\text{ex}\,(U) \subset \text{cl}_Y(U)$ holds and $\text{ex}\,(U)$ is the largest open set in Y whose trace on X is U.*
(b) $\text{ex}\,(\emptyset) = \emptyset$ *and* $\text{ex}\,(X) = Y$.
(c) $\text{ex}\,(U \cap V) = \text{ex}\,(U) \cap \text{ex}\,(V)$.
(d) $\text{ex}\,(U \cup V) \supset \text{ex}\,(U) \cup \text{ex}\,(V)$.
(e) $\text{B}_Y(\text{ex}\,(U)) \cap X = \text{B}_X(U)$.
(f) *If Z is an extension of Y, then $\text{ex}_Z(U) \cap Y = \text{ex}_Y(U)$.*
(g) *When a metric for Y is restricted to X, the diameters of U and $\text{ex}\,(U)$ are equal.*

Proof. We have $W \cap (X \setminus U) = \emptyset$ for any an open set W of Y with $W \cap X \subset U$. So $W \cap \text{cl}_Y(X \setminus U) = \emptyset$ follows. Now we have $W \subset \text{ex}\,(U)$ and (a) is proved. The formulas in (b), (c) and (d) are verified by straightforward computations. Formula (e) is verified by the following computation.

$$\begin{aligned} \mathrm{B}_Y(\mathrm{ex}(U)) \cap X &= \big(\mathrm{cl}_Y(\mathrm{ex}(U)) \setminus \mathrm{ex}(U)\big) \cap X \\ &= \big(\mathrm{cl}_Y(U) \setminus \mathrm{ex}(U)\big) \cap X \\ &= \mathrm{cl}_X(U) \setminus U = \mathrm{B}_X(U). \end{aligned}$$

For the proof of formula (f), observe that $\mathrm{ex}_Z(U) \cap Y$ is an open set in Y whose trace on X is U. By (a) we find $\mathrm{ex}_Z(U) \cap Y \subset \mathrm{ex}_Y(U)$. If W is any open set in Z with $W \cap Y = \mathrm{ex}_Y(U)$, then $W \cap X = U$, whence $W \subset \mathrm{ex}_Z(U)$. So we have $\mathrm{ex}_Z(U) \cap Y \supset W \cap Y = \mathrm{ex}_Y(U)$. The last statement follows from the formula in (a).

We shall discuss various extension theorems. The following is the key lemma.

2.11. Lemma. *Suppose that the space Y is an extension of the space X. Let $\mathcal{B}$ be a σ-locally finite base for the open sets of X. Then there exists a G_δ-set Z of Y such that $X \subset Z \subset Y$ and $\{\,\mathrm{ex}_Z(U) : U \in \mathcal{B}\,\}$ is a σ-locally finite base for the open sets of Z.*

Proof. Write $\mathcal{B} = \{\,U_\gamma : \gamma \in \Gamma_i,\ i \in \mathbb{N}\,\}$ with $\{\,U_\gamma : \gamma \in \Gamma_i\,\}$ being locally finite and the indexing by Γ_i being bijective for each i. First, we find for each i in $\mathbb{N}$ an open set V_i of Y such that $X \subset V_i \subset Y$ and $\{\,\mathrm{ex}_Y(U_\gamma) : \gamma \in \Gamma_i\,\}$ is locally finite in V_i. This is done as follows. For each x in X there is an open neighborhood B_x in X such that B_x meets U_γ for only finitely many γ in Γ_i. By (c) of the previous lemma we have $\mathrm{ex}_Y(B_x) \cap \mathrm{ex}_Y(U_\gamma) \neq \emptyset$ for only finitely many γ in Γ_i. Then $V_i = \bigcup\{\,\mathrm{ex}_Y(B_x) : x \in X\,\}$ is the required open set. Next, for each i in $\mathbb{N}$ we let W_i be the set of all points y in Y with the property that y is contained in $\mathrm{ex}_Y(U)$ for some U in $\mathcal{B}$ with diameter less than $\frac{1}{i+1}$. As $\mathcal{B}$ is a base it easily follows that $X \subset W_i$. Finally, define Z by

$$Z = \big(\bigcap\{\,V_i : i \in \mathbb{N}\,\}\big) \cap \big(\bigcap\{\,W_i : i \in \mathbb{N}\,\}\big)$$

and the result easily follows from Lemma 2.10.(f).

The first extension theorem that will be discussed is Tumarkin's extension theorem which has been stated for separable metrizable spaces as Theorem I.7.6. To place emphasis on the role of the basic inductive dimension we shall use the basic inductive dimension in the statement of the theorem even though we know from Theorem 1.12 that the basic inductive dimension and the large inductive dimension coincide in the universe of metrizable spaces.

2.12. Theorem. *Suppose that the space X is a subspace of the space Y and that n is in $\mathbb{N}$. If* Bind $X \leq n$, *then there exists a G_δ-set Z in Y such that $X \subset Z$ and* Bind $Z \leq n$. *In particular, every space X has a complete extension Z with the same basic inductive dimension as X.*

Proof. By replacing Y with $\mathrm{cl}_Y(X)$ if necessary, we may assume that X is dense in Y. As Bind $X \leq n$, there exists a σ-locally finite base for the open sets of X such that Bind $\mathrm{B}_X(U) \leq n-1$ for each U in $\mathcal{B}$. By the previous lemma there exists a G_δ-set Z_0 in Y with $X \subset Z_0$ such that $\{\mathrm{ex}_{Z_0}(U) : U \in \mathcal{B}\}$ is a σ-locally finite base for the open sets of Z_0. By the induction hypothesis, for each U in $\mathcal{B}$ there exists a G_δ-set Z_U such that

$$\mathrm{B}_X(U) \subset Z_U \subset \mathrm{B}_{Z_0}(\mathrm{ex}_{Z_0}(U)) \quad \text{and} \quad \text{Bind}\, Z_U \leq n-1.$$

Let $E = \bigcup\{\mathrm{B}_{Z_0}(\mathrm{ex}_{Z_0}(U)) \setminus Z_U : U \in \mathcal{B}\}$. As E is a subset of $Z_0 \setminus X$ and as $\mathrm{B}_{Z_0}(\mathrm{ex}_{Z_0}(U)) \setminus Z_U$ is an F_σ-set in $\mathrm{B}_{Z_0}(\mathrm{ex}_{Z_0}(U))$ for each U in $\mathcal{B}$, it will follow from Corollary II.9.4 that E is an F_σ-set in Z_0. Finally we define $Z = Z_0 \setminus E$. Obviously Z is a G_δ-set in Y such that $X \subset Z$. Also $\{\mathrm{ex}_{Z_0}(U) \cap Z : U \in \mathcal{B}\}$ is a σ-locally finite base for the open sets of Z. It is easily verified that for each U in $\mathcal{B}$ we have

$$\mathrm{B}_Z(\mathrm{ex}_{Z_0}(U) \cap Z) \subset \mathrm{B}_{Z_0}(\mathrm{ex}_{Z_0}(U)) \cap Z \subset Z_U.$$

As Bind $Z_U \leq n-1$ for each U in $\mathcal{B}$, we have Bind $Z \leq n$.

The second extension theorem will generalize a key result of Section I.7 concerning the completeness degree Icd to general metrizable spaces.

2.13. Theorem. *Suppose that X is a subset of a complete space Y and let n be in $\mathbb{N}$. If $\mathcal{C}$-*Bind $X \leq n$, *then there exists a complete subspace Z of Y such that* Bind $(Z \setminus X) \leq n$ *and Z is an extension of X.*

Proof. Because $\mathcal{C}$-Bind$X \leq n$, there exists a σ-locally finite base $\mathcal{B}$ for the open sets of X such that $\mathcal{C}$-Bind $\mathrm{B}_X(U) \leq n-1$ for each U

in $\mathcal{B}$. As in the proof of the previous theorem we may assume that X is dense in Y and that $\{\mathrm{ex}_Y(U) : U \in \mathcal{B}\}$ is a σ-locally finite base for Y. For each U in $\mathcal{B}$, $\mathrm{B}_Y(\mathrm{ex}_Y(U))$ is a complete space. By the induction hypothesis there exists a complete space Z_U such that $\mathrm{B}_X(U) \subset Z_U \subset \mathrm{B}_Y(\mathrm{ex}_Y(U))$ and $\mathrm{Bind}\,(Z_U \setminus \mathrm{B}_X(U)) \leq n-1$. For each U in $\mathcal{B}$, the set $\mathrm{B}_Y(\mathrm{ex}_Y(U)) \setminus Z_U$ is an F_σ-set in Y. We define $E = \bigcup\{\mathrm{B}_Y(\mathrm{ex}_Y(U)) \setminus Z_U : U \in \mathcal{B}\}$. From Corollary II.9.4 it follows that E is an F_σ-set in Y. Finally we define $Z = Y \setminus E$. As Z is a G_δ-set in Y, it is a complete extension of X. The collection $\{\mathrm{ex}_Y(U) \cap (Z \setminus X) : U \in \mathcal{B}\}$ is a σ-locally finite base of $Z \setminus X$. An easy computation will show that, for each U in $\mathcal{B}$,

$$\mathrm{B}_{Z\setminus X}(\mathrm{ex}_Y(U) \cap (Z \setminus X)) \subset \mathrm{B}_Y(\mathrm{ex}_Y(U)) \cap (Z \setminus X) \subset Z_U.$$

It follows that $\mathrm{Bind}\,(Z \setminus X) \leq n$.

Following the pattern of Section I.7, we first prove a coincidence theorem for the completeness dimension functions.

2.14. Theorem (Main theorem for completeness degree). *For every space X,*

$$\mathcal{C}\text{-Bind}\,X = \mathrm{Icd}\,X = \mathcal{C}\text{-Def}\,X.$$

Proof. Recall that $\mathrm{Icd} = \mathcal{C}\text{-Ind}$. The theorem will follow from Theorem 1.10, Proposition II.8.7 and Theorem 2.13.

Now that the main theorem has been established one can prove various adaptations of the results for icd from Section I.7. As the proofs of the Theorems 2.15 through 2.19 are simple modifications of the proofs in Section I.7, they will be left for the reader to complete.

2.15. Theorem. *For every G_δ-set Y in a space X,*

$$\mathrm{Icd}\,Y \leq \mathrm{Icd}\,X.$$

2.16. Theorem (Addition theorem). *If $X = Y \cup Z$, then*

$$\mathrm{Icd}\,X \leq \mathrm{Icd}\,Y + \mathrm{Icd}\,Z + 1.$$

In particular, the large inductive completeness degree Icd *of a space cannot be increased by the adjunction of a complete space or a point.*

In contrast with this result for Icd, note that icd can be raised by the addition of one point as we have shown in Section III.1.

2.17. Theorem (Intersection theorem). *If Y and Z are subsets of a space X, then*

$$\operatorname{Icd}(Y \cap Z) \leq \operatorname{Icd} Y + \operatorname{Icd} Z + 1.$$

2.18. Theorem (Locally finite sum theorem). *Let $\boldsymbol{F}$ be a locally finite closed cover of a space X. Then*

$$\operatorname{Icd} X = \sup \{ \operatorname{Icd} F : F \in \boldsymbol{F} \}.$$

2.19. Theorem (Structure theorem). *For every space X and every natural number n, $\operatorname{Icd} X \leq n$ if and only if there exists a complete extension Y of X such that the set X can be represented as the intersection of $n+1$ subsets Y_i of Y, $i = 0, 1, \ldots, n$, with $\operatorname{Icd} Y_i \leq 0$ for every i.*

The results of Theorems 2.13 and 2.14 will be generalized by replacing $\mathcal{C}$ with the multiplicative absolute Borel classes $\mathsf{M}(\alpha)$, $\alpha \geq 2$.

2.20. Theorem. *Let α be an ordinal number with $\alpha \geq 2$. Suppose that X is a subset of a complete space Y and that n is in $\mathbb{N}$. If $\mathsf{M}(\alpha)$-$\operatorname{Bind} X \leq n$, then there exist a subspace Z of Y such that $X \subset Z \subset Y$, $Z \in \mathsf{M}(\alpha)$ and $\operatorname{Bind}(Z \setminus X) \leq n$.*

Proof. Because $\mathsf{M}(\alpha)$-$\operatorname{Bind} X \leq n$, there exists a σ-locally finite base $\mathcal{B}$ for the open sets of X such that $\mathsf{M}(\alpha)$-$\operatorname{Bind} \mathrm{B}_X(U) \leq n-1$ for each U in $\mathcal{B}$. In view of Lemma 2.11 we may assume that X is dense in Y and that $\{ \operatorname{ex}_Y(U) : U \in \mathcal{B} \}$ is a σ-locally finite base for Y. For each U in $\mathcal{B}$ the space $\mathrm{B}_Y(\operatorname{ex}_Y(U))$ is complete. By the induction hypothesis there exists a space Z_U with $\mathrm{B}_X(U) \subset Z_U \subset \mathrm{B}_Y(\operatorname{ex}_Y(U))$, $Z_U \in \mathsf{M}(\alpha)$ and $\operatorname{Bind}\big(Z_U \setminus \mathrm{B}_X(U)\big) \leq n-1$. For each U in $\mathcal{B}$ the set $\mathrm{B}_Y(\operatorname{ex}_Y(U)) \setminus Z_U$ belongs to the additive Borel class α by Theorem II.9.1. Let $E = \bigcup \{ \mathrm{B}_Y(\operatorname{ex}_Y(U)) \setminus Z_U : U \in \mathcal{B} \}$. From Corollary II.9.4 it follows that E is a member of the additive class α in Y. Finally we define $Z = Y \setminus E$. As Z is a member of the multiplicative class α in the complete space Y, we have $Z \in \mathsf{M}(\alpha)$ by Theorem II.9.6. The collection $\{ \operatorname{ex}_Y(U) \cap (Z \setminus X) : U \in \mathcal{B} \}$ is a σ-locally finite base for $Z \setminus X$. An easy computation will show that for each U in $\mathcal{B}$

$$\mathrm{B}_{Z \setminus X}(\operatorname{ex}_Y(U) \cap (Z \setminus X)) \subset \mathrm{B}_Y(\operatorname{ex}_Y(U)) \cap (Z \setminus X) \subset Z_U.$$

So Bind $\big(\mathrm{ex}_Y(U) \cap (Z \setminus X)\big) \leq n-1$ by Theorems 1.12 and 1.14, whence Bind $(Z \setminus X) \leq n$.

The natural generalization of Theorem 2.14 reads as follows.

2.21. Theorem. *For every space X and every ordinal $\alpha \geq 1$,*

$$\mathsf{M}(\alpha)\text{-Bind}\, X = \mathsf{M}(\alpha)\text{-Ind}\, X = \mathsf{M}(\alpha)\text{-Def}\, X.$$

The proof of the following result is an application of Theorem 2.12.

2.22. Lemma. *Suppose that the class $\mathcal{P}$ is F_σ-monotone. Then*

$$\mathcal{P}\text{-Sur} \leq \mathcal{P}\text{-Def}.$$

It is to be noted that the classes $\mathsf{A}(\alpha)$ are F_σ-monotone for each α and that classes $\mathsf{M}(\alpha)$ are F_σ-monotone when $\alpha \geq 2$. These properties follow from Theorems II.9.1 and II.9.6.

The concept of complementarity of dimension functions that is illustrated by Proposition II.10.2 can be developed a little further.

2.23. Definition. A space Z is said to be *ambiguous relative to the classes of spaces $\mathcal{P}$ and $\mathcal{Q}$* provided that $X \in \mathcal{P}$ if and only if $Y \in \mathcal{Q}$ whenever X and Y are *complementary* in Z, that is, whenever $Z = X \cup Y$ and $X \cap Y = \emptyset$. For classes of spaces $\mathcal{P}$, $\mathcal{Q}$ and $\mathcal{R}$, the classes $\mathcal{P}$ and $\mathcal{Q}$ are called *complementary with respect to $\mathcal{R}$* if every Z in $\mathcal{R}$ is ambiguous relative to $\mathcal{P}$ and $\mathcal{Q}$.

In Section II.9 we have seen that the additive Borel class $\mathsf{A}(\alpha)$ and the multiplicative Borel class $\mathsf{M}(\alpha)$ are complementary with respect to the class $\mathcal{C}$ of all complete spaces when $\alpha \geq 2$. We shall make a small digression from our discussion of absolute Borel classes to collect some results concerning complementary classes. The first result follows directly from the definitions.

2.24. Lemma. *Suppose that a space Z is ambiguous relative to the classes $\mathcal{P}$ and $\mathcal{Q}$. Then*

$$\mathcal{Q}\text{-Def}\, X \leq \mathcal{P}\text{-Sur}\, Y,$$

whenever X and Y are complementary in Z.

By combining the Lemmas 2.22 and 2.24, we get the following result.

2.25. Theorem. *Let $\mathcal{P}$ and $\mathcal{Q}$ be F_σ-monotone classes. Suppose that the space Z is ambiguous relative to $\mathcal{P}$ and $\mathcal{Q}$. Then*

$$\mathcal{P}\text{-Def}\, X = \mathcal{P}\text{-Sur}\, X = \mathcal{Q}\text{-Def}\, Y = \mathcal{Q}\text{-Sur}\, Y$$

whenever X and Y are complementary in Z.

With the help of this theorem we find for each ordinal α with $\alpha \geq 2$ the equalities

$$\mathsf{A}(\alpha)\text{-Sur} = \mathsf{A}(\alpha)\text{-Def}, \quad \mathsf{M}(\alpha)\text{-Sur} = \mathsf{M}(\alpha)\text{-Def}$$

and

$$\big(\mathsf{M}(\alpha) \cap \mathsf{A}(\alpha)\big)\text{-Sur} = \big(\mathsf{M}(\alpha) \cap \mathsf{A}(\alpha)\big)\text{-Def}.$$

For the covering dimension functions modulo a class there is the following surprising theorem.

2.26. Theorem. *Suppose that the space Z is ambiguous relative to $\mathcal{P}$ and $\mathcal{Q}$. Then*

$$\mathcal{P}\text{-Dim}\, X = \mathcal{Q}\text{-Dim}\, Y \quad \text{and} \quad \mathcal{P}\text{-dim}\, X = \mathcal{Q}\text{-dim}\, Y$$

whenever X and Y are complementary in Z.

Proof. The proof will easily follow from the following observation. If X and Y are complementary sets in Z and if $\mathbf{V}$ is an open collection in Z, then the trace of $\mathbf{V}$ on X is a $\mathcal{P}$-border cover of X if and only the trace of $\mathbf{V}$ on Y is a $\mathcal{Q}$-border cover of Y.

Let us return to the discussion of the basic inductive dimensions and complementarity.

2.27. Theorem. *Suppose that the classes $\mathcal{P}$ and $\mathcal{Q}$ are complementary with respect to a class $\mathcal{R}$. If X and Y are complementary subspaces in a space Z in $\mathcal{R}$, then*

$$\mathcal{Q}\text{-Bind}\, Y \leq \mathcal{P}\text{-Ind}\, X.$$

Proof. The proof is by induction. Let n be a natural number and assume $\mathcal{P}\text{-Ind}\, X \leq n$. By Proposition 1.9, the space Z has a framework $\big(\{\, U_\gamma : \gamma \in \Gamma \,\}, \{\, F_\gamma : \gamma \in \Gamma \,\}\big)$ of a σ-locally finite base for the open sets. As $\mathcal{P}$-Ind $X \leq n$, by Proposition I.4.6 there exists for each γ in Γ an open set V_γ such that $\mathcal{P}$-Ind$\big(\mathrm{B}_Z(V_\gamma) \cap X\big) \leq n-1$ and $F_\gamma \subset V_\gamma \subset U_\gamma$. Observe that $\mathrm{B}_Z(V_\gamma)$ is in $\mathcal{R}$. By the induction hypothesis, we have $\mathcal{Q}$-Bind$\big(\mathrm{B}_Z(V_\gamma) \cap Y\big) \leq n-1$. It now follows that $\mathcal{Q}$-Bind $Y \leq n$ because $\mathrm{B}_Y(V_\gamma \cap Y) \subset \mathrm{B}_Z(V_\gamma) \cap Y$.

Returning to Theorems 2.13 and 2.14, we can generalize these theorems along other lines. The following is a typical example.

2.28. Theorem. *Let X be a space and let Y be a complete space. Suppose that $\boldsymbol{F}$ and $\boldsymbol{G}$ are σ-locally finite collections of closed subsets of X and let $\{ f_i : i \in \mathrm{N} \}$ be a countable family of continuous maps of X to Y. Then there exists a complete extension $\widetilde{X}$ of X with the properties:*

(a) $\bigcap\{ \mathrm{cl}_{\widetilde{X}}(F_i) : i = 1,\dots,m \} = \mathrm{cl}_{\widetilde{X}}(\bigcap\{ F_i : i = 1,\dots,m \})$ *for all finite subcollections $F_1, \dots, F_m$ of $\boldsymbol{F}$.*
(b) Bind $\mathrm{cl}_{\widetilde{X}}(G) =$ Bind G *for each G in $\boldsymbol{G}$.*
(c) Bind $\big(\mathrm{cl}_{\widetilde{X}}(G) \setminus X\big) = \mathcal{C}$-Bind G *for each G in $\boldsymbol{G}$.*
(d) *Each f_i has a continuous extension $\widetilde{f}_i \colon \widetilde{X} \to Y$.*

Proof. Let Z be a complete extension of X. The space $\widetilde{X}$ will be obtained by trimming F_σ-sets off Z.

As $\boldsymbol{F}$ and $\boldsymbol{G}$ are σ-locally finite closed collections of X, we can find a G_δ-set in Z containing X such that $\boldsymbol{F}$ and $\boldsymbol{G}$ are σ-locally finite collections in this G_δ-set. Thus we may assume that Z is an extension of X such that the collections $\boldsymbol{F}$ and $\boldsymbol{G}$ are also σ-locally finite in Z. For each finite subset $\{ F_1, \dots, F_n \}$ of $\boldsymbol{F}$ the set

$$D = \bigcap\{ \mathrm{cl}_Z(F_i) : i = 1,\dots,m \} \setminus \mathrm{cl}_Z(\bigcap\{ F_i : i = 1,\dots,m \})$$

is an F_σ-set in Z that is contained in $Z \setminus X$. The family $\boldsymbol{D}$ of all sets D that can be obtained in this way is σ-locally finite. By Corollary II.9.4 the set $\bigcup \boldsymbol{D}$ is an F_σ-set in Z. For each G in $\boldsymbol{G}$, by Tumarkin's theorem (Theorem 2.12) there exists a G_δ-set E_G in $\mathrm{cl}_Z(G)$ such that $G \subset E_G$ and Bind $E_G =$ Bind G. Denote the collection $\{ \mathrm{cl}_Z(G) \setminus E_G : G \in \boldsymbol{G} \}$ by $\boldsymbol{E}$. As $\mathrm{cl}_Z(G) \setminus E_G$ is an F_σ-set in $\mathrm{cl}_Z(G)$, by Corollary II.9.4 the set $\bigcup \boldsymbol{E}$ is an F_σ-set in Z. Furthermore, by Theorem 2.13, for each G in $\boldsymbol{G}$ there exists a G_δ extension Z_G of G such that $Z_G \subset \mathrm{cl}_Z(G)$ and Bind $(Z_G \setminus X) = \mathcal{C}$-Bind G. (The equality follows because the set G is closed in X.) Let $\boldsymbol{Z}$ be the collection $\{\operatorname{cl}_Z(G) \setminus Z_G : G \in \mathbf{G}\}$. By Lemma I.7.2, for each i in $\mathbb{N}$ there is a set C_i and there is a continuous map $\widetilde{f}_i : C_i \to Y$ such that C_i is a G_δ-set in Z containing X and $\widetilde{f}_i$ is an extension of f_i. Now the space $\widetilde{X}$ is defined by

$$\widetilde{X} = \bigcap \{ C_i : i \in \mathbb{N} \} \setminus ((\bigcup \boldsymbol{D}) \cup (\bigcup \boldsymbol{E}) \cup (\bigcup \boldsymbol{Z})).$$

The verification of properties (a) through (d) is straightforward.

3. The order dimension

The order dimension function modulo the class $\mathcal{P}$, $\mathcal{P}$-Odim, will be studied in this section. The function Skl, which was introduced in Section I.6, is of this type and will be used in Section VI.5 to characterize $\mathcal{K}$-def. Also the order dimension $\mathcal{P}$-Odim will play a key role in the proof of the inequality $\mathcal{P}$-Bind $\leq \mathcal{P}$-Dim.

The definition of $\mathcal{P}$-Odim for general metrizable spaces is a natural generalization of the definitions of $\mathcal{C}$-Odim (I.7.19) and Skl (I.6.8). As in the definition of $\mathcal{P}$-Bind given in Section 1.1, the metrizability of the space will be part of the definition of $\mathcal{P}$-Odim.

3.1. Definition. Let $\mathcal{P}$ be a class of spaces and let X be a (metrizable) space. For $n = -1$ or $n \in \mathbb{N}$ the space X is said to have $\mathcal{P}$-Odim $\leq n$ if there exists a σ-locally finite base $\{ U_\gamma : \gamma \in \Gamma \}$ for the open sets of X such that $\mathrm{B}(U_{\gamma_0}) \cap \cdots \cap \mathrm{B}(U_{\gamma_n})$ belongs to $\mathcal{P}$ for any $n+1$ different indices $\gamma_0, \ldots, \gamma_n$ from Γ. (Of course, this gives $X \in \mathcal{P}$ for $n = -1$.)

We shall start by collecting some fairly simple properties.

3.2. Proposition. *If $\mathcal{P}$ and $\mathcal{Q}$ are classes with $\mathcal{P} \supset \mathcal{Q}$, then $\mathcal{P}$-Odim $\leq \mathcal{Q}$-Odim. In particular, $\mathcal{P}$-Odim $\leq$ Odim.*

Recall that for any subspace Y of a space X and for every subset U of X we have by Proposition II.4.2.B that $\mathrm{B}_Y(U \cap Y) \subset \mathrm{B}_X(U)$. So, easy inductive proofs will give the following propositions.

3.3. Proposition. *For every closed subspace Y of a space X, $\mathcal{P}$-Odim $Y \leq \mathcal{P}$-Odim X.*

3.4. Proposition. *The class $\mathcal{P}$ is monotone if and only if the function $\mathcal{P}$-*Odim *is monotone, i.e., for every subspace Y of a space X,*

$$\mathcal{P}\text{-Odim}\, Y \leq \mathcal{P}\text{-Odim}\, X.$$

The corresponding statements hold for F_σ-, G_δ- and open-monotone classes.

The next open subspace theorem is similar to Theorem II.2.17.

3.5. Proposition. *Let X be such that $0 \leq \mathcal{P}\text{-Odim}\, X$. For every open subspace Y of X, $\mathcal{P}\text{-Odim}\, Y \leq \mathcal{P}\text{-Odim}\, X$.*

Our next goal is to show the relation between $\mathcal{P}$-Dim and $\mathcal{P}$-Odim. The relation between $\mathcal{P}$-dim and $\mathcal{P}$-Odim for separable spaces will follow as a corollary. But first we need to extend Lemma I.9.4 to infinite collections. Notice that in the extended theorem the hypothesis that X be metrizable can be weakened to hereditarily paracompact. The corresponding result for $\mathcal{P}$-dim has been stated for hereditarily normal spaces in Theorem II.5.9.

3.6. Theorem. *For a natural number n let X be a metrizable space with $\mathcal{P}\text{-Dim}\, X \leq n$. Suppose that G is a closed $\mathcal{P}$-kernel of X and suppose that $\{ F_\gamma : \gamma \in \Gamma \}$ and $\{ U_\gamma : \gamma \in \Gamma \}$ are locally finite collections of closed and open sets of X respectively such that $F_\gamma \subset U_\gamma \mod G$, $\gamma \in \Gamma$. Then there exists a closed $\mathcal{P}$-kernel H of X and two open collections $\{ V_\gamma : \gamma \in \Gamma \}$ and $\{ W_\gamma : \gamma \in \Gamma \}$ with*

$$G \subset H,$$
$$F_\gamma \subset V_\gamma \subset \mathrm{cl}_X(V_\gamma) \subset W_\gamma \subset U_\gamma \mod H, \qquad \gamma \in \Gamma,$$
$$\mathrm{ord}\, \{ W_\gamma \setminus \mathrm{cl}_X(V_\gamma) : \gamma \in \Gamma \} \leq n \mod H.$$

Moreover, the conditions

$$V_\gamma \subset X \setminus H \text{ and } W_\gamma \subset X \setminus H, \qquad \gamma \in \Gamma,$$

may also be imposed.

Proof. We may assume that the indexing by Γ is one-to-one. Consider the $\mathcal{P}$-border cover

$$\mathbf{D} = \bigwedge \{ \{ U_\gamma \setminus G, (X \setminus F_\gamma) \setminus G \} : \gamma \in \Gamma \}.$$

This border cover is locally finite because $\{ U_\gamma : \gamma \in \Gamma \}$ is locally finite. Let $\boldsymbol{E}$ be an open cover of the subspace $X \setminus G$ such that each member of $\boldsymbol{E}$ meets U_γ for only finitely many indices γ from Γ. We may further require that $\boldsymbol{E}$ be locally finite. By Proposition II.5.5.C, there exists a locally finite $\mathcal{P}$-border cover $\boldsymbol{L} = \{ L_\alpha : \alpha \in A \}$ with enclosure H such that $\boldsymbol{L}$ shrinks $\boldsymbol{E}$ modulo H and $\operatorname{ord} \boldsymbol{L} \leq n+1$. Clearly the $\mathcal{P}$-kernel H contains G and is closed. The locally finite open collection $\boldsymbol{L}$ in the subspace $X \setminus H$ has a closed (in $X \setminus H$) shrinking $\{ K_\alpha : \alpha \in A \}$. As $\boldsymbol{L}$ is a refinement of $\boldsymbol{E}$, for each α in A there are only finitely many γ in Γ such that $L_\alpha \cap F_\gamma \neq \emptyset$. We list the indices γ for which $L_\alpha \cap F_\gamma \neq \emptyset$ in a one-to-one way as $\gamma(0,\alpha), \dots, \gamma(n(\alpha),\alpha)$. Using the normality of the subspace $X \setminus H$, we can find for each α in A a collection $\{ K^i_\alpha : i = 0, 1, \dots, n(\alpha)+1 \}$ of open sets such that

$$(1) \qquad K_\alpha \subset K^i_\alpha \subset \operatorname{cl}_{X\setminus H}(K^i_\alpha) \subset L_\alpha, \qquad i = 0, 1, \dots, n(\alpha)+1,$$

and

$$(2) \qquad \operatorname{cl}_{X\setminus H}(K^i_\alpha) \subset K^{i+1}_\alpha, \qquad i = 0, 1, \dots, n(\alpha).$$

It should be noted that from (1) and (2) we have

$$(3) \quad \big(K^{i+1}_\alpha \setminus \operatorname{cl}_{X\setminus H}(K^i_\alpha)\big) \cap \big(K^{i'+1}_\alpha \setminus \operatorname{cl}_{X\setminus H}(K^{i'}_\alpha)\big) = \emptyset, \qquad i < i'.$$

Now for γ in Γ we define

$$V_\gamma = \bigcup \{ K^i_\alpha : \gamma = \gamma(i,\alpha) \},$$
$$W_\gamma = \bigcup \{ K^{i+1}_\alpha : \gamma = \gamma(i,\alpha) \}.$$

Because $\boldsymbol{L} = \{ L_\alpha : \alpha \in A \}$ is a cover of $X \setminus H$ and $\boldsymbol{L}$ is a refinement of $\boldsymbol{E}$, one readily proves the inclusions

$$F_\gamma \subset V_\gamma \subset \operatorname{cl}(V_\gamma) \subset W_\gamma \subset U_\gamma \mod H, \qquad \gamma \in \Gamma.$$

Thus the first two formulas have been established.

Now, by way of a contradiction, we shall prove the last of the formulas. Assume for some p in $X \setminus H$ and some distinct indices γ_m, $m = 0, 1, \dots, n$, that

$$p \in \bigcap \{ W_{\gamma_m} \setminus \operatorname{cl}_{X\setminus H}(V_{\gamma_m}) : m = 0, 1, \dots, n \}.$$

Note that we have

$$W_\gamma \setminus \mathrm{cl}_{X\setminus H}(V_\gamma) \subset \bigcup\{\, K_\alpha^{i+1} \setminus \mathrm{cl}_{X\setminus H}(K_\alpha^i) \ : \gamma = \gamma(i,\alpha)\,\}, \qquad \gamma \in \Gamma.$$

Thus for suitable pairs (i_m, α_m) we have

$$p \in \bigcap\{\, K_{\alpha_m}^{i_m+1} \setminus \mathrm{cl}_{X\setminus H}(K_{\alpha_m}^{i_m}) \ : m = 0,1,\ldots,n\,\}.$$

It follows from (3) and the definition of $\gamma(k,\alpha)$, $k = 0,1,\ldots,n(\alpha)$, that the indices α_m must be distinct. As $\{\, K_\alpha : \alpha \in A\,\}$ covers X, we have $p \in K_\eta$ for some η in A. From (1) it follows that η is distinct from each of the α_m's. Hence

$$p \in L_\eta \cap L_{\alpha_0} \cap \cdots \cap L_{\alpha_n},$$

that is,

$$\mathrm{ord}_p\, \mathbf{L} \geq n+2.$$

This is a contradiction.

Now let us disclose the relation between $\mathcal{P}$-Odim and $\mathcal{P}$-Dim.

3.7. Theorem. *For every space X,*

$$\mathcal{P}\text{-Odim}\, X \leq \mathcal{P}\text{-Dim}\, X.$$

Proof. Let n be a natural number and suppose $\mathcal{P}$-Dim $X \leq n$. Let $(\{\, U_\gamma : \gamma \in \Gamma\,\}, \{\, H_\gamma : \gamma \in \Gamma\,\})$ be the framework of a base of X. We may assume that the indexing of the collection $\{\, U_\gamma : \gamma \in \Gamma\,\}$ by Γ is one-to-one. For each γ in Γ we select an open set C_γ such that

$$H_\gamma \subset C_\gamma \subset F_\gamma \subset U_\gamma,$$

where $F_\gamma = \mathrm{cl}\,(C_\gamma)$. We write $\Gamma = \bigcup\{\, \Gamma_i : i \in \mathbb{N}\,\}$ in such a way that $\{\, U_\gamma : \gamma \in \Gamma_i\,\}$ is locally finite for each i. Observe that for each j the collection $\{\, U_\gamma : \gamma \in \Gamma_0 \cup \cdots \cup \Gamma_j\,\}$ is locally finite. Inductively on j, we shall define a closed $\mathcal{P}$-kernel G_j and, for each γ in $\Gamma_0 \cup \cdots \cup \Gamma_j$, open sets V_γ^j and W_γ^j possessing the properties:

For each j and each γ in $\Gamma_0 \cup \cdots \cup \Gamma_j$,

$$\begin{gathered} V_\gamma^j \subset X \setminus G_j, \quad W_\gamma^j \subset X \setminus G_j, \\ F_\gamma \subset V_\gamma^j \subset \operatorname{cl}(V_\gamma^j) \subset W_\gamma^j \subset U_\gamma \mod G_j \end{gathered} \tag{4}$$

and

$$\begin{gathered} G_j \subset G_{j+1}, \\ \operatorname{cl}(V_\gamma^j) \subset V_\gamma^{j+1} \mod G_{j+1}, \quad \operatorname{cl}(W_\gamma^{j+1}) \subset W_\gamma^j \mod G_{j+1}; \end{gathered} \tag{5}$$

and, for each j,

$$\operatorname{ord}\{ W_\gamma^j \setminus \operatorname{cl}(V_\gamma^j) : \gamma \in \Gamma_0 \cup \cdots \cup \Gamma_j \} \le n \mod G_j. \tag{6}$$

As the first step of the definition is similar to the inductive step, we shall describe only the inductive one. To this end, suppose that s is such that there have been defined for each j a closed $\mathcal{P}$-kernel G_j and open sets V_γ^j and W_γ^j for each γ in $\Gamma_0 \cup \cdots \cup \Gamma_j$ so that (4) and (6) hold for $j \le s$ and (5) holds for $j < s$. To define G_{s+1} and V_γ^{s+1} and W_γ^{s+1} for each γ in $\Gamma_0 \cup \cdots \cup \Gamma_{s+1}$, we apply the preceding theorem to the $\mathcal{P}$-kernel G_s and the collections

$$\{ \operatorname{cl}(V_\gamma^s) \setminus G_s : \gamma \in \Gamma_0 \cup \cdots \cup \Gamma_s \} \cup \{ F_\gamma \setminus G_s : \gamma \in \Gamma_{s+1} \}$$

and

$$\{ W_\gamma^s \setminus G_s : \gamma \in \Gamma_0 \cup \cdots \cup \Gamma_s \} \cup \{ U_\gamma \setminus G_s : \gamma \in \Gamma_{s+1} \}$$

so as to obtain the closed $\mathcal{P}$-kernel G_{s+1} and the open collections

$$\{ V_\gamma^{s+1} : \gamma \in \Gamma_0 \cup \cdots \cup \Gamma_{s+1} \} \text{ and } \{ W_\gamma^{s+1} : \gamma \in \Gamma_0 \cup \cdots \cup \Gamma_{s+1} \}$$

satisfying (4) and (6) for $j = s + 1$ and (5) for $j = s$, thereby completing the inductive step.

From the inclusion

$$H_\gamma \subset C_\gamma \subset V_\gamma^i \cup (U_\gamma \cap G_i), \qquad \gamma \in \Gamma_i,$$

we find that the interior of $V_\gamma^i \cup (U_\gamma \cap G_i)$, denoted by V_γ^{i*}, satisfies

$$H_\gamma \subset C_\gamma \subset V_\gamma^{i*} \subset U_\gamma, \qquad \gamma \in \Gamma_i, \tag{7}$$

and

$$V_\gamma^i = V_\gamma^{i*} \setminus G_i, \qquad \gamma \in \Gamma_i.$$

for each i in $\mathbb{N}$. Moreover, when $i \le k \le j$ we have from (4) and (5) that $V_\gamma^j \subset W_\gamma^k$ modulo G_j for each γ in $\Gamma_0 \cup \cdots \cup \Gamma_j$, and consequently

$$V_\gamma^j \setminus G_j \subset W_\gamma^k \subset U_\gamma.$$

Finally we define $\mathcal{B} = \{ V_\gamma : \gamma \in \Gamma \}$ by

$$V_\gamma = V_\gamma^{i*} \cup \big(\bigcup \{ V_\gamma^j \setminus G_j : j > i \} \big), \qquad \gamma \in \Gamma_i.$$

As the pairs (H_γ, U_γ) are from the framework of the base, we have from (7) that $\mathcal{B}$ is a base for the open sets. Notice

$$\mathrm{B}(V_\gamma) \subset \big(W_\gamma^k \setminus \mathrm{cl}(V_\gamma^k) \big) \cup G_{k+1}, \qquad k \ge i.$$

Thus it follows that

$$\mathrm{ord} \{ \mathrm{B}(V_\gamma) : \gamma \in \Gamma_0 \cup \cdots \cup \Gamma_k \} \le n \mod G_{k+1}, \qquad k \in \mathbb{N}.$$

We have shown for any $n+1$ different indices $\gamma_0, \ldots, \gamma_n$ that the intersection $\mathrm{B}(V_{\gamma_0}) \cap \cdots \cap \mathrm{B}(V_{\gamma_n})$ is contained in some G_k. It follows that the intersection is a member of $\mathcal{P}$. Thereby, $\mathcal{P}$-Odim $X \le n$.

The following corollary of the above proof is easily derived. For, in that proof, Theorem 3.6 can be replaced with Theorem II.5.9 (b) and the framework of the base for X used there can be chosen in such a way that Γ_i is the collection $\{ i \}$. See also the second part of the proof of Theorem I.9.5.

3.8. Corollary. *For every separable space X,*

$$\mathcal{P}\text{-Odim}\, X \le \mathcal{P}\text{-dim}\, X.$$

Applying this corollary to the class $\mathcal{K}$ of all compact spaces, we get the following important result.

3.9. Corollary. *For every separable space X,*

$$\mathrm{Skl}\, X \le \mathcal{K}\text{-dim}\, X.$$

The next theorem compares $\mathcal{P}$-Odim and $\mathcal{P}$-Bind.

3.10. Theorem. *For every space X,*

$$\mathcal{P}\text{-Bind}\, X \le \mathcal{P}\text{-Odim}\, X.$$

Proof. The proof is by induction on $\mathcal{P}$-Odim X. As the inequality is obvious for $\mathcal{P}$-Odim $X = 0$, let us go to the inductive step. Let $n > 0$ and assume $\mathcal{P}$-Odim $X \le n$. There exists a σ-locally finite base $\mathcal{B} = \{ V_\gamma : \gamma \in \Gamma \}$ such that $\text{ord}\{ \mathrm{B}(V_\gamma) : \gamma \in \Gamma \} \le n$. For each η in Γ the collection $\mathcal{B}_\eta = \{ V_\gamma \cap \mathrm{B}(V_\eta) : \gamma \in \Gamma \setminus \{ \eta \} \}$ is a σ-locally finite base for the open sets of $\mathrm{B}(V_\eta)$. From the fact that $\mathrm{B}_{\mathrm{B}(V_\eta)}(V_\gamma \cap \mathrm{B}(V_\eta))$ is a closed subset of $\mathrm{B}(V_\gamma) \cap \mathrm{B}(V_\eta)$ we easily see that $\mathcal{P}$-Odim $\mathrm{B}(V_\eta) \le n - 1$ for each η in Γ. By the induction hypothesis we have $\mathcal{P}$-Bind $\mathrm{B}(V_\eta) \le n - 1$, whence $\mathcal{P}$-Bind $X \le n$.

It is to be noted that the Theorems 3.7 and 3.10 have been proved under the minimal requirement that the class $\mathcal{P}$ be closed-monotone. When the class $\mathcal{P}$ is a cosmic family, a nice lining up of the dimension functions modulo $\mathcal{P}$ will result.

3.11. Theorem. *Suppose that $\mathcal{P}$ is a strongly closed-additive cosmic family. Then for every space X*

$$\mathcal{P}\text{-Bind}\, X \le \mathcal{P}\text{-Odim}\, X \le \mathcal{P}\text{-dim}\, X = \mathcal{P}\text{-Dim}\, X \le \mathcal{P}\text{-Ind}\, X.$$

Proof. The equality of the covering dimensions has been established in Theorem IV.2.7. The last inequality in the theorem follows from Theorem IV.4.2. The other inequalities follow from the previous theorems.

3.12. Theorem. *Suppose that $\mathcal{P}$ is a normal family or that $\mathcal{P}$ is a multiplicative Borel class $\mathsf{M}(\alpha)$, $\alpha > 0$. Then for every space X*

$$\mathcal{P}\text{-Bind}\, X = \mathcal{P}\text{-Odim}\, X = \mathcal{P}\text{-dim}\, X = \mathcal{P}\text{-Dim}\, X = \mathcal{P}\text{-Ind}\, X.$$

Proof. Note that $\mathcal{P}$ is a cosmic family. The theorem follows from the previous result and Theorems 2.4, 2.14 and 2.21.

3.13. Corollary. *With the hypothesis of the theorem, for every separable space X,*

$$\mathcal{P}\text{-ind}\, X = \mathcal{P}\text{-Ind}\, X.$$

The previous theorem includes the coincidence theorem for the ordinary dimension functions.

3.14. Theorem (Coincidence theorem). *For every space X,*

$$\operatorname{Bind} X = \operatorname{Odim} X = \dim X = \operatorname{Dim} X = \operatorname{Ind} X.$$

3.15. Remark. We have found in the preceding theorem that $\dim X = \operatorname{Ind} X$ for every X in $\mathcal{M}$. We can use Theorems III.3.25 and IV.3.5 to get the same equality for every X in $\mathsf{N}[\mathcal{M} : \mathcal{N}_H]$, the normal extension of $\mathcal{M}$ in $\mathcal{N}_H$. As $\mathcal{M}$ is contained in the class $\mathcal{E}$ of strongly hereditarily normal spaces, we have that $\mathsf{N}[\mathcal{M} : \mathcal{N}_H]$ is contained in the Dowker universe $\mathcal{D}$. It is obvious that $\mathsf{N}[\mathcal{M} : \mathcal{N}_H]$ properly contains $\mathcal{M}$.

3.16. Corollary. *For every class $\mathcal{P}$,*

$$\mathcal{P}\text{-sur} = \mathcal{P}\text{-Sur} \quad \textit{and} \quad \mathcal{P}\text{-def} = \mathcal{P}\text{-Def}.$$

4. The mixed inductive dimension

The purpose of this short section is to place the compactness dimension function Cmp into a proper perspective by presenting the natural generalization of the function Cmp, namely the function $\mathcal{P}$-Mind called the mixed inductive dimension modulo a class $\mathcal{P}$. Most of the theory about this dimension function is yet to be developed.

The function Cmp is defined as a large inductive variation of cmp which agrees with cmp in the values -1 and 0. The function $\mathcal{P}$-Mind will be defined in a completely analogous manner.

4.1. Definition. Let $\mathcal{P}$ be a class of spaces. To every space X one assigns the *mixed inductive dimension*, $\mathcal{P}\text{-Mind}\, X$, as follows.

(i) For $n = -1$ or 0, $\mathcal{P}\text{-Mind}\, X = n$ if and only if $\mathcal{P}\text{-ind}\, X = n$.

(ij) For each positive natural number n, $\mathcal{P}\text{-Mind}\, X \le n$ if for each pair of disjoint closed sets F and G there is a partition S between F and G such that $\mathcal{P}\text{-Mind}\, S \le n - 1$.

Obviously the mixed inductive dimension modulo a class $\mathcal{P}$ is a topological invariant which coincides with Cmp when $\mathcal{P} = \mathcal{K}$. Although the mixed inductive dimension has been defined for all topological spaces, a theory for non-metrizable spaces does not seem very promising at this time. But the following proposition can be established by an easy inductive proof which will be left to the reader.

4.2. Proposition. *For each closed subspace Y of a space X,*

$$\mathcal{P}\text{-Mind}\, Y \leq \mathcal{P}\text{-Mind}\, X.$$

As we shall see in the next theorem, the proof of Theorem I.12.1 can be carried out in a more general setting. The theorem concerns the existence of certain partitions between closed sets. In Theorem IV.3.9 we have already encountered a partitioning theorem in terms of the relative dimension with respect to $\mathcal{P}$-border covers.

4.3. Theorem. *Suppose that $\mathcal{P}$ is closed-additive in the universe of separable metrizable spaces. Let X be a space with $\mathcal{P}$-Odim $X \leq n$ where $n \geq 1$. Then between any two disjoint closed sets of X there is a partition S such that $\mathcal{P}$-Odim $S \leq n-1$.*

Proof. The proof is by induction on $\mathcal{P}$-Odim X. To prove the inductive step let us suppose that X is a space with $\mathcal{P}$-Odim $X \leq n$ where $n \geq 1$. Let F and G be disjoint closed sets of X. Only the case where both F and G are nonempty need be considered. We shall construct first a partition S.

From the definition of $\mathcal{P}$-Odim $X \leq n$ there exists a countable base $\mathcal{B} = \{ U_i : i \in \mathbb{N} \}$ for the open sets of X such that the intersection $\mathrm{B}(U_{i_0}) \cap \cdots \cap \mathrm{B}(U_{i_n})$ belongs to $\mathcal{P}$ for any $n+1$ different indices $i_0, \ldots, i_n$ in $\mathbb{N}$. Consider the collection $\{ (C_k, D_k) : k \in \mathbb{N} \}$ of all pairs of elements of $\mathcal{B}$ with $\mathrm{cl}(C_k) \subset D_k$ such that $\mathrm{cl}(D_k) \cap F = \emptyset$ or $\mathrm{cl}(D_k) \cap G = \emptyset$. Observe that the indexing for this collection of pairs is unrelated to that of $\mathcal{B}$. Now for each k in $\mathbb{N}$ let

$$V_k = D_k \setminus \bigcup \{ \mathrm{cl}(C_j) : j = 0, 1, \ldots, k-1 \}.$$

The collection $\mathbf{V} = \{ V_k : k \in \mathbb{N} \}$ is a locally finite cover of X, and

$$\mathrm{B}(V_k) \subset \mathrm{B}(C_0) \cup \cdots \cup \mathrm{B}(C_{k-1}) \cup \mathrm{B}(D_k)$$

for each k in $\mathbb{N}$. Let $W = \bigcup \{ V_k : \mathrm{cl}(D_k) \cap G = \emptyset \}$. Obviously the set W is open and $F \subset W$. From local finiteness of $\mathbf{V}$ we have $\mathrm{cl}(W) \cap G = \emptyset$ and $\mathrm{B}(W) \subset \bigcup \{ \mathrm{B}(V_k) : \mathrm{cl}(D_k) \cap G = \emptyset \}$. It follows that $\mathrm{B}(W)$ is a partition between F and G.

Let us show $\mathcal{P}$-Odim $\mathrm{B}(W) \leq n-1$. As the inequality is obvious when $\mathrm{B}(W) = \emptyset$ holds, we assume $\mathrm{B}(W) \neq \emptyset$. Observe that each $\mathrm{B}(V_k)$ is the union of a finite collection of closed subsets of boundaries of elements of $\mathcal{B}$. It follows that $\mathrm{B}(W)$ is the union of a locally finite collection $\{E_j : j \in N_0\}$ where N_0 is a subset of $\mathbb{N}$ and each E_j is a nonempty closed subset of $\mathrm{B}(U_j)$. From here on we shall be returning to the original indexing of $\mathcal{B}$. Next we shall select a subcollection $\mathcal{B}'$ of $\mathcal{B}$ such that $\mathcal{B}'$ is still a base for X and has the added property that the collection $\{U \cap \mathrm{B}(W) : U \in \mathcal{B}'\}$ is a base for the open sets of $\mathrm{B}(W)$ witnessing the fact $\mathcal{P}$-Odim $\mathrm{B}(W) \leq n-1$. This will be done in two steps. The first step follows. For each j in N_0 a point p_j from E_j is selected. Let $P = \{p_j : j \in N_0\}$ and observe that P is closed. With $N_1 = \{i \in \mathbb{N} : U_i \in \mathcal{B}, \mathrm{B}(U_i) \cap P = \emptyset\}$ we form the subcollection $\mathcal{B}^* = \{U_i : i \in N_1\}$ of $\mathcal{B}$. It is easily seen that $\mathcal{B}^*$ is a base for the open sets of X and that the boundary $\mathrm{B}(U)$ of each U in $\mathcal{B}^*$ is distinct from the boundary $\mathrm{B}(U_j)$ for any j in N_0. So, in particular, $N_0 \cap N_1 = \emptyset$. For the second step let $\mathbf{G}$ be an open cover of X such that each element of $\mathbf{G}$ meets at most finitely many members of the locally finite collection $\{E_j : j \in N_0\}$. Let $\mathcal{B}' = \{U_i : i \in N_2\}$ be the set of all U_i in $\mathcal{B}^*$ such that $\mathrm{cl}(U) \subset G$ for some G in $\mathbf{G}$. Obviously $N_2 \subset N_1$ holds and $\mathcal{B}'$ is also a base for X. Let us show that the base $\{U \cap \mathrm{B}(W) : U \in \mathcal{B}'\}$ for the open sets of $\mathrm{B}(W)$ witnesses the fact $\mathcal{P}$-Odim $\mathrm{B}(W) \leq n-1$. To this end, let $i_0, \ldots, i_{n-1}$ be n distinct indices from N_2 and, with ∂ denoting the boundary operator in $\mathrm{B}(W)$, let

$$S = \partial\left(U_{i_0} \cap \mathrm{B}(W)\right) \cap \cdots \cap \partial\left(U_{i_{n-1}} \cap \mathrm{B}(W)\right).$$

Clearly S is a closed subset of $R = \mathrm{B}(U_{i_0}) \cap \cdots \cap \mathrm{B}(U_{i_{n-1}}) \cap \mathrm{B}(W)$. As each $\mathrm{B}(U_{i_j})$ is a subset of some G in $\mathbf{G}$ and each member of $\mathbf{G}$ meets E_j for at most finitely many j in N_0, it follows that the equality $R = \bigcup\{R \cap E_k : k \in N'\}$ will hold for some finite subset N' of N_0. For each j in N', the set $R \cap E_j$ is a closed subset of $\mathrm{B}(U_j)$. Now $R \cap \mathrm{B}(U_k)$ belongs to $\mathcal{P}$ because the indices $j, i_0, \ldots, i_{n-1}$ are distinct. Because $\mathcal{P}$ is closed-additive, it follows that R and, consequently, S belong to $\mathcal{P}$. We have shown $\mathcal{P}$-Odim $\mathrm{B}(W) \leq n-1$.

Remark. The set P in the above construction was introduced to avoid the indexing pitfall mentioned in Example I.6.13. In Section VI.5 we shall introduce another method to serve the same purpose.

With this result the next statement follows by an easy inductive proof.

4.4. Corollary. *Suppose that $\mathcal{P}$ is closed-additive in the universe of separable metrizable spaces. For every space X,*

$$\mathcal{P}\text{-ind}\, X = \mathcal{P}\text{-Bind}\, X \leq \mathcal{P}\text{-Mind}\, X \leq \mathcal{P}\text{-Odim}\, X.$$

Observe that the closed-additivity of $\mathcal{P}$ is only required to establish the last inequality. Obviously the class $\mathcal{K}$ is closed-additive. In addition to the class $\mathcal{K}$, the class $\mathcal{L}$ of locally compact spaces is closed-additive. Indeed, $\mathcal{L}$ is a cosmic family.

For later use we state a variation of Theorem 4.3 in which the condition $\mathcal{P}$-Odim $X \leq n$ has been replaced by a weaker condition. The proof is essentially the same.

4.5. Theorem. *Let $\mathcal{P}$ be closed-additive in the universe of separable metrizable spaces. Suppose that X is a space for which there exists a collection $\mathbf{V}$ of open sets satisfying the following conditions.*

(a) $\{V : V \in \mathbf{V}$ *or* $(X \setminus \text{cl}\,(V)) \in \mathbf{V}\}$ *is a base for the open sets.*
(b) $\text{B}\,(V_{i_0}) \cap \cdots \cap \text{B}\,(V_{i_n})$ *is in $\mathcal{P}$ for any $n+1$ different indices $i_0, \ldots, i_n$ in $\mathbb{N}$.*

Then $\mathcal{P}$-Mind $X \leq n$.

5. Historical comments and unsolved problems

Several of the results in this chapter are forerunners of other results in this book. For example, the proof of Theorem 2.13 is almost the same as the proof of the main result in Aarts [1968] which appeared early in history of the subject. Other examples come from the notion of basic inductive dimension which had been only marginally discussed in the [1973] paper by Aarts and Nishiura and many earlier versions of these results had been found by using this dimension function. The excision theorem (Theorem 2.3) was established in Aarts and Nishiura [1973] in essentially the same way as in Section 2. The first extension theorem can be found in Aarts [1968]. Results like Theorem 2.24 can be found in Aarts [1971a], Nagami [1965] and Wenner [1969], [1970]. The proofs in Section 3 are new and follow the pattern of the proofs of corresponding results in Nagata [1965].

Unsolved problems

1. This problem concerns an ind analogue of Tumarkin's theorem for Ind. Suppose that X is a subset of a metrizable space Y and that $\operatorname{ind} X \leq n$ holds.

Does there exist a G_δ-set Z of Y with $X \subset Z$ and $\operatorname{ind} Z \leq n$?

2. Does there exist an example of a strongly closed-additive cosmic family $\mathcal{P}$ for which $\mathcal{P}$-$\operatorname{Bind} X < \mathcal{P}$-$\operatorname{Ind} X$ for some space X?

3. Observe that Theorem 3.12 is not applicable to the absolute Borel class $\mathsf{M}(0)$, that is, the class $= \mathcal{K} \cap \mathcal{M}$. Also Kimura's theorem yields $\operatorname{def} X = \operatorname{Skl} X$ only for separable metrizable space X.

Is there a meaningful generalization of the theory of Skl for general metrizable spaces?

Chart 2. Compactness dimension functions

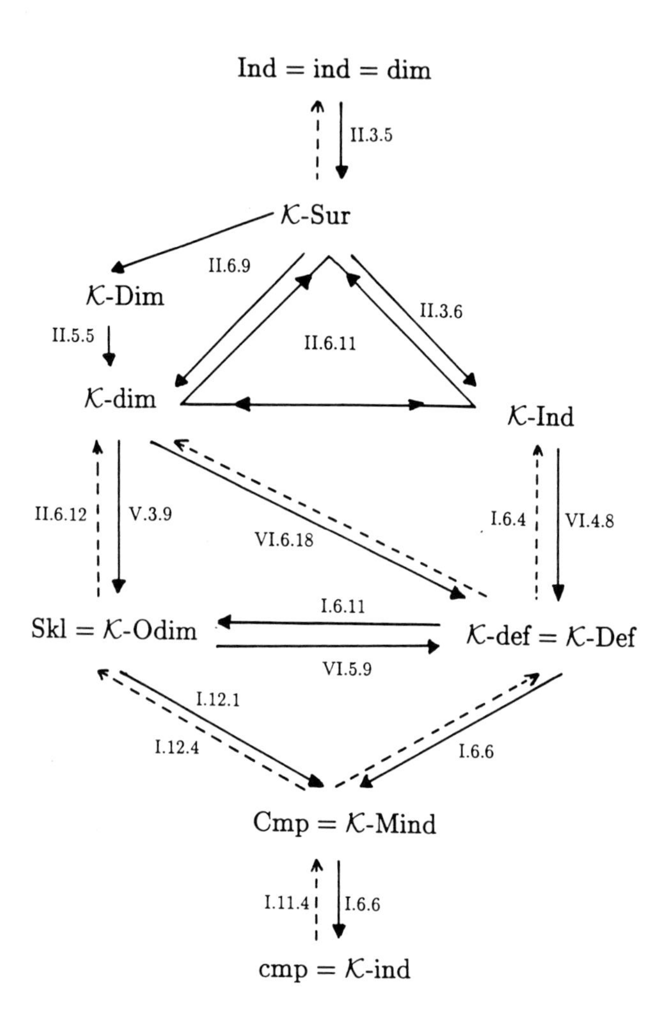

The chart summarizes the relations among the compactness dimension functions in the universe $\mathcal{M}_0$ of separable metrizable spaces. The solid arrows indicate inequalities and the dashed arrows indicate counterexamples. For example, the solid arrow from $\mathcal{K}$-Ind to $\mathcal{K}$-def indicates that $\mathcal{K}\text{-def}\, X \leq \mathcal{K}\text{-Ind}\, X$ holds for every separable metrizable space X, and the number next to the arrow shows where the proof is found. The dashed arrow from cmp to Cmp indicates that there exists a space X such that $\operatorname{cmp} X < \operatorname{Cmp} X$.

CHAPTER VI

COMPACTIFICATIONS

The sole topic that remains is the compactness deficiency of a space. Thus the central theme of this chapter will be compactifications. The chart on the opposite page gives an overview of the relationship among the various compactness dimension functions in the universe $\mathcal{M}_0$ of all separable metrizable spaces. By now, most of the arrows have been validated and all counterexamples have been presented. The following arrows are the ones that still require validation.

(A) Skl $\longrightarrow$ $\mathcal{K}$-def, that is, $\mathcal{K}\text{-def}\, X \leq \text{Skl}\, X$ for every separable metrizable space X.
(B) $\mathcal{K}$-dim $\longrightarrow$ $\mathcal{K}$-def.
(C) $\mathcal{K}$-Ind $\longrightarrow$ $\mathcal{K}$-Def.

Each of these arrows requires a proof of the existence of a compactification.

It is to be noted that the equalities def $=$ $\mathcal{K}$-def $=$ $\mathcal{K}$-Def hold in the universe $\mathcal{M}_0$. So statement (A), which will be proved in Section 5, is the essential remaining part of Kimura's theorem. Observe that (B) and (C) will follow from (A) because Corollary V.3.9 and the two equalities Skl $=$ $\mathcal{K}$-def and $\mathcal{K}$-dim $=$ $\mathcal{K}$-Ind are valid for separable metrizable spaces. Nonetheless, the statements (B) and (C), or proper modifications of them, can be proved in a universe that is much larger than the universe $\mathcal{M}_0$ of separable metrizable spaces. This will be done for (B) in Section 6 and for (C) in Section 4.

The proofs will require rather delicate constructions. In advance of the presentation of these proofs, a discussion of some of the simpler results in compactification theory will be given. The first section will deal with the Wallman compactification. This compactification method is especially well suited for relating various covering properties, including that of dimension and homology, of a space and its compactification. In the second section a modification of the Wallman compactification and several theorems about dimension and weight preserving compactifications will be discussed. This modification will be used in the next section for a presentation of the Freudenthal compactification. Included in this presentation is de Groot's Theorem I.5.3 which is central to this book.

Agreement. *The universe is the class of T_1-spaces.*

1. Wallman compactifications

A self-contained development of lattice representations of topological spaces will be given. The theory of the representation of topological spaces by lattices was introduced by Wallman in [1938]. His approach was similar to that of the representation of zero-dimensional spaces by Boolean rings as developed by Stone in [1937]. The main result of our development is the construction of the Wallman compactification ωX for each space X. It will be shown that for every normal space X the dimensions, both Ind and dim, of X and ωX are the same.

We shall begin with a discussion of some simple notions in lattice theory. The notation and terminology of Birkhoff [1973] will be used.

1.1. Definition. A *lattice* is a nonempty set L with a reflexive partial ordering $\leq$ such that for each pair (x, y) of elements of L there is a unique smallest element $x \vee y$, called the *join* of x and y, satisfying $x \leq x \vee y$ and $y \leq x \vee y$ and there is a unique largest element $x \wedge y$, called the *meet* of x and y, satisfying $x \wedge y \leq x$ and $x \wedge y \leq y$.

A lattice L is *distributive* provided $x \vee (y \wedge z) = (x \vee y) \wedge (x \vee z)$ and $x \wedge (y \vee z) = (x \wedge y) \vee (x \wedge z)$ hold for all x, y and z.

The collection of all subsets of a given set X ordered by inclusion is an example of a lattice, the join and meet are the union and intersection respectively. The same holds true for any collection $\mathcal{F}$ of subsets of X that is closed under the formation of finite unions and finite intersections. It might be useful to keep these examples in mind in what follows.

Various properties can be deduced easily from the definition. We shall mention just a few. The easy proofs will be left to the reader.

1.2. Proposition. *For all x, y and z of L,*

(a) $x \wedge x = x, \qquad x \vee x = x,$
(b) $x \wedge y = y \wedge x, \qquad x \vee y = y \vee x,$
(c) $(x \wedge y) \wedge z = x \wedge (y \wedge z), \qquad (x \vee y) \vee z = x \vee (y \vee z),$
(d) $x \wedge (x \vee y) = x \vee (x \wedge y) = x,$
(e) $x \leq y$ *if and only if* $x \wedge y = x$,
(f) $x \leq y$ *if and only if* $x \vee y = y$.

1.3. Definition. The lattices L and L' are said to be *isomorphic* if there exists a bijection $\vartheta: L \to L'$ such that for all x and y in L

(i) $\vartheta(x \wedge y) = \vartheta(x) \wedge \vartheta(y)$,
(ij) $\vartheta(x \vee y) = \vartheta(x) \vee \vartheta(y)$.

Such a bijection is called an *isomorphism.* A function h from L to L' that satisfies conditions (i) and (ij) is called a *homomorphism.*

In view of property (e) above, for an isomorphism ϑ we have $x \leq y$ if and only if $\vartheta(x) \leq \vartheta(y)$.

Our first goal is to construct a topological space out of a lattice. The points of the space will be generated by the maximal dual ideals which we shall now define.

1.4. Definition. A nonempty subset D of a lattice L is called a *dual ideal* if the following hold for all x and y:

(i) If $x \in \mathsf{D}$ and $y \in \mathsf{D}$, then $x \wedge y \in \mathsf{D}$.
(ij) If $x \in \mathsf{D}$ and $x \leq y$, then $y \in \mathsf{D}$.

A dual ideal D in a lattice L is said to be *maximal* if $\mathsf{D} \neq L$ and either $\mathsf{D} = \mathsf{E}$ or $\mathsf{E} = L$ for each dual ideal E with $\mathsf{D} \subset \mathsf{E}$.

Concerning the conditions (i) and (ij), one can readily show that each set D_0 satisfying the condition (i) is contained in the dual ideal $\mathsf{D} = \{\, y : x \leq y$ for some x in $\mathsf{D}_0 \,\}$. As a special case, for each a in L the set $\{\, y : y \in L,\ a \leq y \,\}$ is a dual ideal. We shall see that dual ideals in the lattice L correspond to closed filters in the space generated by L. Let us collect some properties of dual ideals.

If a lattice has a smallest element (that is, an element a for which $a \leq x$ for every x in L), this element must be unique. When it exists, the smallest element will be denoted by 0. Dually, when it exists, the largest element of a lattice is unique and is denoted by 1. For a dual ideal E in a lattice L with 0, it is easily seen that $\mathsf{E} = L$ if and only if $0 \in \mathsf{E}$.

1.5. Lemma. *Let L be a lattice with 0. Every dual ideal D in L with $\mathsf{D} \neq L$ is contained in a maximal dual ideal.*

Proof. Let $\boldsymbol{E}$ be the collection of all dual ideals E with the property that $\mathsf{D} \subset \mathsf{E}$ and $\mathsf{E} \neq L$. The collection $\boldsymbol{E}$ is partially ordered by inclusion. Each chain in $\boldsymbol{E}$ has an upper bound, namely the union of the chain. By Zorn's lemma, $\boldsymbol{E}$ has a maximal element which is the maximal dual ideal containing D.

There is the following criterion for maximality of dual ideals.

1.6. Lemma. *Let L be a lattice with 0. A dual ideal D of L is a maximal dual ideal if and only if the condition*

$(*)$ $\quad x \in \mathsf{D}$ *if and only if* $z \wedge x \neq 0$ *for every* z *in* D

holds for each x in L.

Proof. To prove sufficiency suppose that D is a dual ideal for which $(*)$ holds. Clearly, from $(*)$ we have $\mathsf{D} \neq L$. Let E be a dual ideal such that $\mathsf{D} \subset \mathsf{E}$ and $0 \notin \mathsf{E}$ and let $x \in \mathsf{E}$. As $0 \notin \mathsf{E}$, we have $x \wedge z \neq 0$ for each z in D. By $(*)$ we have $x \in \mathsf{D}$ thereby showing $\mathsf{D} = \mathsf{E}$. That is, D is maximal.

For the proof of the necessity, let us assume for some x in L that $x \wedge z \neq 0$ for every z in D. Define the subset E of L by

$$\mathsf{E} = \{\, y : x \wedge z \leq y \text{ for some } z \text{ in } D \,\}.$$

Then E is a dual ideal such that $\mathsf{D} \subset \mathsf{E}$, $x \in \mathsf{E}$ and $0 \notin \mathsf{E}$. As D is maximal, we have $\mathsf{D} = \mathsf{E}$ and thus $x \in \mathsf{D}$. The rest of the proof is obvious.

The condition in the next lemma is the lattice counterpart of a well-known property of ultrafilters.

1.7. Lemma. *Suppose that D is a maximal dual ideal in a distributive lattice L with 0. Then the condition*

$$(x \vee y) \in \mathsf{D} \quad \textit{if and only if} \quad x \in \mathsf{D} \textit{ or } y \in \mathsf{D}$$

holds for every x and y in L.

Proof. The sufficiency part of the condition is obvious. For the necessity suppose that $x \vee y \in \mathsf{D}$ and $x \notin \mathsf{D}$. Then by Lemma 1.6 there is a z in D such that $z \wedge x = 0$. Now from $y \geq z \wedge y = z \wedge (x \vee y)$ and $z \wedge (x \vee y) \in \mathsf{D}$ we have $y \in \mathsf{D}$.

There is yet another property of lattices that will play a role in the representation of lattices by topological spaces.

1.8. Definition. Suppose that L is a lattice with 0. Then L is said to have the *disjunction property* if for all x and y in L with $x \neq y$ there exists a z in L with the property that one of the meets $x \wedge z$ and $y \wedge z$ is 0 and the other is not.

Having collected the relevant parts of lattice theory, we now can discuss the representation of lattices by spaces. In the preceding chapters we have always used bases $\mathcal{B}$ for the open sets of a space X. For the purposes of the theory of Wallman representation and compactification it is much more convenient to work with bases for the closed sets of a space.

1.9. Definition. A collection $\mathcal{F}$ of closed sets of a topological space X is called a *base for the closed sets* of X if each closed subset of X is the intersection of some subcollection of $\mathcal{F}$.

Note that a collection $\mathcal{F}$ of closed sets of a topological space X will be a base for the closed sets of X if $X \in \mathcal{F}$ and if for each closed subset G of X and for each point p of X with $p \notin G$ there is an F in $\mathcal{F}$ such that $p \notin F$ and $G \subset F$. Also, observe that a collection $\mathcal{F}$ of subsets of a set X will be a base for the closed sets of a uniquely determined topology for the set X if $X \in \mathcal{F}$ and $\mathcal{F}$ has the property that for all F and G in $\mathcal{F}$ the set $F \cup G$ is the intersection of some subcollection of $\mathcal{F}$. In particular, $\mathcal{F}$ will be a base when $\mathcal{F}$ is a lattice of subsets of X and $X \in \mathcal{F}$.

We can now formulate the representation theorem.

1.10. Theorem. *Suppose that L is a distributive lattice with* 0 *and* 1, *and suppose that L has the disjunction property. Then there exists a compact T_1-space ωL and there exists a base $\mathcal{F}$ for the closed sets of ωL such that $\mathcal{F}$ is a lattice that is isomorphic with L.*

The space ωL is called the *Wallman representation of* L. Of course, the proof will consist of two parts, where the first consists of the constructions of the pair ωL and $\mathcal{F}$ and the isomorphism $\vartheta\colon L \to \mathcal{F}$ and the second consists of verifying that $\mathcal{F}$ generates a compact T_1-space topology on ωL. Since the need to refer to the construction will occur often, this part of the proof will be set off from the other.

1.11. Construction of the Wallman representation. Consider the collection of all maximal dual ideals of L. Let this collection be indexed in a bijective fashion and denote the indexing set by ωL. Then the collection of maximal dual ideals can be written as $\{ \mathsf{D}_x : x \in \omega L \}$. For each t in L define the subset B_t of ωL to be

$$B_t = \{ x : t \in \mathsf{D}_x \},$$

and define the collection $\mathcal{F}$ to be

$$\mathcal{F} = \{ B_t : t \in L \}.$$

It will be shown that the collection $\mathcal{F}$ possesses the properties

(1) if u and v are distinct members of L, then $B_u \neq B_v$,

(2) $B_u \cup B_v = B_{u \vee v}$ and $B_u \cap B_v = B_{u \wedge v}$, $u, v \in L$.

That is, $\mathcal{F}$ is a lattice. Observe that the natural map $\vartheta \colon L \to \mathcal{F}$ defined by $\vartheta(t) = B_t$ is an isomorphism.

Let us verify property (1). Suppose that u and v are distinct members of L. Because L has the disjunction property, there exists a z in L such that one of $z \wedge u$ and $z \wedge v$ is 0 and the other is not. We may assume that $z \wedge u = 0$ and $z \wedge v \neq 0$. By Lemma 1.5 there is a maximal dual ideal D_y such that $(z \wedge v) \in \mathsf{D}_y$. It follows that $v \in \mathsf{D}_y$ and $u \notin \mathsf{D}_y$, whence $y \in B_v \setminus B_u$. Thereby property (1) is verified. Turn now to property (2). We shall prove only the first formula since the second formula is proved in a similar way. Clearly $B_u \cup B_v \subset B_{u \vee v}$ holds. So, let $x \in B_{u \vee v}$. By definition $(u \vee v) \in \mathsf{D}_x$ holds, whence $u \in \mathsf{D}_x$ or $v \in \mathsf{D}_x$ must hold by Lemma 1.7. Thereby property (2) is verified.

Proof of Theorem 1.10. Because $\mathcal{F}$ is a lattice, it will generate the closed sets of a topology on the set ωL. We shall prove that ωL with this topology is compact. Let $\mathbf{G}$ be a collection of closed subsets of ωL with the finite intersection property. We need to show $\bigcap \mathbf{G} \neq \emptyset$. We may assume that $\mathbf{G}$ is closed under finite intersections. Define the dual ideal H of L by

$$\mathsf{H} = \{ t \in L : \text{there exists a } G \text{ in } \mathbf{G} \text{ with } B_t \supset G \}.$$

Because $\{ B_t : t \in L \}$ is a base for the closed sets, we have

$$\bigcap \{ B_t : t \in \mathsf{H} \} = \bigcap \mathbf{G}.$$

Now choose a maximal dual ideal D_y such that $\mathsf{H} \subset \mathsf{D}_y$. Then for every t in H we have $t \in \mathsf{D}_y$. So $y \in B_t$ for all t in H and hence $y \in \bigcap\{ B_t : t \in \mathsf{H} \} = \bigcap \boldsymbol{G}$.

Next suppose that x and y are distinct points of ωL. Then D_x and D_y are distinct and, moreover, neither D_x nor D_y contains the other. By Lemma 1.6 there are s and t from L such that $s \in \mathsf{D}_x$, $t \in \mathsf{D}_y$ and $t \wedge s = 0$. It follows that $x \in B_s$, $y \in B_t$ and $B_s \cap B_t = \emptyset$, that is, ωL is a T_1-space.

The following theorem is very important because it provides a lattice criterion for the Wallman representation to be Hausdorff.

1.12. Theorem. *The Wallman representation ωL of a distributive lattice L with 0 and 1 is a Hausdorff space if and only if $s \wedge t = 0$ implies the existence of u and v in L such that the condition*

$$(*) \qquad s \wedge v = 0, \quad u \wedge t = 0 \quad \text{and} \quad u \vee v = 1$$

holds.

It should be noted that $s \leq u$ follows from the condition $(*)$ because $s = s \wedge 1 = s \wedge (u \vee v) = (s \wedge u) \vee (s \wedge v) = s \wedge u$. By symmetry, $t \leq v$ also holds.

Proof. To prove the sufficiency of the condition let x and y be distinct points of ωL. In view of Lemma 1.6 there are s and t in L such that $x \in B_s$, $y \in B_t$ and $s \wedge t = 0$. With u and v as in $(*)$, from the isomorphism between L and $\{ B_t : t \in L \}$ we get

$$B_t \cap B_u = \emptyset, \quad B_s \cap B_v = \emptyset \quad \text{and} \quad B_u \cup B_v = \omega L.$$

It follows that $\omega L \setminus B_v$ and $\omega L \setminus B_u$ are disjoint neighborhoods of x and y respectively.

To prove the necessity we assume that ωL is a Hausdorff space. From the compactness of ωL it follows that ωL is normal. Suppose that s and t satisfy $s \wedge t = 0$. Then B_s and B_t are disjoint closed subsets of ωL. Let V and U be disjoint open neighborhoods of B_s and B_t respectively. Consider the collection $\mathsf{H} = \{ B_r : \omega L \setminus U \subset B_r \}$. By way of contradiction we shall show that there is a B_u in H such that $B_u \cap B_t = \emptyset$. If $B_r \cap B_t \neq \emptyset$ for every B_r in H, then by the compactness of ωL the set $B_t \cap \bigcap \mathsf{H}$ is not empty. Let x be a point in this set. As $\{ B_r : r \in L \}$ is a base for the closed sets, there is a B_q in H such that $x \notin B_q$. This is a contraction. Similarly, there is a v in L such that $\omega L \setminus V \subset B_v$ and $B_v \cap B_s = \emptyset$. Obviously, $B_u \cup B_v = \omega L$. As the lattice L and $\{ B_t : t \in L \}$ are isomorphic, the condition $(*)$ holds.

Let us now turn our attention to a topological space X. There is always a natural distributive lattice with 0 and 1 associated with its topology, namely the lattice of all closed subsets of X. As X is a T_1-space, this lattice has the disjunction property because singleton sets are closed. Often we shall be dealing with sublattices L of the lattice of all closed subsets of X. Under such a setting the collection $\mathsf{D}_x = \{ F : F \in L,\, x \in F \}$ is a dual ideal but it may not be a maximal dual ideal of L. For the moment let us assume that D_x is a maximal dual ideal and that $\bigcap \{ F \in L : x \in F \} = \{ x \}$ for each x in X. Then there is a natural injective map

$$(3) \qquad \varphi\colon X \to \omega L \quad \text{such that} \quad \mathsf{D}_{\varphi(x)} = \{ F \in L : x \in F \}.$$

Obviously the existence of φ will be established by the equality $\bigcap \{ F \in L : x \in F \} = \{ x \}$. This assumption is valid when L is the lattice of all closed subsets of a space X.

We shall define next the Wallman compactification of X.

1.13. Definition. Let X be a space and let L be the lattice of all closed subsets of X. The Wallman representation ωL of L is called the *Wallman compactification* of X. The Wallman compactification of X will also be denoted by ωX.

That ωX is a compactification of X will follow from the next theorem whose proof relies only on the theory that has been developed thus far.

1.14. Theorem. *Let X be a space. Then the Wallman representation ωX of the lattice $L = \{ F : F$ is a closed subset of $X \}$ is a compactification of X. Also, there is a canonical embedding φ of X into ωX with the following properties.*

(a) $\mathsf{D}_{\varphi(x)}$ *is the maximal dual ideal* $\{ F : F \in L,\, x \in F \}$.

(b) *The collection* $\mathcal{F}_L = \{ \operatorname{cl}_{\omega X}(\varphi[F]) : F \in L \}$ *is a base for the closed sets of* ωX.

(c) *The function* ϑ *defined by* $\vartheta(F) = \operatorname{cl}_{\omega X}(\varphi[F])$, $F \in L$, *is a bijection from* L *to* $\mathcal{F}_L$.

(d) *When F and G are in L,*

$$\operatorname{cl}_{\omega X}(\varphi[F]) \cap \operatorname{cl}_{\omega X}(\varphi[G]) = \operatorname{cl}_{\omega X}(\varphi[F \cap G]),$$
$$\operatorname{cl}_{\omega X}(\varphi[F]) \cup \operatorname{cl}_{\omega X}(\varphi[G]) = \operatorname{cl}_{\omega X}(\varphi[F \cup G]).$$

Consequently ϑ is a lattice isomorphism.

Moreover, ωX is the Čech-Stone compactification of X whenever X is normal.

The bijective function ϑ will be called the canonical isomorphism.

Proof. We shall employ the construction of the Wallman representation found in Section 1.11 above. Because $\{ B_G : G \in L \}$ is a base for the closed sets of ωX, we have for all F in L

$$\operatorname{cl}_{\omega X}(\varphi[F]) = \bigcap\{ B_G : B_G \supset \varphi[F] \} = \bigcap\{ B_G : G \supset F \} = B_F.$$

The properties (a)–(d) are now obvious.

For the proof of the final statement of the theorem, we begin by noting that the lattice L satisfies the condition $(*)$ of Theorem 1.12 when X is normal. Hence ωX is Hausdorff. Recall that a *zero-set* of X is a set of the form $\{ x : f(x) = 0 \}$ for some continuous real-valued function f. As disjoint zero-sets, like any disjoint closed subsets of X, have disjoint closures in ωX, from the characterization given by Čech in [1937] it follows that ωX coincides with the maximal Hausdorff compactification of X (see also Gillman and Jerison [1960], in particular Section 6.5).

A nice feature of the Wallman compactification is that it carries over to closed subspaces.

1.15. Theorem. *Suppose that Z is a closed subset of X and let $\varphi\colon X \to \omega X$ be the canonical embedding of X onto the Wallman compactification ωX of X. Then there is an embedding ψ of ωZ into ωX that is an extension of the inclusion map $i\colon Z \to X$. And the image $\psi[\omega Z]$ coincides with $\operatorname{cl}_{\omega X}(\varphi[Z])$.*

Proof. We shall use the notation of the preceding proof. There L denoted the lattice of all closed subsets of X. The canonical embedding of Z into its Wallman compactification ωZ will be denoted by φ_1. Let D be a maximal dual ideal of L_Z, the lattice of all closed subsets of Z. Then define $\mathsf{E} = \{ G \in L : G \cap Z \in \mathsf{D} \}$. It is easily seen that E is maximal dual ideal of L. Thus for a unique ζ in ωZ and a unique ξ in ωX we have $\mathsf{D} = D_\zeta$ and $\mathsf{E} = \mathsf{E}_\xi$. The required map ψ is defined by $\xi = \psi(\zeta)$. We shall show that the following diagram is commutative, where $i \colon Z \to X$ is the inclusion map.

$$(4) \qquad \begin{array}{ccc} Z & \xrightarrow{\varphi_1} & \omega Z \\ {\scriptstyle i}\downarrow & & \downarrow{\scriptstyle \psi} \\ X & \xrightarrow{\varphi} & \omega X \end{array}$$

Let z be a point in Z. Then we have $\mathsf{D}_{\varphi_1(z)} = \{ F \in L_Z : z \in F \}$ and $\mathsf{E}_{\psi(\varphi_1(z))} = \{ G \in L : i(z) \in G \} = \mathsf{E}_{\varphi(i(z))}$, which shows commutativity. Let us show that ψ is a bijection between $\mathrm{cl}_{\omega X}(\varphi_1[Z])$ and $\mathrm{cl}_{\omega X}(\varphi[\, i[Z]\,])$. This will follow from the fact that Z is a member of a maximal dual ideal E of L if and only if E is $\{ G \in L : G \cap Z \in \mathsf{D} \}$ for some maximal dual ideal D of L_Z. In particular, we find that the image $\psi[\omega Z]$ coincides with $\mathrm{cl}_{\omega X}(\varphi[\, i[Z]\,])$.

Remark. With very few exceptions, we shall be identifying the space X via the canonical embedding φ with its image $\varphi[X]$ thus making X a subset of ωX. This will occur for the first time in the next lemma.

A virtue of the Wallman compactification is the existence of a simple relation between dimension (and homology) of a space and the dimension (and homology) of its Wallman compactification. The closure, expansion and boundary operators behave rather nicely in the Wallman compactification. (See Definition V.2.9 for the definition of the expansion operator.)

1.16. Lemma. *If A is a subset of X and if U and V are open subsets of X, then the following are true.*

(a) $\mathrm{cl}_{\omega X}(A) = \mathrm{cl}_{\omega X}(\mathrm{cl}_X(A))$.
(b) $\mathrm{ex}_{\omega X}(U \cup V) = \mathrm{ex}_{\omega X}(U) \cup \mathrm{ex}_{\omega X}(V)$.
(c) $\mathrm{B}_{\omega X}(\mathrm{ex}_{\omega X}(U)) = \mathrm{cl}_{\omega X}(\mathrm{B}_X(U))$.

Proof. The equality (a) is obvious. To prove (b) let $F = X \setminus U$ and $G = X \setminus V$. Then

$$\mathrm{ex}_{\omega X}(U \cup V) = \omega X \setminus \mathrm{cl}_{\omega X}(X \setminus (U \cup V))$$

$$
\begin{aligned}
&= \omega X \setminus \mathrm{cl}_{\omega X}(F \cap G) \\
&= \omega X \setminus \big(\mathrm{cl}_{\omega X}(F) \cap \mathrm{cl}_{\omega X}(G)\big) \\
&= \mathrm{ex}_{\omega X}(U) \cup \mathrm{ex}_{\omega X}(V).
\end{aligned}
$$

For (c) consider the closed sets $X \setminus U$ and $\mathrm{cl}_X(U)$ of X. As X is dense in ωX, we have $U \subset \mathrm{ex}_{\omega X}(U) \subset \mathrm{cl}_{\omega X}(\mathrm{cl}_X(U))$. From (a) we infer $\mathrm{cl}_{\omega X}(\mathrm{ex}_{\omega X}(U)) = \mathrm{cl}_{\omega X}(\mathrm{cl}_X(U))$. Now we have

$$
\begin{aligned}
\mathrm{cl}_{\omega X}(\mathrm{B}_X(U)) &= \mathrm{cl}_{\omega X}((X \setminus U) \cap \mathrm{cl}_X(U)) \\
&= \mathrm{cl}_{\omega X}((X \setminus U)) \cap \mathrm{cl}_{\omega X}(\mathrm{cl}_X(U)) \\
&= \big(\omega X \setminus \mathrm{ex}_{\omega X}(U)\big) \cap \mathrm{cl}_{\omega X}(\mathrm{ex}_{\omega X}(U)) \\
&= \mathrm{B}_{\omega X}(\mathrm{ex}_{\omega X}(U)).
\end{aligned}
$$

The maximality of the Wallman compactification, which follows from the next theorem, is the most important property of these compactifications.

1.17. Theorem. *Suppose that $f\colon X \to K$ is a continuous map from a space X into a compact Hausdorff space K. Then f can be extended to a continuous map $\widetilde{f}\colon \omega X \to K$.*

The proof of this theorem, which is left to the reader, is essentially the same as that of Theorem 2.5 below since $K = \omega K$ holds for every compact space K.

To conclude this section we shall show that the Wallman compactification of a normal space X has the same large inductive dimension and the same covering dimension as X.

1.18. Theorem. *For every normal space X,*

$$
\operatorname{Ind} X = \operatorname{Ind} \omega X \quad \textit{and} \quad \dim X = \dim \omega X.
$$

Proof. To establish the equalities we shall prove four inequalities by induction. As $\omega\emptyset$ is the empty space $\emptyset$, the initial step of the induction is trivial. Only the inductive steps remains to be shown.

$\operatorname{Ind} X \le \operatorname{Ind} \omega X$: Assume $\operatorname{Ind} \omega X \le n$ and let F and G be disjoint closed subsets of X. By Theorem 1.14, $\mathrm{cl}_{\omega X}(F)$ and $\mathrm{cl}_{\omega X}(G)$ are disjoint. There is a partition S between $\mathrm{cl}_{\omega X}(F)$ and $\mathrm{cl}_{\omega X}(G)$

in ωX with $\operatorname{Ind} S \leq n-1$. We have $\omega(S \cap X) = \operatorname{cl}_{\omega X}(S \cap X) \subset S$ by Theorem 1.15. In view of Proposition I.4.3, $\operatorname{Ind} \omega(S \cap X) \leq n-1$, whence $\operatorname{Ind}(S \cap X) \leq n-1$ by the induction hypothesis. Thus we have $\operatorname{Ind} X \leq n$.

$\operatorname{Ind} X \geq \operatorname{Ind} \omega X$: Assume $\operatorname{Ind} X \leq n$ and let F and G be disjoint closed sets of ωX. Let U_1 and V_1 be respective open neighborhoods of F and G with disjoint closures. We infer from the definition of Ind the existence of a partition S between $X \cap U_1$ and $X \cap V_1$ in X with $\operatorname{Ind} S \leq n-1$. Select disjoint open sets U and V of X such that $X \setminus S = U \cup V$, $X \cap U_1 \subset U$ and $X \cap V_1 \subset V$. By Lemma V.2.10 we have $F \subset U_1 \subset \operatorname{ex}_{\omega X}(U)$, $G \subset V_1 \subset \operatorname{ex}_{\omega X}(V)$ and $\operatorname{ex}_{\omega X}(U) \cap \operatorname{ex}_{\omega X}(V) = \emptyset$. And from Lemma 1.16 and the definition of $\operatorname{ex}_{\omega X}$ we have $\omega X \setminus \operatorname{cl}_{\omega X}(S) = \operatorname{ex}_{\omega X}(U) \cup \operatorname{ex}_{\omega X}(V)$. Consequently, $\operatorname{cl}_{\omega X}(S)$ is a partition between F and G in ωX. By the induction hypothesis, $\operatorname{Ind} \omega S \leq n-1$. Hence $\operatorname{Ind} \operatorname{cl}_{\omega X}(S) \leq n-1$ by Theorem 1.15. Thereby, $\operatorname{Ind} \omega X \leq n$.

$\dim X \leq \dim \omega X$: Assume $\dim \omega X \leq n$ and let $\boldsymbol{U}$ be a finite open cover of X. By Lemma 1.16, $\{ \operatorname{ex}_{\omega X}(U) : U \in \boldsymbol{U} \}$ is a finite open cover of ωX. This cover has an open refinement $\boldsymbol{V}$ such that $\operatorname{ord} \boldsymbol{V} \leq n+1$. The trace of $\boldsymbol{V}$ on X is an open refinement of $\boldsymbol{U}$ of order less than or equal to $n+1$. It follows that $\dim X \leq n$.

$\dim X \geq \dim \omega X$: Assume $\dim X \leq n$ and consider a finite open cover $\boldsymbol{U} = \{ U_1, \ldots, U_k \}$ of ωX. As ωX is compact, we may assume that each U_i is regularly open in ωX. Then we infer from Lemma V.2.10 that $U_i = \operatorname{ex}_{\omega X}(U_i)$ holds for each i. There is an open shrinking $\boldsymbol{W}' = \{ W_1, \ldots, W_k \}$ of the cover $\{ U_1 \cap X, \ldots, U_k \cap X \}$ of X with $\operatorname{ord} \boldsymbol{W}' \leq n+1$. By Lemmas 1.15 and V.2.10 we have that $\boldsymbol{W} = \{ \operatorname{ex}_{\omega X}(W_1), \ldots, \operatorname{ex}_{\omega X}(W_k) \}$ is an open cover of ωX such that $\operatorname{ex}_{\omega X}(W_i) \subset U_i$ and $\operatorname{ord} \boldsymbol{W} \leq n+1$. So, $\dim \omega X \leq n$.

2. Dimension preserving compactifications

It was indicated by Frink in [1964] that Wallman's theory has a wide range of applications in the theory of compactifications. The short discussion of the Wallman compactification given in the first section illustrated the lattice approach of Wallman. The Wallman-type (or more simply, Wallman) compactifications will be the topic of this section. This method of compactification will prove to be very useful for constructing special compactifications. Indeed, many of the compactifications presented in this chapter will be constructed by means of Wallman compactifications. The section will include theorems on dimension and weight preserving compactifications and on extensions of mappings to such compactifications.

Basic to the theory of Wallman compactifications is the replacement of the lattice of all closed sets by lattices formed from suitably chosen bases for the closed sets of a space. Following Frink, we shall introduce the notion of a normal base.

2.1. Definition. A base $\mathcal{F}$ for the closed sets of a space X is called a *normal base* if the following conditions are satisfied.

(i) $\mathcal{F}$ is a *ring*: $\mathcal{F}$ is closed under the formation of finite unions and finite intersections.

(ij) $\mathcal{F}$ is *disjunctive*: If G is a closed set and x is a point not in G, then there is an F in $\mathcal{F}$ such that $x \in F$ and $F \cap G = \emptyset$.

(iij) $\mathcal{F}$ is *base-normal*: If J and K are disjoint members of $\mathcal{F}$, then there exist G and H in $\mathcal{F}$ such that

$$J \cap H = \emptyset, \quad G \cap K = \emptyset \quad \text{and} \quad G \cup H = X.$$

The pair (G, H) is called a *screening* of (J, K).

A space X is called *base-normal* if it has a normal base for the closed sets.

The name "base normality" reflects the requirement that disjoint members of $\mathcal{F}$ are contained in disjoint open sets whose complements are in $\mathcal{F}$. As we shall see shortly, the base-normal spaces are precisely the completely regular spaces. Consequently base normality can be regarded as a separation property that naturally fits in between regularity and normality. In passing, we remark that Frink used in [1964] the name "semi-normal" for our "base-normal".

2.2. Theorem. *Suppose that X is a base-normal space with a normal base $\mathcal{F}$. Then the collection $\mathcal{F}$ with the partial order given by inclusion is a lattice with 0 and 1 that is distributive and has the disjunction property. The Wallman representation $\omega(\mathcal{F}, X)$ of the lattice $\mathcal{F}$ is a Hausdorff compactification of X. There is a canonical embedding φ of X into $\omega(\mathcal{F}, X)$ with the following properties.*

(a) $\mathsf{D}_{\varphi(x)}$ *is the maximal dual ideal* $\{ F : F \in \mathcal{F}, x \in F \}$.

(b) *The collection* $\{ \mathrm{cl}_{\omega(\mathcal{F},X)}(\varphi[F]) : F \in \mathcal{F} \}$ *is a base for the closed sets of* $\omega(\mathcal{F}, X)$.

(c) *The function* ϑ *defined by* $\vartheta(F) = \mathrm{cl}_{\omega(\mathcal{F},X)}(\varphi[F])$, $F \in \mathcal{F}$, *is a bijection from* $\mathcal{F}$ *to* $\{\, \mathrm{cl}_{\omega(\mathcal{F},X)}(\varphi[F]) : F \in \mathcal{F}\,\}$.

(d) *When* F *and* G *are in* $\mathcal{F}$,

$$\mathrm{cl}_{\omega(\mathcal{F},X)}(\varphi[F]) \cap \mathrm{cl}_{\omega(\mathcal{F},X)}(\varphi[G]) = \mathrm{cl}_{\omega(\mathcal{F},X)}(\varphi[F \cap G]),$$
$$\mathrm{cl}_{\omega(\mathcal{F},X)}(\varphi[F]) \cup \mathrm{cl}_{\omega(\mathcal{F},X)}(\varphi[G]) = \mathrm{cl}_{\omega(\mathcal{F},X)}(\varphi[F \cup G]).$$

The compactification $\omega(\mathcal{F}, X)$ in the above theorem is called the *Wallman compactification of* X *with respect to the base* $\mathcal{F}$.

Remark. Just as we have remarked after Theorem 1.15, the space X will usually be considered to be a subset of $\omega(\mathcal{F}, X)$ via its identification with $\varphi[X]$ that is provided by the canonical embedding φ. Observe that the weight of $\omega(\mathcal{F}, X)$ has the cardinality of $\mathcal{F}$ as an upper bound when X is an infinite set.

The following corollary will provide a characterization of complete regularity by means of separation properties of closed sets. Its proof is obvious.

2.3. Corollary. *A space* X *is completely regular if and only if it is base-normal.*

Proof of Theorem 2.2. The construction of the Wallman representation discussed in the last section will be used. Let $L = \mathcal{F}$. Obviously $\emptyset$ and X are in $\mathcal{F}$ by the ring property of the normal base $\mathcal{F}$. Hence L is a distributive lattice with 0 and 1. To show that L has the disjunction property we let F and G be distinct members of $\mathcal{F}$ such that $F \subset G$ and then choose an x in $G \setminus F$. From the disjunctive property of the normal base $\mathcal{F}$ there is an H in $\mathcal{F}$ such that $x \in H$ and $H \cap F = \emptyset$. Obviously, $H \cap G \neq \emptyset$. It follows that the lattice L has the disjunction property. Consequently the representation space $\omega(\mathcal{F}, X)$ exists. The remainder of the proof is the same as that of Theorem 1.14 as only the lattice properties of L were used there. Because $\mathcal{F}$ is base-normal, the lattice $\mathcal{F}$ has property $(*)$ of Theorem 1.12. Consequently the space $\omega(\mathcal{F}, X)$ is Hausdorff.

2.4. Examples. It is not difficult to see that the collection $\mathcal{Z}$ of all zero-sets of a completely regular space X is a normal base. The compactification $\omega(\mathcal{Z}, X)$ coincides with the Čech-Stone compactification βX. This will follow from the argument found in the proof of Theorem 1.14.

It is also quite easy to show that the one-point compactification αX of a locally compact Hausdorff space X is a Wallman compactification for an appropriate normal base.

Another exploitation of the lattice structure of $\mathcal{F}$ is the following theorem on the extension of continuous mappings. This theorem will be used many times.

2.5. Theorem. *Suppose that X and Y are base-normal spaces with respective normal bases $\mathcal{F}$ and $\mathcal{G}$. Suppose that $f\colon X \to Y$ is a continuous mapping such that $f^{-1}[G] \in \mathcal{F}$ holds for each G in $\mathcal{G}$. Then f can be extended to a continuous map $\widetilde{f}\colon \omega(\mathcal{F},X) \to \omega(\mathcal{G},Y)$.*

Proof. We shall use the notation of Theorem 2.2. For notational convenience we shall also use $\widetilde{X} = \omega(\mathcal{F},X)$ and $\widetilde{Y} = \omega(\mathcal{G},Y)$ and use φ_X and φ_Y as the respective canonical embeddings of X and Y. The maximal dual ideals in $\mathcal{F}$ will be denoted by D's with subscripts and the maximal dual ideals in $\mathcal{G}$ will be denoted by E's. Let us define the map $\widetilde{f}$ that will make the following diagram commutative.

$$
\begin{array}{ccc}
X & \xrightarrow{\varphi_X} & \widetilde{X} = \omega(\mathcal{F},X) \\
{\scriptstyle f}\big\downarrow & & \big\downarrow{\scriptstyle \widetilde{f}} \\
Y & \xrightarrow{\varphi_Y} & \widetilde{Y} = \omega(\mathcal{G},Y)
\end{array}
\tag{5}
$$

Let ξ be a member of $\widetilde{X}$. Then ξ is the index of a maximal dual ideal D_ξ of the lattice $\mathcal{F}$. Let $\mathsf{H} = \{\, G \in \mathcal{G} : f^{-1}[G] \in \mathsf{D}_\xi \,\}$. It is easily seen that H is a dual ideal in the lattice $\mathcal{G}$. By Lemma 1.5, H is contained in a maximal dual ideal E. We shall show by way of a contradiction that there is only one such maximal dual ideal. Assume that E' is another maximal ideal of $\mathcal{G}$ containing H. By Lemma 1.6, there is a G in E and there is a G' in E' such that $G \cap G' = \emptyset$. Let (H, H') be a screening of (G, G'). Observe that $H \notin \mathsf{E}'$ holds because $H \cap G' = \emptyset$. Similarly, $H' \notin \mathsf{E}$. By Lemma 1.7 we have that $f^{-1}[H]$ or $f^{-1}[H']$ is in D_ξ, whence H or H' is in H. As the collection H is contained in $\mathsf{E} \cap \mathsf{E}'$, a contradiction will result. Let η be the unique member of $\widetilde{Y}$ which corresponds to the maximal dual ideal that contains H and define $\widetilde{f}(\xi)$ to be η. It is easily verified that $\widetilde{f}$ is an extension of f and thereby the diagram is commutative.

To show that $\widetilde{f}$ is continuous, we shall prove for every G in $\mathcal{G}$ that $\widetilde{f}^{-1}[\widetilde{Y} \setminus \mathrm{cl}_{\widetilde{Y}}(\varphi_Y[G])]$ is open in $\widetilde{X}$. Suppose that ξ is a member of $\widetilde{f}^{-1}[\widetilde{Y} \setminus \text{cl}_{\widetilde{Y}}(\varphi_Y[G])]$ and write $\eta = \widetilde{f}(\xi)$. Then $\eta \notin \text{cl}_{\widetilde{Y}}(\varphi_Y[G])$. It follows that G is not in E_η, the maximal dual ideal of $\mathcal{G}$ with index η. By Lemma 1.6, we have $G \cap H = \emptyset$ for some H in E_η. The base normality of $\mathcal{G}$ yields a screening (J, K) of (G, H). Let us show that the open set $U = \widetilde{X} \setminus \text{cl}_{\widetilde{X}}(f^{-1}[J])$ in $\widetilde{X}$ is a neighborhood of ξ with $\widetilde{f}[U] \subset \widetilde{Y} \setminus \text{cl}_{\widetilde{Y}}(\varphi_Y[G])$. As $H \cap J = \emptyset$, the sets $\text{cl}_{\widetilde{Y}}(H)$ and $\text{cl}_{\widetilde{Y}}(J)$ are disjoint. So $\eta \notin \text{cl}_{\widetilde{Y}}(f^{-1}[J])$ follows, whence $\xi \in U$. For ξ' in U let $\eta' = \widetilde{f}(\xi')$. As $\xi' \notin \text{cl}_{\widetilde{X}}(f^{-1}[J])$ holds, we have that $\xi' \in \text{cl}_{\widetilde{X}}(f^{-1}[K])$ also holds by Lemma 1.7, and consequently $f^{-1}[K]$ is a member of $\mathsf{D}_{\xi'}$. From the definition of $\widetilde{f}$ we get $K \in \mathsf{E}_{\eta'}$. Thus we have $G \notin \mathsf{D}_{\eta'}$, or equivalently $\eta' \notin \text{cl}_{\widetilde{Y}}(\varphi_Y[G])$. The continuity of $\widetilde{f}$ has been established.

We shall now discuss compactifications that preserve both the weight and the dimension. In particular, dimension preserving metrizable compactifications of separable metrizable spaces, which were used in the proofs of Theorems I.5.11 and III.4.3, will be obtained.

The following is a preparatory lemma. Let us recall that the *weight* $w(X)$ of the space X is the minimal cardinality of a base for the open sets of X.

2.6. Lemma. *Suppose that X is a regular space with infinite weight. Then there is a disjunctive base $\mathcal{F}$ for the closed sets such that $|\mathcal{F}| = w(X)$.*

Proof. Let $\{ U_\alpha : \alpha \in A \}$ be a base for the open sets with $|A|$ equal to $w(X)$. Then $\mathcal{F} = \{ \text{cl}(U_\alpha) : \alpha \in A \} \cup \{ X \setminus U_\alpha : \alpha \in A \}$ is a disjunctive base for the closed sets with $|\mathcal{F}| = w(X)$.

Observe that the above disjunctive base $\mathcal{F}$ is not a ring. We shall often be faced with such a situation. To rectify this we shall employ certain operations to form rings from $\mathcal{F}$. The following notations will be adopted for these operations.

Notations. Let X be a space and $\mathcal{F}$ be a nonempty collection of subsets of X. Then the operations $\wedge$, $\vee$, r, c and $\perp$ on the collection $\mathcal{F}$ are defined as follows.

(i) $\mathcal{F}^\wedge$ is the collection of all sets that are intersections of finite subsets of $\mathcal{F}$.

(ij) $\mathcal{F}^\vee$ is the collection of all sets that are unions of finite subsets of $\mathcal{F}$.

(iij) $\mathcal{F}^r = \mathcal{F}^{\vee\wedge} = \mathcal{F}^{\wedge\vee}$.
(iv) $\mathcal{F}^c = \{ X \setminus F : F \in \mathcal{F} \}$.
(v) $\mathcal{F}^\perp = \{ \mathrm{cl}(F) : F \in \mathcal{F}^c \}$.

Observe that if $|\mathcal{F}| \geq \aleph_0$, then $|\mathcal{F}| = |\mathcal{F}^r| = |\mathcal{F}^c|$.

The first construction will be for the large inductive dimension in the universe of hereditarily normal spaces. A bonus will be the simultaneous dimension preserving compactification for a large prescribed collection of subsets of a space.

2.7. Theorem. *Let X be a hereditarily normal space such that* $\mathrm{Ind}\, X < \infty$. *Suppose that* $\mathbf{E}$ *is a collection of closed subsets of X with* $|\mathbf{E}| \leq w(X)$. *Then there exists a Hausdorff compactification Y of X with $w(X) = w(Y)$ and such that* $\mathrm{Ind}\, Y \leq \mathrm{Ind}\, X$ *and* $\mathrm{Ind}\, \mathrm{cl}_Y(E) \leq \mathrm{Ind}\, E$ *for each E in* $\mathbf{E}$.

Note that the usual Wallman compactification ωX will do if the requirement $w(X) = w(Y)$ is dropped.

Proof. We may assume that $w(X)$ is infinite and $X \in \mathbf{E}$. Our task is to construct an appropriate normal base $\mathcal{F}$. Observe that any base for the closed sets that contains a disjunctive one is also disjunctive. In light of Lemma 2.6 there is a disjuntive base $\mathcal{F}_0$ for the closed sets such that $\mathbf{E} \subset \mathcal{F}_0$ and $|\mathcal{F}_0| = w(X)$. Let us inductively define a sequence of collections $\mathcal{F}_i$, $i = 0, 1, \ldots$. Assume for $0 \leq i$ that $\mathcal{F}_i$ has been defined. With the aid of Proposition I.4.6, for each nonempty set F in $\mathcal{F}_i$ and for each pair (A, B) of disjoint members of $\mathcal{F}_i$ we choose a partition S between A and B such that $\mathrm{Ind}\,(S \cap F) \leq \mathrm{Ind}\, F - 1$. For this partition S let U and V be disjoint open sets with $A \subset U$ and $B \subset V$ such that $X \setminus S = U \cup V$. Then, with $C = S \cup U$ and $D = S \cup V$, the pair (C, D) is a screening of (A, B) for which $\mathrm{Ind}\,(C \cap D \cap F) \leq \mathrm{Ind}\, F - 1$. Let $\mathcal{S}_i$ be the collection of all sets C and D that are obtained in this way and define $\mathcal{F}_{i+1} = (\mathcal{F}_i)^r \cup \mathcal{S}_i$. As $\mathbf{E} \subset \mathcal{F}_0 \subset \mathcal{F}_i \subset \mathcal{F}_{i+1}$ for all i, the collection $\mathcal{F} = \bigcup\{ \mathcal{F}_i : i = 0, 1, 2, \ldots \}$ is a normal base with $|\mathcal{F}| = w(X)$. Let Y be the Wallman compactification $\omega(\mathcal{F}, X)$ with repect to $\mathcal{F}$.

To complete the proof we shall show that $\mathrm{Ind}\, \mathrm{cl}_Y(F) \leq \mathrm{Ind}\, F$ for each F in $\mathcal{F}$. The proof is by induction on $\mathrm{Ind}\, F$. Only the inductive step will be given. Let $\mathrm{Ind}\, F \leq k$. Suppose that G and H are disjoint closed subsets of Y. Because Y is compact and $\{ \mathrm{cl}_Y(F) : F \in \mathcal{F} \}$ is a base for the closed sets of Y, there are disjoint members A

and B of $\mathcal{F}$ such that $G \subset \mathrm{cl}_Y(A)$ and $H \subset \mathrm{cl}_Y(B)$. Let (C, D) be a screening of (A, B) with C and D in $\mathcal{F}$ and $\mathrm{Ind}\,(F \cap C \cap D) \le k - 1$. Then $\mathrm{cl}_Y(C) \cap \mathrm{cl}_Y(D)$ is a partition between G and H and

$$\mathrm{cl}_Y(C) \cap \mathrm{cl}_Y(D) \cap \mathrm{cl}_Y(F) = \mathrm{cl}_Y(C \cap D \cap F).$$

The inequality $\mathrm{Ind}\,\big(\mathrm{cl}_Y(C) \cap \mathrm{cl}_Y(D) \cap \mathrm{cl}_Y(F)\big) \le k - 1$ holds by the induction hypothesis. Thus we have shown $\mathrm{Ind}\,\mathrm{cl}_Y(F) \le k$.

For the covering dimension we shall use the following lemma which converts the condition $\dim X \le n$ into a condition expressed in terms of closed sets. The straightforward proof which is based on Proposition I.8.6 and the De Morgan formulas will be left to the reader.

2.8. Lemma. *Let X be a normal space and let n be in $\mathbb{N}$. Then $\dim X \le n$ if and only if for each finite collection $\boldsymbol{F}$ of closed sets such that $\bigcap \boldsymbol{F} = \emptyset$ there is a finite collection $\mathbf{G}$ of closed sets such that*

(a) *for each G in $\mathbf{G}$ there is an F in $\boldsymbol{F}$ such that $F \subset G$,*
(b) $\bigcap \mathbf{G} = \emptyset$,
(c) *each subcollection of $\mathbf{G}$ with more than $n + 1$ distinct members is a cover of X.*

2.9. Theorem. *Let X be a normal space with $\dim X < \infty$. Then there exists a Hausdorff compactification Y of X such that $w(Y) = w(X)$ and $\dim Y \le \dim X$.*

Proof. Let n be a natural number such that $\dim X \le n$. We may assume that $w(X)$ is infinite. As in the proof of the previous theorem we can find a disjunctive base $\mathcal{F}_0$ for the closed sets such that $|\mathcal{F}_0| = w(X)$ and $X \in \mathcal{F}_0$. We shall inductively define a sequence of closed collections $\mathcal{F}_i$, $i = 0, 1, 2, \ldots$. Assume that $\mathcal{F}_i$ has been defined. For each finite collection $\boldsymbol{F}$ of elements of $\mathcal{F}_i$ with $\bigcap \boldsymbol{F} = \emptyset$ we select exactly one collection $\mathbf{G}$ of closed sets satisfying (a), (b) and (c) of Lemma 2.8. Let $\mathcal{H}_i$ be the union of all such collections that are obtained in this way. For each pair (A, B) of disjoint members of $\mathcal{F}_i$ we choose a screening (C, D) of (A, B) and use the resulting collection $\mathcal{S}_i$ of all sets C and D chosen in this way for the screening collection. We let $\mathcal{F}_{i+1} = (\mathcal{F}_i)^r \cup \mathcal{S}_i \cup \mathcal{H}_i$ and the induction is completed. The collection $\mathcal{F} = \bigcup\{\, \mathcal{F}_i : i = 0, 1, 2, \ldots\}$ is a normal base with $|\mathcal{F}| = w(X)$. Let $Y = \omega(\mathcal{F}, X)$. Obviously $w(Y)$ is equal to $w(X)$. That $\dim Y \leq n$ is verified with the aid of Lemma 2.8 in the following way. Suppose that $\boldsymbol{F}'$ is a finite family of closed subsets of Y with $\bigcap \boldsymbol{F}' = \emptyset$. In view of the compactness of Y we may assume that $\boldsymbol{F}'$ is a subcollection of the base $\{\operatorname{cl}_{\omega X}(F) : F \in \mathcal{F}\}$. That is, $\boldsymbol{F}' = \{\operatorname{cl}_{\omega X}(F_j) : j = 0, 1, \ldots, k\}$. It follows that there is an i such that $\boldsymbol{F} = \{F_j : j = 0, 1, \ldots, k\} \subset \mathcal{F}_i$. Then there is a subcollection $\boldsymbol{G}$ of $\mathcal{H}_i$ that corresponds to the collection $\boldsymbol{F}$. Finally we have that the collection $\boldsymbol{G}' = \{\operatorname{cl}_{\omega X}(G) : G \in \boldsymbol{G}\}$ satisfies (a), (b) and (c) of Lemma 2.8. It follows that $\dim Y \leq n$.

There are many possible variations and combinations of the above compactification theorems. We shall present two, one for inductive dimension and the other for covering dimension.

2.10. Theorem. *Suppose that X is a hereditarily normal space. Let Φ be a collection of continuous maps of X into a compact space Y with $|\Phi| \leq w(X)$ and $w(Y) \leq w(X)$. Then there is a Hausdorff compactification $\widetilde{X}$ of X with $w(\widetilde{X}) = w(X)$ such that $\operatorname{Ind} \widetilde{X} \leq \operatorname{Ind} X$ holds and each φ in Φ can be extended to a continuous map $\widetilde{\varphi}$ of $\widetilde{X}$ into Y.*

Proof. We may assume $w(X) \geq \aleph_0$ and $\operatorname{Ind} X < \infty$. Let $\mathcal{G}$ be a normal base for the closed sets of Y with $|\mathcal{G}| = w(Y)$. As Y is a compact space, we have $Y = \omega(\mathcal{G}, Y)$. We shall use the normal base $\mathcal{F}$ constructed in the proof of Theorem 2.7 with $\boldsymbol{E}$ being the collection $\{\varphi^{-1}[G] : G \in \mathcal{G}, \varphi \in \Phi\}$ whose cardinal number is at most $w(X)$. Then we may apply Theorem 2.5 to each φ in Φ. The theorem will follow easily.

2.11. Theorem. *Suppose that X is a normal space. Let Φ be a collection of homeomorphisms of X into X with $|\Phi| \leq w(X)$. Then there is a Hausdorff compactification $\widetilde{X}$ of X with $w(\widetilde{X}) = w(X)$ such that $\dim \widetilde{X} \leq \dim X$ holds and each φ in Φ can be extended to a homeomorphism $\widetilde{\varphi}\colon \widetilde{X} \to \widetilde{X}$.*

Proof. We may assume that $w(X) \geq \aleph_0$ and $\dim X < \infty$. We may further assume that $\varphi^{-1} \in \Phi$ whenever $\varphi \in \Phi$. The proof of Theorem 2.9 is modified in the following way. Let $\mathcal{F}_0$ be chosen as in the proof of Theorem 2.9. For $i \geq 0$ let

$$\begin{aligned} \mathcal{J}_i &= \{\varphi^{-1}[F] : F \in \mathcal{F}_i, \varphi \in \Phi\}, \\ \mathcal{F}_{i+1} &= (\mathcal{F}_i)^r \cup \mathcal{S}_i \cup \mathcal{H}_i \cup \mathcal{J}_i, \end{aligned}$$

where the collections $\mathcal{S}_i$ and $\mathcal{H}_i$ are defined in the same manner as in the proof of Theorem 2.9. Then $\widetilde{X} = \omega(\mathcal{F}, X)$ satisfies the weight and dimension conditions of the theorem. The extensions of the homeomorphisms follow from Theorem 2.5.

One begins to recognize common features in the above constructions of normal bases. The collection $(\mathcal{F}_i)^r$ is used to induce the ring structure and the collection $\mathcal{S}_i$ is used to induce the base-normal structure. We shall call $\mathcal{S}_i$ the *screening collection* in all subsequent constructions of this type.

3. The Freudenthal compactification

Is there an ideal compactification for a topological space? In the forties many people, most notably Freudenthal in [1942], [1951] and [1952], tried to provide a definitive answer to this question. For locally compact spaces the one-point compactification has many natural features, but is it natural? For example, which is the natural compactification for the complex plane, the Riemann sphere or the Poincaré disc for the hyperbolic geometry? Such questions are difficult to answer, but through the years the ideas have converged to the following definition given by de Groot in [1942].

A compactification $\widetilde{X}$ of a space X is called *ideal* if

(i) the dimension of $\widetilde{X} \setminus X$ is less than or equal to zero, where the dimension function is still to be specified,
(ij) $\widetilde{X}$ is maximal with respect to the property in (i).

With the usual dimension functions in mind, in view of (ij), the two-point compactification $[-1, 1]$ of the open interval $(-1, 1)$ is ideal and the one-point compactification is not. But, in view of (i), the Riemann sphere is the ideal compactification of the complex plane and the Poincaré disc is not.

In [1942] and [1951] Freudenthal constructed ideal compactifications for rim-compact spaces. In this section we shall present the theory of Freudenthal compactifications and its ramifications. The fundamental result of de Groot that originated the theory in this book will be included in the discussion. At the end of the section we shall prove the theorem of Zippin that inspired some of de Groot's work, notably his [1942] thesis.

Recall that a set F in a space X is said to be *regularly closed* if it is the closure of its interior. Similarly, a set U is said to be *regularly open* if it is the interior of its closure. The complement of a regularly closed set is regularly open and conversely. Also the closure of an open set is regularly closed. The union of two regularly closed sets is again regularly closed but the intersection need not be.

Throughout this section the symbol $\mathcal{G}$ will be reserved for the collection of all regularly closed sets with compact boundaries. We shall use the notations adopted in the previous section. The collection of regularly open sets with compact boundaries is clearly the collection $\mathcal{G}^c = \{X \setminus G : G \in \mathcal{G}\}$. Observe that $\mathcal{G} = \mathcal{G}^\perp$. The collection $\mathcal{G}^\wedge$ is the collection of all sets that are finite intersections of members of $\mathcal{G}$. As $\mathcal{G} = \mathcal{G}^\vee$, we have $\mathcal{G}^r = \mathcal{G}^\wedge$. If X is a rim-compact space, then the ring $\mathcal{G}^\wedge$ will be a base for the closed sets of X.

3.1. Lemma. *Let X be a rim-compact space. Then $\mathcal{G}^\wedge$ is a normal base. Moreover, if D is a subset of X with compact boundary and if V is an open neighborhood of D, then there is a member G of $\mathcal{G}$ such that $D \cap G = \emptyset$ and $X \setminus V \subset G$.*

Proof. It has already been observed that $\mathcal{G}^\wedge$ is a ring. To see that $\mathcal{G}^\wedge$ is disjunctive, we observe that each point of X has arbitrarily small open neighborhoods V with compact boundaries. Since X is a regular space and since $\mathrm{cl}\,(V)$ is a regularly closed set it follows that each point has arbitraily small regularly closed neighborhoods with compact boundaries. To complete the proof of the first statement of the theorem we observe that the base-normal property will follow from the second statement because $\mathrm{cl}\,(X \setminus G) \in \mathcal{G}$ whenever $G \in \mathcal{G}$.

To prove the second statement consider the open cover of $\mathrm{B}\,(D)$ consisting of all open sets U with $\mathrm{cl}\,(U) \subset V$ such that $\mathrm{B}\,(U)$ is compact. As $\mathrm{B}\,(D)$ is compact, there is a finite subcover $\boldsymbol{U}$ of this cover. It is easily seen that $W = D \cup \bigcup \boldsymbol{U}$ is an open neighborhood of D such that $\mathrm{cl}\,(W) \in \mathcal{G}$ and $\mathrm{cl}\,(W) \subset V$. The regularly closed set $G = \mathrm{cl}\,(X \setminus \mathrm{cl}\,(W))$ satisfies the required property.

3.2. Definition. The *Freudenthal compactification* $\digamma X$ of a rim-compact space X is the Wallman compactification of X with respect to the base $\mathcal{G}^\wedge$, that is, $\digamma X = \omega(\mathcal{G}^\wedge, X)$. The base $\mathcal{G}^\wedge$ will be called *the Freudenthal base.*

The following theorem contains the key properties of the Freudenthal compactification.

3.3. Theorem. *Suppose that X is a rim-compact space. Let U and V be regularly open sets with compact boundaries. Then*

(a) $\mathrm{ex}_{\mathcal{F}X}(U \cup V) = \mathrm{ex}_{\mathcal{F}X}(U) \cup \mathrm{ex}_{\mathcal{F}X}(V)$,

(b) $\mathrm{B}_{\mathcal{F}X}(\mathrm{ex}_{\mathcal{F}X}(U)) = \mathrm{B}_X(U)$.

Proof. It is to be observed that the complements of regularly open subsets U and V of X with compact boundaries are members of the Freudenthal base $\mathcal{G}^\wedge$. Using the lattice $\mathcal{G}^\wedge$, we can copy the proof of Lemma 1.16 to get a proof of (a) and (b), where the second formula will require the additional fact that $\mathrm{B}_X(U)$ is compact.

Our first task will be to show that the Freudenthal compactification of a rim-compact space is an ideal compactification. To this end, we shall define yet another dimensional property. In contrast with the usual dimension functions this property will not be an absolute property but will be related to ambient spaces.

3.4. Definition. A subset S of a space Y is *zero-dimensionally embedded* in Y if Y has a base $\mathcal{B}$ for the open sets such that $\mathrm{B}_Y(U)$ is disjoint from S for each U in $\mathcal{B}$.

Observe that $\operatorname{ind} S \leq 0$ holds whenever S is zero-dimensionally embedded in Y. But the converse is not true as the following example will show.

3.5. Example. In Example III.1.2 we have exhibited a metrizable space X with $\operatorname{ind} X = 1$ and $\operatorname{ind}(X \setminus \{p\}) = 0$ for some point p in X. Clearly all sufficiently small neighborhoods U of p have non-empty boundaries $\mathrm{B}_X(U)$. It follows that $X \setminus \{p\}$ is not zero-dimensionally embedded in X.

There is also a compact space with a zero-dimensional subset that is not zero-dimensionally embedded in the compact space. Consider the compact space $\beta\Delta$ where Δ is Roy's example. Each point x of the subspace Δ has arbitrarily small open-and-closed neighborhoods U. For each such U we have that $\mathrm{cl}_{\beta\Delta}(U)$ is an open-and-closed neighborhood of x in $\beta\Delta$. Thus each point x in Δ has a base for its neighborhoods in $\beta\Delta$ such that each member of the base has empty boundary, whence disjoint from Δ. This property is somewhat stronger than the property of $\operatorname{ind}\Delta = 0$. We shall show the existence of a point p in $\beta\Delta \setminus \Delta$ such that $\operatorname{ind}(\Delta \cup \{p\}) = 1$. It will then follow that Δ is not zero-dimensionally embedded in $\beta\Delta$. Let A and B

be disjoint closed subsets of Δ for which the empty set is not a partition between A and B in Δ. Observe that $\mathrm{cl}_{\beta\Delta}(A) \cap \mathrm{cl}_{\beta\Delta}(B) = \emptyset$. There must exist a point p in $\mathrm{cl}_{\beta\Delta}(A)$ such that $\mathrm{B}_{\beta\Delta}(U) \cap \Delta \neq \emptyset$ for each of its neighborhoods U with $\mathrm{cl}_{\beta\Delta}(U) \cap \mathrm{cl}_{\beta\Delta}(B) = \emptyset$. If this were not the case, then from the compactness of $\mathrm{cl}_{\beta\Delta}(A)$ we would have the contradiction that the empty set is a partition between A and B in Δ.

There is an intimate relation between zero-dimensionally embedded subsets of a compact space and rim-compact spaces.

3.6. Theorem. *A space X is rim-compact if and only if for some Hausdorff compactification Y of X the set $Y \setminus X$ is zero-dimensionally embedded in Y.*

Proof. Let Y be a compactification of X such that $Y \setminus X$ is zero-dimensionally embedded in Y. Choose a base $\mathcal{B}$ for the open sets of Y such that $\mathrm{B}_Y(U)$ is disjoint from $Y \setminus X$ for each U in $\mathcal{B}$. The trace of $\mathcal{B}$ on X witnesses the fact that X is rim-compact.

For the converse let X be rim-compact and consider its Freudenthal compactification $\mathcal{F}X$. We infer from Lemma 3.1 that the collection $\{\, \mathrm{ex}_{\mathcal{F}X}(U) : U \in \mathcal{G}^c \,\}$ is a base for the open sets of $\mathcal{F}X$. By Theorem 3.3 we have $\mathrm{B}_{\mathcal{F}X}(\mathrm{ex}_{\mathcal{F}X}(U)) = \mathrm{B}_X(U) \subset X$ for each U in $\mathcal{G}^c$. This proves that $\mathcal{F}X \setminus X$ is zero-dimensionally embedded in $\mathcal{F}X$.

Now we can show that the Freudenthal compactification is an ideal compactification. The following theorem will also provide a characterization of the Freudenthal compactification.

3.7. Theorem. *The Freudenthal compactification $\mathcal{F}X$ of a rim-compact space X has the following properties:*

(a) *$\mathcal{F}X \setminus X$ is zero-dimensionally embedded in $\mathcal{F}X$.*
(b) *$\mathcal{F}X$ is maximal with respect to* (a). *That is, if γX is a compactification of X such that $\gamma X \setminus X$ is zero-dimensionally embedded in γX, then the identity map of X has a continuous extension from $\mathcal{F}X$ to γX.*

Proof. As (a) has already been proved in the previous theorem, only (b) requires a proof.

Let $\mathcal{H}$ be the collection of all regularly closed subsets H of γX with $\mathrm{B}_{\gamma X}(H) \subset X$. Since $\gamma X \setminus X$ is zero-dimensionally embedded in γX, the collection $\mathcal{H}$ is a base for the closed sets of γX. Moreover, $\mathcal{H}^\wedge$ is a ring because $\mathcal{H}$ is closed under finite unions. We assert that this ring is a normal base for γX. This assertion will follow from the analogue of Lemma 3.1 which is the subject of the next paragraph.

Let us prove that if A is a subset of γX such that $\mathrm{B}_{\gamma X}(A)$ is compact and if V is any open neighborhood of A, then there is a member H of $\mathcal{H}$ such that $A \cap H = \emptyset$ and $\gamma X \setminus V \subset H$. To this end, let $\boldsymbol{U}$ be the collection consisting of all regularly open sets U of γX such that $\mathrm{cl}_{\gamma X}(U) \subset V$ and $\mathrm{B}_{\gamma X}(U) \subset X$. Since $\gamma X \setminus X$ is zero-dimensionally embedded in γX we have that $\boldsymbol{U}$ is a cover of $\mathrm{B}_{\gamma X}(A)$. The remainder of the proof is the same as the one for Lemma 3.1.

Consider the collection $\mathcal{H}^* = \{ H \cap X : H \in \mathcal{H}^\wedge \}$. Because the boundaries of the members of $\mathcal{H}^*$ are contained in X, we have for H_1 and H_2 in $\mathcal{H}$ that $H_1 \cap H_2 \cap X = \emptyset$ if and only if $H_1 \cap H_2 = \emptyset$. It will follow that $\mathcal{H}^*$ is a normal base for X. To complete the proof we shall show that γX is canonically homeomorphic with $\omega(\mathcal{H}^*, X)$. Once this is established, the existence of a continuous extension of the identity map is a direct consequence of Theorem 2.5. We define a map $\psi : \gamma X \to \omega(\mathcal{H}^*, X)$ in the following way. The notation of Theorems 1.10 and 1.14 will be used. Let ξ be in γX. Define the collection D by $\mathsf{D} = \{ H \cap X : H \in \mathcal{H}, \xi \in H \}$. Because $\mathcal{H}^*$ is disjunctive, by Lemma 1.6 the collection D is a maximal dual ideal of $\mathcal{H}^*$. Thus $\mathsf{D} = \mathsf{D}_\eta$ for some unique η in $\omega(\mathcal{H}^*, X)$. Define $\eta = \psi(\xi)$. It is easily seen that ψ is injective. Let $H \in \mathcal{H}^*$. The basic closed set in $\omega(\mathcal{H}^*, X)$ corresponding to H is $B_{H \cap X}$. It is not difficult to verify

$$\psi^{-1}[B_{H \cap X}] = \{ \xi \in \gamma X : \psi(\xi) \in B_{H \cap X} \} = \{ \xi \in \gamma X : H \cap X \in \mathsf{D}_{\psi(\xi)} \} = H.$$

It follows that ψ is a homeomorphism.

Theorem 3.7 characterizes the Freudenthal compactification by its maximality. We shall briefly discuss another characterization that uses perfect compactifications.

3.8. Definition. A compactification Y of a space X is called *perfect* if for each point y of $Y \setminus X$ and for each neighborhood U of y

in Y the set $U \cap X$ is not the disjoint union of two open subsets V and W of $U \cap X$ such that $y \in \mathrm{cl}_Y(V) \cap \mathrm{cl}_Y(W)$.

It is well known that the Čech-Stone compactification βX is a perfect compactification of X. There is the following characterization of $\mathcal{F}X$.

3.9. Theorem. *Suppose that γX is a compactification of a space X such that $\gamma X \setminus X$ is zero-dimensionally embedded in γX. Then $\gamma X = \mathcal{F}X$ if and only if γX is a perfect compactification of X.*

Proof. Let us show that $\mathcal{F}X$ is perfect. Assume for some point y of $\mathcal{F}X \setminus X$ that there is an open neighborhood U in $\mathcal{F}X$ and there are disjoint open subsets V and W of X such that $U \cap X = V \cup W$ and y is a member of $\mathrm{cl}_{\mathcal{F}X}(V) \cap \mathrm{cl}_{\mathcal{F}X}(W)$. One can easily show that each open neighborhood of y contained in U also has the same property. So we may assume that U is the open set $\mathcal{F}X \setminus \mathrm{cl}_{\mathcal{F}X}(G)$ for some G in $\mathcal{G}$. As V and W are disjoint open sets of X whose union is $X \setminus G$, we have that $\mathrm{cl}_X(V)$ and $\mathrm{cl}_X(W)$ are in $\mathcal{G}$. We infer from $G \in \mathcal{G}$ that $\mathrm{cl}_X(V)$ and $\mathrm{cl}_X(W)$ are disjoint. Theorem 2.2 implies $\mathrm{cl}_{\mathcal{F}X}(V) \cap \mathrm{cl}_{\mathcal{F}X}(W) = \emptyset$. This leads to the contradiction $y \in \emptyset$.

Conversely, suppose that γX is a perfect compactification of X. As $\mathcal{F}X$ is maximal, there is a continuous map $\varphi\colon \mathcal{F}X \to \gamma X$ which extends the identity map. Let us show that φ is injective. Assume for some point z in γX that the set $\varphi^{-1}(z)$ consists of more than one point. As $\mathcal{F}X \setminus X$ is zero-dimensionally embedded in $\mathcal{F}X$, we have $\mathrm{Ind}\, \varphi^{-1}(z) = 0$. So the set $\varphi^{-1}(z)$ is the disjoint union of two nonempty closed subsets, say A and B. Let V and W be disjoint open sets such that $A \subset V$ and $B \subset W$. The set $\varphi[V \cup W]$ is a neighborhood of z, and for every open neighborhood U of z with $U \subset \varphi[V \cup W]$ the set $U \cap X$ admits a splitting into the disjoint open sets $U \cap X \cap V$ and $U \cap X \cap W$. But this is a contradiction to $z \in \mathrm{cl}_{\gamma X}(U \cap X \cap V) \cap \mathrm{cl}_{\gamma X}(U \cap X \cap W)$, a contradiction.

In the introductory discussion of ideal compactifications we did not specify the dimension function. Let us consider ind. Related to Example 3.5 is the following quite natural question. Suppose that a space X has a compactification $\widetilde{X}$ such that $\mathrm{ind}\,(\widetilde{X} \setminus X) \leq 0$. Under what condition does this imply that X is rim-compact? Theorem 3.12 below will provide a partial answer to this question.

3.10. Definition. A completely regular space X is said to be *Lindelöf at infinity* if each compact subset F of X is contained in a compact subset K with a countable base for its neighborhood, that is, there is a countable family of open sets containing K such that each open set containing K necessarily contains an open set from the countable family.

The next lemma will explain the name Lindelöf at infinity.

3.11. Lemma. *Let X be a completely regular space. Then X is Lindelöf at infinity if and only if the subspace $Y \setminus X$ of Y is Lindelöf for every compactification Y of X.*

Proof. Assume that X is Lindelöf at infinity. For a compactification Y of X let $\mathbf{V}$ be an open cover of $Y \setminus X$. The set $F = Y \setminus \bigcup \mathbf{V}$ is a compact subset of X. Let K be a compact subset of X containing F for which a countable base $\boldsymbol{U} = \{ U_i : i \in \mathbb{N} \}$ for the neighborhoods in X of K exists. For each i the set $\mathrm{cl}_Y(X \setminus U_i)$ is a compact subset of $\bigcup \mathbf{V}$ and therefore is covered by finitely many members of $\mathbf{V}$. Because $\boldsymbol{U}$ is a neighborhood base for K, the set $Y \setminus X$ is covered by $\{ \mathrm{cl}_Y(X \setminus U_i) : i \in \mathbb{N} \}$. It will follow that $Y \setminus X$ is a Lindelöf space.

For the converse suppose that X has a compactification Y such that $Y \setminus X$ is a Lindelöf space. Let F be a compact subset of X. Consider the open cover $\mathbf{V}$ of $Y \setminus X$ consisting of cozero-sets V of Y with $V \cap F = \emptyset$. As $Y \setminus X$ is Lindelöf, there is a countable subcover $\mathbf{V}' = \{ V_j : j \in \mathbb{N} \}$ of $\mathbf{V}$. Let K be the compact set $Y \setminus \bigcup \mathbf{V}'$. Clearly, $F \subset K = X \setminus \bigcup \mathbf{V}'$. Let us exhibit a countable base for the open neighborhoods of K. As V_j is a cozero-set, there exists a sequence V_{jk}, $k = 0, 1, 2, \ldots$, of open subsets of Y with $\mathrm{cl}_Y(V_{jk}) \subset V_j$ for every k and $V_j = \bigcup \{ V_{jk} : k \in \mathbb{N} \}$. For each i define U_i to be the open set $Y \setminus \bigcup \{ \mathrm{cl}_Y(V_{jk}) : j \leq i,\, k \leq i \}$. Clearly $\boldsymbol{U}' = \{ U_i : i \in \mathbb{N} \}$ is a collection of neighborhoods of K whose intersection is K. From the fact that $\{ V_{jk} : j \in \mathbb{N},\, k \in \mathbb{N} \}$ is an open cover of $Y \setminus K$ we infer that $\boldsymbol{U} = \{ U_i \cap X : i \in \mathbb{N} \}$ is the required base for the neighborhoods in X of the compact set K. Thereby the converse is proved. Observe that $Y \setminus K$ is σ-compact.

In addition to justifying the name Lindelöf at infinity, the above characterization will be useful in the proof of the next theorem which characterizes rim-compactness for these spaces.

3.12. Theorem. *Suppose that the space X is Lindelöf at infinity. Then X is rim-compact if and only if there is a compactification Y such that* $\operatorname{ind}(Y \setminus X) \leq 0$.

Proof. We have already shown that $\operatorname{ind}(\digamma X \setminus X) \leq 0$ when X is rim-compact.

Let Y be a compactification of X with $\operatorname{ind}(Y \setminus X) = 0$. We have that $Y \setminus X$ is a Lindelöf space from Lemma 3.11. In view of Theorem 3.6, it is sufficient to prove that for each point p of Y and for each neighborhood U of p there is an open set V such that $p \in V \subset U$ and $\mathrm{B}_Y(V) \cap (Y \setminus X) = \emptyset$. To this end, it would be useful to have $\operatorname{Ind}(Y \setminus X) = 0$. This will be so by the assertion in the next paragraph.

We assert that the equivalence $\operatorname{ind} Z = 0$ if and only if $\operatorname{Ind} Z = 0$ holds for Lindelöf spaces Z. We only need to prove that $\operatorname{ind} Z = 0$ implies $\operatorname{Ind} Z = 0$ when Z is a Lindelöf space. Let F and G be disjoint closed sets of Z. Consider the open cover of Z consisting of all open-and-closed sets U such that $U \cap F = \emptyset$ or $U \cap G = \emptyset$. Such a cover exists because $\operatorname{ind} Z = 0$. This cover has a countable subcover $\{ U_i : i \in \mathrm{N} \}$. Let $V_0 = U_0$ and let $V_n = U_n \setminus \bigcup \{ U_i : i < n \}$ when $n \geq 1$. The collection $\mathbf{V} = \{ V_n : n \in \mathrm{N} \}$ is a cover of Z consisting of mutually disjoint open-and-closed sets. It is easy to see that $W = \bigcup \{ V_n : V_n \in \mathbf{V}, V_n \cap F \neq \emptyset \}$ is an open-and-closed set with $F \subset W$ and $W \cap G = \emptyset$.

Returning to the proof of the theorem, we have $\operatorname{Ind}(Y \setminus X) = 0$. Let p and U be given. The space $Z = (Y \setminus X) \cup \{ p \}$ is Lindelöf. By Theorem I.4.12 we have $\operatorname{Ind} Z = 0$. It follows that Z is a disjoint union of open-and-closed subsets A and B of Z such that $p \in A \subset U$. Let $F = \mathrm{cl}_Y(A) \cap \mathrm{cl}_Y(B)$. Since Z is a Lindelöf space we have by Lemma 3.11 that $Y \setminus Z$ is Lindelöf at infinity. As $F \subset Y \setminus Z$, there is a compact subset K of $Y \setminus Z$ such that $F \subset K$ holds and K has a countable base for its neighborhoods in $Y \setminus Z$. Moreover, $Y \setminus K$ may be assumed to be σ-compact, whence Lindelöf and therefore normal. As the set $\mathrm{cl}_Y(A) \setminus K$ and $\mathrm{cl}_Y(B) \setminus K$ are disjoint in $Y \setminus K$, they have disjoint open neighborhoods V' and W' in Y respectively. Let V be the set $V' \cap U$. It is easy to see that $A \subset V$ and $B \subset W'$. Consequently, $\mathrm{B}_Y(V) \cap (Y \setminus X) = \emptyset$ and the proof is now completed.

In Section I.5 we announced the result that a separable metrizable space X is rim-compact if and only if there is a metrizable compactification Y of X with $\operatorname{ind}(Y \setminus X) \leq 0$. The if part of the statement follows from Theorem I.5.8 or Theorem 3.12. But the converse will not follow from the Freudenthal compactification because the Freudenthal compactification need not be weight preserving. In particular, the space $\mathcal{F}X$ need not be metrizable for a separable metrizable space X. For example, the Freudenthal compactification $\mathcal{F}\mathbb{N}$ for the space $\mathbb{N}$ coincides with $\beta\mathbb{N}$ and $\aleph_0 = w(\mathbb{N}) < w(\beta\mathbb{N}) = \mathfrak{c}$. A necessary condition for the Freudenthal compactification to be weight preserving will be presented. But at this point we shall make a brief digression to prove the following weight preserving modification of the Freudenthal compactification. The theorem includes de Groot's result as a special case (see Theorem I.5.7). This is the first of two proofs of de Groot's theorem; the second one is found in Subsection 3.20.

3.13. Theorem. *Suppose that X is a rim-compact space. Then there is a compactification Y of X such that Y and X have the same weight and $Y \setminus X$ is zero-dimensionally embedded in Y.*

Proof. The proof is a modification of some of the previous ones. We may assume that $w(X)$ is infinite. Let $\mathcal{B} = \{ U_\alpha : \alpha \in A \}$ be a base for the open sets of X with $|A| = w(X)$ and consider all pairs (U_α, U_β) of members of $\mathcal{B}$ with $\operatorname{cl}_X(U_\alpha) \subset U_\beta$. When (U_α, U_β) is such that there exists a regularly open set G with compact boundary and $\operatorname{cl}_X(U_\alpha) \subset G \subset \operatorname{cl}_X(G) \subset U_\beta$, choose one such G and denote it by $G_{\alpha,\beta}$. Otherwise, choose $G_{\alpha,\beta} = \emptyset$. The collection $\mathcal{F}_0 = \{ G_{\alpha,\beta} : \alpha \in A,\, \beta \in A \}$ is a disjunctive base with cardinality $w(X)$ that is contained in the Freudenthal base $\mathcal{G}^\wedge$. We may assume that X is in $\mathcal{F}_0$. We shall define a sequence $\mathcal{F}_i$, $i = 0, 1, 2, \ldots$, of subcollections of the Freudenthal base $\mathcal{G}^\wedge$ with cardinality $w(X)$. Assume that $\mathcal{F}_i$ has already been defined. With the aid of Lemma 3.1 we form a screening collection by selecting for each pair (A, B) of disjoint members of $\mathcal{F}_i$ a screening (C, D) of (A, B) by regularly closed sets C and D with compact boundaries. As usual, $\mathcal{S}_i$ is the collection of all such sets C and D that were selected in this way. Define the collection $\mathcal{F}_{i+1}$ to be the subcollection $(\mathcal{F}_i)^r \cup (\mathcal{F}_i)^\perp \cup \mathcal{S}_i$ of $\mathcal{G}^\wedge$. Clearly, $|\mathcal{F}_{i+1}| = w(X)$. The collection $\mathcal{F} = \bigcup\{ \mathcal{F}_i : i = 0, 1, 2, \ldots\}$ is a normal base with $|\mathcal{F}| = w(X)$ and $\mathcal{F} \subset \mathcal{G}^\wedge$. Let Y be the Wallman compactification $\omega(\mathcal{F}, X)$ of X with respect to $\mathcal{F}$.

The verification of the fact that $Y \setminus X$ is zero-dimensionally embedded will be done in the exact same way as in Theorem 3.6. We infer from the way the collections $\mathcal{S}_i$ were constructed that the collection $\{\operatorname{ex}_Y(X \setminus G) : G \in (\mathcal{G} \cap \mathcal{F})\}$ is a base for the open sets of Y. Moreover, if U denotes any $\operatorname{ex}_Y(X \setminus G)$ in this base, then we also have $\mathrm{B}_Y(U) = \mathrm{B}_X(X \setminus G)$. Consequently, $Y \setminus X$ is zero-dimensionally embedded in Y.

Let us return to the discussion of weight preserving Freudenthal compactifications. Whether or not the Freudenthal compactification of a space is weight preserving depends on the collection of open-and-closed sets of the space. To make this more precise we shall briefly discuss the notion of the quasi-component space. Suppose that X is a space. The *quasi-component* of a point x of X is the intersection of all open-and-closed sets that contain x. The collection of quasi-components of a space is pairwise disjoint, and each quasi-component is closed. The *quasi-component space* $Q(X)$ is the set of all quasi-components of X endowed with the topology generated by the open-and-closed subsets of X. This topology need not be the quotient topology. But it will coincide with the quotient topology when X is compact. The projection map π_X will induce a bijection between the respective collections of open-and-closed sets of X and $Q(X)$. It is not difficult to show that $Q(X)$ is a totally disconnected Hausdorff space. The quasi-component space can be a rather powerful tool, especially when it is compact. Some properties are collected in the following theorem.

3.14. Theorem. *Let X be a space with infinite weight and compact quasi-component space $Q(X)$. Then $w(Q(X)) \leq w(X)$ and the cardinality of the family of open-and-closed subsets of X is less than or equal to $w(X)$. Moreover, if Y is a closed subset of X such that $\mathrm{B}_X(Y)$ is compact, then $Q(Y)$ is also compact.*

Proof. Let $\mathcal{B} = \{U_\alpha : \alpha \in A\}$ be a base for the open sets such that $|A| = w(X)$. If for a pair (U_α, U_β) of disjoint elements of $\mathcal{B}$ there exists an open-and-closed set H such that $U_\alpha \subset H \subset X \setminus \mathrm{cl}_X(U_\beta)$, we choose one such set and denote it by $H_{\alpha,\beta}$. If no such set exists, we choose $H_{\alpha,\beta}$ to be $\emptyset$. The collection $\mathcal{Q}$ of finite intersections and finite unions of all finite subfamilies of $\{\pi_X[H_{\alpha,\beta}] : \alpha, \beta \in A\}$, where π_X is the projection map, is a base for the topology of $Q(X)$ because of the compactness of $Q(X)$. Consequently we have $w\big(Q(X)\big) \leq w(X)$. Let W be an open-and-closed subset of $Q(X)$. Then W is compact, and therefore W can be written as a finite union of basic open sets from $\mathcal{Q}$. Thus the cardinality of the family of all open-and-closed subsets of $Q(X)$ is finite or equals $w(Q(X))$ and the same holds for the cardinality of the family of all open-and-closed subsets of X.

For the proof of the last statement of the theorem, suppose that Y is a closed subset of X such that $\mathrm{B}_X(Y)$ is compact. Let us show that if $\mathcal{R} = \{ R_\alpha : \alpha \in A \}$ is a collection of closed subsets of $Q(Y)$ with the finite intersection property, then $\bigcap \mathcal{R}$ is not empty. Denote the projection map of Y onto $Q(Y)$ by π_Y. From the compactness of $\mathrm{B}_X(Y)$ it will follow that $\bigcap \mathcal{R} \neq \emptyset$ when $\pi_Y^{-1}[R_\alpha] \cap \mathrm{B}_X(Y) \neq \emptyset$ for every α. So suppose that $\pi_Y^{-1}[R_\beta] \cap \mathrm{B}_X(Y) = \emptyset$ for some β in A. It is easily verified that if z is in R_β, then $\pi_Y^{-1}[z]$ is not only a quasi-component of Y but also a quasi-component of X. Consequently the set $\pi_Y^{-1}[R_\alpha \cap R_\beta]$ is a closed subset of X that is the union of quasi-components of X for each α. By the compactness of $Q(X)$ the intersection of $\{ \pi_X[\pi_Y^{-1}[R_\beta \cap R_\alpha]] : \alpha \in A \}$ is nonempty. Consequently $\mathcal{R}$ has a nonempty intersection. Thereby $Q(Y)$ is compact.

It will turn out that the Freudenthal compactification is weight preserving if the quasi-component space is compact.

3.15. Theorem. *Let X be a rim-compact space such that $Q(X)$ is compact. Then $w(\digamma X) = w(X)$.*

Proof. We may assume that $w(X)$ is infinite. Choose a base $\mathcal{B} = \{ U_\alpha : \alpha \in A \}$ for the open sets such that $|A| = w(X)$. We may assume that $\mathcal{B}$ is closed under finite unions. If for the pair (U_α, U_β) of disjoint elements of $\mathcal{B}$ there exists a regularly closed set H with compact boundary such that $U_\alpha \subset H \subset X \setminus U_\beta$, we select one such set and denote it by $H_{\alpha,\beta}$. If no such set exists, we select $H_{\alpha,\beta}$ to be $\emptyset$. Define

$$\mathcal{F}_0 = \{ H_{\alpha,\beta} : \alpha \in A,\ \beta \in A \} \cup \{ \mathrm{cl}_X(X \setminus H_{\alpha,\beta}) : \alpha \in A,\ \beta \in A \}.$$

Let $H \in \mathcal{F}_0$. The quasi-component space $Q(H)$ is compact by Theorem 3.14. By the same theorem the cardinality of the family of all open-and-closed subsets of H does not exceed $w(H)$. We define $\mathcal{F}_1$ to be the collection of all open-and-closed subsets of elements of $\mathcal{F}_0$. Let $\mathcal{F}_2 = \mathcal{F}_1{}^{\vee}$, the collection of all finite unions of elements of $\mathcal{F}_1$. Obviously, $|\mathcal{F}_2| \leq w(X)$. Recall that $\mathcal{G}$ denotes the collection of all regularly closed subsets of X and note that $\mathcal{F}_2 \subset \mathcal{G}$.

Let us first prove the assertion: *For any two disjoint elements F and G of $\mathcal{G}$ there exists an H in $\mathcal{F}_2$ such that $F \subset H \subset X \setminus G$.* For a proof we first note that by Lemma 3.1 one can select from the collection $\mathcal{G}$ a pair of disjoint closed neighborhoods F' and G' of F and G respectively. As $\mathrm{B}_X(F)$ is compact, there is an U in $\mathcal{B}$ such that $\mathrm{B}_X(F) \subset U \subset F'$. Similarly there is a V in $\mathcal{B}$ such that $\mathrm{B}_X(G) \subset V \subset G'$. So there is a J in $\mathcal{F}_0$ with $U \subset J \subset X \setminus V$. Letting $K = \mathrm{cl}_X(X \setminus J)$, we have $V \subset K \subset X \setminus U$. As $\mathrm{B}_X(G) \subset V$ and $V \cap J = \emptyset$, both $J \cap G$ and $J \setminus G$ are open-and-closed subsets of J. Similarly $K \cap F$ and $K \setminus F$ are open-and-closed subsets of K. It follows that $J \cap G$, $J \setminus G$, $K \cap F$ and $K \setminus F$ are members of $\mathcal{F}_1$. Define $H = (J \setminus G) \cup (K \cap F)$. Then $H \in \mathcal{F}_2$ and $F \subset H \subset X \setminus G$. Thereby the assertion is proved.

Since $\mathcal{G}^\wedge$ is a normal base we can easily derive from the assertion that the collection $\mathcal{F}_2{}^\wedge$ is a normal base. By Theorem 2.5 there is a continuous map $\varphi\colon \mathcal{F}X \to \omega(\mathcal{F}_2{}^\wedge, X)$ that is the extension of the identity mapping of X. To complete the proof of the theorem we only have to show that φ is injective. Let ξ and η be distinct points of $\mathcal{F}X$ and denote their respective maximal dual ideals of $\mathcal{G}^\wedge$ by D_ξ and D_η. There are disjoint sets F and G in $\mathcal{G}$ such that $F \in D_\xi$ and $G \in D_\eta$. By the above assertion there are disjoint sets J and K in $\mathcal{F}_2$ such that $F \subset J$ and $G \subset K$. It will follow that $\varphi(\xi) \in \mathrm{cl}_{\omega(\mathcal{F}_2{}^\wedge, X)}(J)$ and $\varphi(\eta) \in \mathrm{cl}_{\omega(\mathcal{F}_2{}^\wedge, X)}(K)$. As the intersection $\mathrm{cl}_{\omega(\mathcal{F}_2{}^\wedge, X)}(J) \cap \mathrm{cl}_{\omega(\mathcal{F}_2{}^\wedge, X)}(K)$ is empty, φ is injective.

3.16. Corollary. *Let X be a rim-compact, separable metrizable space. Then $\mathcal{F}X$ is metrizable if and only if $Q(X)$ is compact.*

Proof. When $Q(X)$ is compact, we have $w(\mathcal{F}X) = w(X) \leq \aleph_0$. Consequently $\mathcal{F}X$ is metrizable.

Before proceding to the converse we shall have need of the connection between the quasi-components of X and the components of $\mathcal{F}X$. As $\mathcal{F}X$ is compact, we have that the components and quasi-components agree in $\mathcal{F}X$. So each quasi-component of X will yield a component of $\mathcal{F}X$. Conversely, each component of $\mathcal{F}X$ which meets X yields a quasi-component of X.

Now suppose that $\mathcal{F}X$ is metrizable. Denote the quotient map of $\mathcal{F}X$ onto $Q(\mathcal{F}X)$ by π and the natural embedding of X into $\mathcal{F}X$ by φ. The composition $\pi\varphi$ can be factored through the space $Q(X)$. Let g be the factor map from $Q(X)$ into $Q(\mathcal{F}X)$. Then g is injective and the topology of $Q(X)$ makes g into an embedding. As $\mathcal{F}X$ has been assumed to be metrizable, we have that its image under the surjective map π is also metrizable. Suppose that the factor map g is not surjective. Then there exists a point q in $Q(\mathcal{F}X) \setminus Q(X)$. As $Q(X)$ is dense in $Q(\mathcal{F}X)$ and $Q(\mathcal{F}X)$ is a compact zero-dimensional space, there is a sequence U_i, $i \in \mathbb{N}$, of open-and-closed sets of $Q(\mathcal{F}X)$ with $U_{i+1} \subset U_i$ such that $V_i = U_i \setminus U_{i+1}$ is not empty for each i and $\bigcap\{ V_i : i \in \mathbb{N} \} = \{ q \}$. Since $Q(\mathcal{F}X)$ is compact, we have that $\bigcup\{ V_i : i \in N \}$ is open-and-closed in the subspace $Q(\mathcal{F}X) \setminus \{ q \}$ for each subset N of $\mathbb{N}$. So $\pi_X^{-1}\left[g^{-1}[\bigcup\{ V_i : i \in N \}]\right]$, $N \in \mathbb{N}$, is an uncountable collection of open-and-closed subsets of X, where π_X is the projection of X onto $\mathcal{F}(X)$. Thus a contradiction to Theorem 3.14 has appeared. Thereby we have that g is surjective.

The extension of the identity map to the Freudenthal compactification played a role in Theorem 3.7. More general situations are considered in the following two theorems. It will turn out that the class of rim-compact spaces with closed continuous maps whose point inverses have compact boundaries forms a category. The Freudenthal compactification then becomes a reflection of this category to the category of compact Hausdorff spaces with continuous maps.

3.17. Theorem. *Let X and Y be rim-compact spaces. Suppose that $f\colon X \to Y$ is a closed, continuous map such that point inverses have compact boundaries. Then there is a continuous extension $\mathcal{F}f\colon \mathcal{F}X \to \mathcal{F}Y$ for the map f.*

Proof. Let us show that if G is a closed subset of Y with compact boundary, then $f^{-1}[G]$ has a compact boundary. For notational convenience let H be the compact set $\mathrm{B}_X(f^{-1}[G])$ and g be the restriction of f to H. Note that g is closed and continuous. It is easily verified that $g[H] \subset \mathrm{B}_Y(G)$ and $g^{-1}[y] \subset \mathrm{B}_X(f^{-1}[y]) \cap H$ for each y in $\mathrm{B}_Y(G)$. Suppose that $\boldsymbol{U}$ is an open cover of H. Let $y \in \mathrm{B}_Y(G)$. The set $g^{-1}[y]$ is covered by a finite subfamily of $\boldsymbol{U}$. Denote by W_y the union of this finite subfamily and let $V_y = Y \setminus f[X \setminus W_y]$. As f is closed, V_y is a neighborhood of y. And as $g[H]$ is compact, it is covered by finitely many V_y. Consequently $\boldsymbol{U}$ has a finite subcover. The theorem will now follow from Theorem 2.5.

3.18. Theorem. *Suppose that X and Y are metrizable rim-compact spaces. If $f\colon X \to Y$ is a closed continuous map, then f has a continuous extension $\mathcal{F}f\colon \mathcal{F}X \to \mathcal{F}Y$.*

Proof. It is sufficient to show that point inverses have compact boundaries. The fact that closed maps of metrizable spaces have this property has been observed in Lemma IV.4.13.

In our discussion of complete metric extensions, we have shown in Theorem V.2.28 that for a metrizable space X there are metrizable extensions $\widetilde{X}$ that preserve dimension and at the same time satisfy the requirement that $\operatorname{Ind}(\widetilde{X} \setminus X)$ be minimal. More specifically, for every metrizable space X there is a completion $\widetilde{X}$ such that $\operatorname{Ind} \widetilde{X} = \operatorname{Ind} X$ and $\operatorname{Ind}(\widetilde{X} \setminus X) = \operatorname{Icd} X$. The situation for metrizable compactifications is somewhat different as the following example will show.

3.19. Example. We shall present a rim-compact subspace X of $\mathbb{S}^n$ with $n \geq 2$ such that

(a) $\operatorname{ind} X = n - 1$,

(b) $\operatorname{ind} \widetilde{X} = n$ for every metrizable compactification $\widetilde{X}$ of X for which $\operatorname{ind}(\widetilde{X} \setminus X) \leq 0$.

Let us make a preliminary observation. If a subset S of $\mathbb{S}^n$ is a partition between two points x and y, then x has arbitrarily small neighborhoods whose boundaries are homeomorphic with S. Such neighborhoods can be obtained by removing the point y and shrinking $\mathbb{S}^n \setminus \{y\}$. As $\operatorname{ind} \mathbb{S}^n = n$, it will follow that no $(n-2)$-dimensional set can be a partition between any two distinct points.

Let $X = \mathbb{S}^n \setminus Q$ where Q is a countable dense subset of $\mathbb{S}^n$, and note that X is complete. We have $\operatorname{ind} X > n - 1$ by the addition theorem. Though we have not proved it, $\operatorname{ind} X \leq n - 1$ also holds (see, for example, Hurewicz and Wallman [1941], page 44). Let $\widetilde{X}$ be any metrizable compactification of X with $\operatorname{ind}(\widetilde{X} \setminus X) \leq 0$. We shall show $\operatorname{ind} \widetilde{X} = n$. To the contrary, let us assume $\operatorname{ind} \widetilde{X} \leq n - 1$. Let x and y be distinct points of X. As X is complete, $\widetilde{X} \setminus X$ is an F_σ-set of $\widetilde{X}$. By Lemma I.4.6 there is a partition S between x and y such that $\operatorname{ind} S \leq n - 2$ and $\operatorname{ind}\big(S \cap (\widetilde{X} \setminus X)\big) \leq -1$, whence $S \subset X$. Let $X \setminus S = U \cup V$ where U and V are open sets of X with $x \in U$ and $y \in V$ such that $U \cap V = \emptyset$. By the sum theorem, $\operatorname{ind}(S \cup Q) \leq n - 2$. The interior of $S \cup Q$ in $\mathbb{S}^n$ is empty. So $Z = U \cup V$ is a dense subspace of $\mathbb{S}^n$. Applying Lemma V.2.10 to this subspace Z, we have $\operatorname{ex}_{\mathbb{S}^n}(U) \cap \operatorname{ex}_{\mathbb{S}^n}(V) = \emptyset$. Consequently $S' = \mathbb{S}^n \setminus \big(\operatorname{ex}_{\mathbb{S}^n}(U) \cup \operatorname{ex}_{\mathbb{S}^n}(V)\big)$ is a subset of $S \cup Q$. That is, S' is a partition between x and y in $\mathbb{S}^n$ with $\operatorname{ind} S' \le n-2$, a contradiction.

There is another proof of $\operatorname{ind} \widetilde{X} = n$. The preliminary observation will lead to the fact that $\mathbb{S}^n$ is a perfect compactification of X. So by Theorem 3.9, $\mathbb{S}^n = \digamma X$. From Theorem 3.7 we have a continuous map $f\colon \mathbb{S}^n \to \widetilde{X}$ with $\dim f = 0$. Theorem IV.4.5 and the coincidence theorem yield $n = \dim \mathbb{S}^n \le \operatorname{Ind} \widetilde{X} = \operatorname{ind} \widetilde{X}$. The addition theorem gives $\operatorname{ind} \widetilde{X} \le \operatorname{ind} X + 1 = n$.

We shall conclude the section by presenting a construction of a metrizable compactification $\widetilde{X}$ of a separable metrizable rimcompact space X such that $\operatorname{ind}(\widetilde{X} \setminus X) \le 0$. The construction has the additional feature that $\widetilde{X} \setminus X$ is countable whenever X is complete.

3.20. The construction. Let X be a separable metrizable rimcompact space and let d be a metric on X. When X is complete, the metric d will be chosen to be complete as well.

Let us construct a collection $\mathcal{F}$ that is the union of countably many collections $\mathcal{F}_i$, $i \in \mathbb{N}$, where $\mathcal{F}_i$ satisfies the conditions (1) through (6) below for every positive i.

(1) Each member F of $\mathcal{F}_i$ is a regularly closed set with compact boundary.
(2) Each pair of sets F and G in $\mathcal{F}_i$ satisfies the requirement $F \cap G = \mathrm{B}_X(F) \cap \mathrm{B}_X(G)$.
(3) Each collection $\mathcal{F}_i$ is a locally finite, countable, closed cover of X.
(4) The collection $\{ F : F \cap C \ne \emptyset, F \in \mathcal{F}_i \}$ is finite for every compact set C.
(5) $\operatorname{mesh} \mathcal{F}_i \le 2^{-i}$.
(6) $\mathcal{F}_i$ is a refinement of $\mathcal{F}_{i-1}$.

Define $\mathcal{F}_0 = \{ X \}$. The collection $\mathcal{F}_1$ is obtained in the following way. In view of the separability of X, there exists a countable open cover $\{ U_k : k \in \mathbb{N} \}$ of X consisting of regularly open sets U_k with compact boundary and diameter less than 2^{-k}. Define for each j in $\mathbb{N}$ the set

$$F_{1j} = \operatorname{cl}_X(U_j \setminus \bigcup \{ \operatorname{cl}_X(U_k) : k < j \})$$

and let $\mathcal{F}_1 = \{ F_{1j} : j \in \mathbb{N} \}$. It is easily seen that conditions (1) through (6) are satisfied for $i = 1$. Suppose that $\mathcal{F}_0, \ldots, \mathcal{F}_n$ have been constructed. Our construction will now be directed toward the subspace F_{nj}. In the *subspace* F_{nj} there is a countable, closed collection $\mathcal{H}_j$ satisfying the conditions (1) through (6), where X and $\mathcal{F}_i$ have been replaced by F_{nj} and $\mathcal{H}_j$ respectively, and the condition mesh $\mathcal{H}_j \leq 2^{-(i+1)}$. Returning to the space X, we let $\mathcal{F}_{n+1} = \bigcup\{\mathcal{H}_j : j \in \mathbb{N}\}$. Then enumerate $\mathcal{F}_{n+1}$ as $\{F_{n+1,j} : j \in \mathbb{N}\}$. It is not difficult to see that conditions (1) through (6) are satisfied for $i = 1, \ldots, n+1$. Thereby the collection $\mathcal{F}$ has been constructed.

Inductively define a sequence of bases $\mathcal{J}_i$ as follows. Let $\mathcal{J}_0 = \mathcal{F}$. Assume that $\mathcal{J}_i$ has been defined. Then let $\mathcal{J}_{i+1} = \mathcal{J}_i^r \cup \mathcal{J}_i^{\perp}$. Finally we let $\mathcal{J} = \bigcup\{\mathcal{J}_i : i \in \mathbb{N}\}$. By inducting on i, we can show that each member of $\mathcal{J}_i$ has a compact boundary. If A is a member of $\mathcal{J}_i$ and V is a neighborhood of A, then there is an m such that $m \geq 1$ and $2^{-(m-1)}$ is less than the distance between $\mathrm{B}(A)$ and $X \setminus V$. Using (4), one can easily find a screening of $(A, X \setminus V)$ with elements of $\mathcal{J}_m$. It follows that $\mathcal{J}$ is a normal base with $|\mathcal{J}| = \aleph_0$. Finally we let $\widetilde{X} = \omega(\mathcal{J}, X)$. This completes the construction of the special Wallman compactification.

This construction was used by de Groot in [1942] for his proofs of Theorem I.5.7 and Theorem 3.13. The details of this part of these proofs will be left to the reader, but we shall indicate how the construction can be used for a proof of Theorem I.5.2 which states that

> *a separable metric space that is complete and rim-compact can be compactified by the addition of at most countably many points.*

Recall that we have assumed in the construction that the metric d is complete. We shall show that there are at most countably many maximal dual ideals of $\mathcal{J}$ that are not fixed. Let H be a maximal dual ideal in $\mathcal{J}$ that is not fixed. Define $k = \sup\{i : \mathcal{F}_i \cap \mathsf{H} \neq \emptyset\}$. We shall show by way of a contradiction that k is finite. Suppose that $k = \infty$. Then $\mathcal{F}_i \cap \mathsf{H} \neq \emptyset$ for every i. It follows that H contains arbitrarily small members. As X is complete, the set $\bigcap \mathsf{H}$ consists of exactly one point of X and H is fixed. As H is not fixed by assumption, we have a contradiction. So, $k < \infty$. In view of (2) there is exactly one member of $\mathcal{F}_k$ in H. Denote this element by G. For each F in $\mathcal{F}_{k+1}$ with $F \subset G$ we have by Lemma 1.7 that $\mathrm{cl}_X(X \setminus F) \in \mathsf{H}$ because H is maximal and F is not a member of H. It can be seen that there is exactly one point in the set

$\mathrm{cl}_{\tilde{X}}(G) \cap \bigcap\{\mathrm{cl}_{\tilde{X}}(X \setminus F) : F \in \mathcal{F}_{k+1}, F \subset G\}$. Thus we have established a correspondence between the added points and a collection of finite sequences of natural numbers (namely, H corresponds to the indices of the unique members from $\mathcal{F}_i \cap \mathsf{H}$ for $i = 0, 1, \ldots, k$).

Observe that the above construction yields the one-point compactification for the space X formed by removing the end points of the Cantor fan and not $\digamma X$, the Cantor fan.

4. The inequality $\mathcal{K}$-Ind $\geq$ $\mathcal{K}$-Def

The condition that a subset be zero-dimensionally embedded in a space turned out to be pivotal for rim-compact spaces. This condition can be inductively raised to higher dimensions to yield the notion of a subset being ($\leq n$)-Inductionally embedded in a space. There is a strong connection between this notion and $\mathcal{K}$-Ind, where $\mathcal{K}$ is the class of compact spaces. Conditions of this type were introduced by de Vries in [1962]. The first result will tie together the condition $\mathcal{K}\text{-Ind}\, X \leq n$ and the condition that $\omega X \setminus X$ be ($\leq n$)-Inductionally embedded in the Wallman compactification ωX of X. Connections between the relative property that a set S is ($\leq n$)-Inductionally embedded in a space Y and the absolute property $\mathrm{Ind}\, S \leq n$ will be discussed; the picture here is far from complete. We shall end the section by continuing our discussion of the interrelations among the compactness dimension functions by proving the inequality in the title of the section for separable metrizable spaces.

4.1. Definition. Let Z be a subset of a normal space X and let $n \geq -1$. Then the expression *Z is ($\leq n$)-Inductionally embedded in X*, denoted $\mathrm{Ind}\,[Z; X] \leq n$, is inductively defined as follows.

(i) $\mathrm{Ind}\,[Z; X] = -1$ if and only if $Z = \emptyset$.

(ij) For $n \geq 0$, $\mathrm{Ind}\,[Z; X] \leq n$ if for any two disjoint closed sets F and G of X there is a partition S between F and G in X such that $\mathrm{Ind}\,[S \cap Z; S] \leq n - 1$.

It is easily verified that $\mathrm{Ind}\,[Z; X] \leq \mathrm{Ind}\, X$ for every normal space X. It is also not difficult to prove that for a closed subset Z of a normal space X the inequality $\mathrm{Ind}\, Z \leq \mathrm{Ind}\,[Z; X]$ holds.

Let us prove the following lemma by induction.

4.2. Lemma. *Suppose that Y is a closed subspace of a normal space X. Let Z be a subset of X. Then*

$$\operatorname{Ind}[Z \cap Y; Y] \leq \operatorname{Ind}[Z; X].$$

Proof. Let n be $\operatorname{Ind}[Z;X]$. The lemma is obvious when $n = -1$. For the inductive step of the proof we let F and G be two disjoint closed subsets of Y. There is a partition S between F and G in X such that $\operatorname{Ind}[S \cap Z; S] \leq n-1$. By the induction hypothesis we get the inequality $\operatorname{Ind}[S \cap Z \cap Y; S \cap Y] \leq n-1$. As $S \cap Y$ is a partition between F and G in Y, $\operatorname{Ind}[Z \cap Y; Y] \leq n$ will hold.

Our first result characterizes the condition $\mathcal{K}\text{-}\operatorname{Ind} X \leq n$ by the condition that $\omega X \setminus X$ is $(\leq n)$-Inductionally embedded in ωX. The proof should be compared with that of Theorem 1.18.

4.3. Theorem. *Suppose that X is a normal space and let n be in $\mathbb{N}$. Then*

$$\mathcal{K}\text{-}\operatorname{Ind} X \leq n \quad \textit{if and only if} \quad \operatorname{Ind}[\omega X \setminus X; \omega X] \leq n.$$

Proof. The proof will be by induction. As the statement is obviously true for $n = -1$, we shall only discuss the inductive steps.

Suppose $\operatorname{Ind}[\omega X \setminus X; \omega X] \leq n$ and let F and G be disjoint closed subsets of X. By Theorem 1.14, $\operatorname{cl}_{\omega X}(F) \cap \operatorname{cl}_{\omega X}(G) = \emptyset$. In ωX there is a partition S between $\operatorname{cl}_{\omega X}(F)$ and $\operatorname{cl}_{\omega X}(G)$ such that $\operatorname{Ind}[S \cap (\omega X \setminus X); S] \leq n-1$. And $\omega(S \cap X) = \operatorname{cl}_{\omega X}(S \cap X) \subset S$ follows from Theorem 1.15. Thus, from the previous lemma, we have $\operatorname{Ind}[\omega(S \cap X) \cap (\omega X \setminus X); \omega(S \cap X)] \leq n-1$. By the induction hypothesis $\mathcal{K}\text{-}\operatorname{Ind}(S \cap X) \leq n-1$ holds. It follows that $\mathcal{K}\text{-}\operatorname{Ind} X \leq n$.

To prove the inductive step of the other implication we assume $\mathcal{K}\text{-}\operatorname{Ind} X \leq n$ and let F and G be disjoint closed sets of ωX. Let U_1 and V_1 be open neighborhoods of F and G with disjoint closures. There is a partition S in X between $U_1 \cap X$ and $V_1 \cap X$ such that $\mathcal{K}\text{-}\operatorname{Ind} S \leq n-1$. Lemmas V.2.10 and 1.16 imply that $\operatorname{cl}_{\omega X}(S)$ is a partition between F and G in ωX. By Theorem 1.15, $\operatorname{cl}_{\omega X}(S) = \omega S$ and $\omega X \setminus X) \cap \omega S = \omega S \setminus S$ hold. The induction hypothesis gives $\operatorname{Ind}[\omega S \setminus S; \omega S] \leq n-1$. Hence $\operatorname{Ind}[\omega X \setminus X; \omega X] \leq n$.

Recall that $\operatorname{Ind} \omega X = \operatorname{Ind} X$ (Theorem 1.18). The result of the previous theorem is then in contrast with Example 3.19, but similar to that of Theorem V.2.28. In general the Wallman compactification is not weight preserving. But for hereditarily normal spaces there is a modification of Theorem 4.3 in which the compactification is weight preserving.

4.4. Theorem. *Suppose that X is a hereditarily normal space with $\mathcal{K}$-Ind $X \leq n$. Then there exists a weight preserving compactification Y of X with*

$$\operatorname{Ind} Y \leq \operatorname{Ind} X, \quad \operatorname{Ind}[Y \setminus X; Y] \leq n.$$

Proof. We may assume that $w(X)$ is infinite. Let $\mathcal{F}_0$ be a disjunctive base for the closed sets such that $|\mathcal{F}_0| = w(X)$. We shall inductively define a sequence of collections $\mathcal{F}_i$.

Assume that $\mathcal{F}_i$ has already been defined. Let us first define a screening collection $\mathcal{S}_i$ for the invariant $\mathcal{K}$-Ind. If $0 \leq k \leq n$ and if an E in $\mathcal{F}_i$ has $\mathcal{K}$-Ind $E \leq k$, then for any pair (F, G) of elements of $\mathcal{F}_i$ with $F \cap G = \emptyset$ we select a partition S between F and G with $\mathcal{K}$-Ind$(E \cap S) \leq k-1$. Such a partition exists by Theorem II.2.21. Write $X \setminus S = U \cup V$ with disjoint open sets U and V such that $F \subset U$ and $G \subset V$. Let $J = U \cup S$ and $K = V \cup S$. The pair (J, K) is a screening of (F, G) with $\mathcal{K}$-Ind$(J \cap K \cap E) \leq k-1$. Let the screening collection $\mathcal{S}_i$ consist of all sets J and K that are obtained in this way.

Next we define a second screening collection $\mathcal{S}'_i$ for Ind in the corresponding way. That is, by using Proposition I.4.6, for each k in $\mathbb{N}$, each E in $\mathcal{F}_i$ with $\operatorname{Ind} E \leq k$ and each pair (F, G) of disjoint members of $\mathcal{F}_i$ we choose a screening (J, K) of (F, G) such that $\operatorname{Ind}(J \cap K \cap E) \leq k-1$. Let the screening collection $\mathcal{S}'_i$ consist of all sets J and K obtained in this way.

Finally we define $\mathcal{F}_{i+1} = \mathcal{F}_i^r \cup \mathcal{S}_i \cup \mathcal{S}'_i$ to complete the inductive construction. Let $\mathcal{F} = \bigcup\{\mathcal{F}_i : i \in \mathbb{N}\}$. Then the collection $\mathcal{F}$ is a normal base with $|\mathcal{F}| = w(X)$.

Let Y be the compactification $\omega(\mathcal{F}, X)$. Obviously, $w(Y) = w(X)$. As in Theorem 2.7, it can be shown that $\operatorname{Ind} Y \leq \operatorname{Ind} X$. We shall show that $\operatorname{Ind}\left[\operatorname{cl}_Y(E) \setminus E; \operatorname{cl}_Y(E)\right] \leq k$ holds whenever $\mathcal{K}$-Ind $E \leq k$ and $E \in \mathcal{F}$ hold, where $-1 \leq k \leq n$. The proof will be by induction on k. Only the inductive step is discussed. Assume $\mathcal{K}$-Ind $E \leq k$ and $E \in \mathcal{F}$. Then let F and G be disjoint closed subsets of Y.

By the compactness of Y there are disjoint sets A and B in $\mathcal{F}$ such that $F \subset \mathrm{cl}_Y(A)$ and $G \subset \mathrm{cl}_Y(B)$. Let (J, K) be a screening of (A, B) with J and K in $\mathcal{F}$ and $\mathcal{K}$-Ind$\,(J \cap K \cap E) \leq k - 1$. Then $\mathrm{cl}_Y(J) \cap \mathrm{cl}_Y(K)$ is a partition between F and G. We have $\big(\mathrm{cl}_Y(J) \cap \mathrm{cl}_Y(K)\big) \cap \mathrm{cl}_Y(E) = \mathrm{cl}_Y(J \cap K \cap E)$. The induction hypothesis gives

$$\mathrm{Ind}\,\big[\mathrm{cl}_Y(J \cap K \cap E) \setminus (J \cap K \cap E); \mathrm{cl}_Y(J \cap K \cap E)\big] \leq k - 1.$$

It follows that $\mathrm{Ind}\,\big[\mathrm{cl}_Y(E) \setminus E; \mathrm{cl}_Y(E)\big] \leq k$.

The strict inequality $\mathrm{Ind}\, Y < \mathrm{Ind}\, X$ is possible in the last theorem as the next example shows.

4.5. Example. We shall again use Roy's example Δ. Let $\mathcal{F}$ be the collection of all open-and-closed subsets of Δ. As $\mathrm{ind}\,\Delta = 0$, the collection $\mathcal{F}$ is a base for the closed sets of Δ. One readily sees that $\mathcal{F}$ is a normal base. Denote $\omega(\mathcal{F}, \Delta)$ by $\gamma\Delta$. As $\mathcal{F} = \mathcal{F}^c$, we find that $\mathrm{cl}_{\gamma\Delta}(F)$ is an open-and-closed subset of $\gamma\Delta$ for each F in $\mathcal{F}$. It follows that $\mathrm{ind}\,\gamma\Delta = 0$. As $\gamma\Delta$ is compact, we also have $\mathrm{Ind}\,\gamma\Delta = 0$. So $\mathrm{Ind}\,\gamma\Delta < \mathrm{Ind}\,\Delta$. A slight adaptation of this construction will result in a weight preserving compactification with the same property.

Now we shall address the problem of finding relations between the relative notion of $(\leq n)$-Inductionally embedded and the absolute notion of $\mathrm{Ind} \leq n$. The next theorem follows from Proposition I.4.6 by an easy inductive argument.

4.6. Theorem. *Suppose that Z is a subset of a hereditarily normal space X. Then*

$$\mathrm{Ind}\,[Z; X] \leq \mathrm{Ind}\, Z.$$

The reverse inequality need not be true as the following example will show.

4.7. Example. There exists a hereditarily normal space X with the properties that $\mathrm{Ind}\, X = 0$ holds and for every n in $\mathbb{N}$ there exists a subspace Z_n with $\mathrm{Ind}\, Z_n = n$. Such a space was constructed by Pol and Pol in [1979]. From $\mathrm{Ind}\, X = 0$ we have $\mathrm{Ind}\,[Z_n; X] \leq 0$.

It is not known whether Theorem 4.6 holds for normal spaces.

To find the relation between $\mathcal{K}$-Ind and $\mathcal{K}$-Def we shall first prove the following proposition.

4.8. Proposition. *Suppose that Z is a subset of a normal space X. Then*

$$\operatorname{ind} Z \leq \operatorname{Ind}[Z; X].$$

Proof. We prove the inductive step only. Suppose $\operatorname{Ind}[Z; X] \leq n$. Let G be a closed subset of Z and p be in $Z \setminus G$. Then $\{p\}$ and $\operatorname{cl}_X(G)$ are disjoint closed sets of X. There is a partition S between $\{p\}$ and $\operatorname{cl}_X(G)$ such that $\operatorname{Ind}[S \cap Z; S] \leq n - 1$. By the induction hypothesis $\operatorname{ind}(S \cap Z) \leq n - 1$. It follows that $\operatorname{ind} Z \leq n$.

We shall prove two results concerning the relation between $\mathcal{K}$-Ind and $\mathcal{K}$-Def.

4.9. Theorem. *Suppose that X is a separable metrizable space. Then*

$$\mathcal{K}\text{-Ind}\, X \geq \mathcal{K}\text{-Def}\, X.$$

Proof. Suppose $\mathcal{K}\text{-Ind}\, X \leq n$. Theorem 4.4 gives a metrizable compactification Y with $\operatorname{Ind}[Y \setminus X; Y] \leq n$. By the previous result the inequality $\operatorname{ind}(Y \setminus X) \leq n$ holds. The coincidence theorem gives $\operatorname{Ind}(Y \setminus X) \leq n$, whence $\mathcal{K}\text{-Def} \leq n$.

From the proof of the theorem it is clear that the equality of ind and Ind played an important role. It is not clear whether the detour via ind cannot be avoided.

4.10. Theorem. *Let the space X be Lindelöf at infinity. Suppose $\mathcal{K}\text{-Ind}\, X \leq 1$. Then*

$$\mathcal{K}\text{-Ind}\, X \geq \mathcal{K}\text{-Def}\, X.$$

Proof. Suppose that $\mathcal{K}\text{-Ind}\, X = n$ with $n = 0$ or 1. Theorem 4.3 gives $\operatorname{Ind}[\omega X \setminus X; \omega X] \leq n$. By Proposition 4.8, $\operatorname{ind}(\omega X \setminus X) \leq n$. The space $\omega X \setminus X$ is Lindelöf. We have shown in the proof of Theorem 3.12 that for the case $n = 0$ the inequality $\operatorname{Ind}(\omega X \setminus X) \leq 0$ holds, whence $\mathcal{K}\text{-Def}\, X \leq 0$. Suppose that $n = 1$. It will follow from a theorem of Vedenisoff [1939] (see Engelking [1978], Section 2.4) that $\operatorname{Ind}(\omega X \setminus X) \leq 1$ and consequently $\mathcal{K}\text{-Def}\, X \leq 1$.

5. Kimura's characterization of $\mathcal{K}$-def

This section is almost entirely devoted to the proof of Kimura's theorem which states that Skl $= \mathcal{K}$-def in the universe $\mathcal{M}_0$ of separable metrizable spaces. In Section I.6 it has already been proved that Skl $\le$ def $= \mathcal{K}$-def. So it still remains to prove that if $\operatorname{Skl} X \le n$, then there exists a weight preserving compactification Y of X such that $\operatorname{ind}(Y \setminus X) \le n$. Suppose that $\mathcal{B} = \{ U_i : i \in \mathrm{N} \}$ is a base for the open sets of X that witnesses the fact that $\operatorname{Skl} X \le n$, i.e., the intersection $\mathrm{B}(U_{i_0}) \cap \cdots \cap \mathrm{B}(U_{i_n})$ is compact for every $n+1$ indices $i_0 < \cdots < i_n$. Kimura's proof uses the beautiful idea of constructing a completely new base out of $\mathcal{B}$ instead of the instinctive idea of enlarging $\mathcal{B}$ to another base by some means. The intricate construction of this new base yields a countable normal base $\mathcal{F}$ that has a "dense" subcollection $\mathcal{G}$ with the property that every $n+1$ of its boundaries has a compact intersection. It will then follow that $\operatorname{ind}\bigl(\omega(\mathcal{F}, X) \setminus X\bigr) \le n$.

Agreement. *The universe of discourse in this section is* $\mathcal{M}_0$.

The nature of the problem under consideration requires the introduction of a substantial number of special notations.

5.1. Notation. Suppose that X is a space and that $\mathcal{H}$ is some collection of its subsets.

Let us recall the notation given in Section 2 that was used to form the following new collections from $\mathcal{H}$: $\mathcal{H}^\wedge$ is the collection of all sets that are finite intersections of elements of $\mathcal{H}$; $\mathcal{H}^\vee$ is the collection of all sets that are finite unions of elements of $\mathcal{H}$; $\mathcal{H}^c$ is the collection $\{ S : X \setminus S \in \mathcal{H} \}$; $\mathcal{H}^\perp$ is the collection of all sets that are closures of elements of $\mathcal{H}^c$. It is to be observed that $\mathcal{H}^{\vee\wedge} = \mathcal{H}^{\wedge\vee}$.

The notation for the topological operators of boundary and closure will be extended to collections $\mathcal{H}$ by

$$\mathrm{B}(\mathcal{H}) = \{ \mathrm{B}(S) : S \in \mathcal{H} \} \quad \text{and} \quad \operatorname{cl}(\mathcal{H}) = \{ \operatorname{cl}(S) : S \in \mathcal{H} \}.$$

We shall assume in the next definitions that $\mathcal{H}$ is countable and has been indexed by the set N as $\mathcal{H} = \{ S_i : i \in \mathrm{N} \}$. For each n define the collection

$$[\mathcal{H}]^n = \{ \{ S_{i_0}, \ldots, S_{i_{n-1}} \} : i_0 < \cdots < i_{n-1} \}.$$

Then write $\kappa(\mathcal{H}) \leq n$ if for each member $\{ S_{i_0}, \ldots, S_{i_n} \}$ of $[\mathcal{H}]^{n+1}$ the intersection $\mathrm{B}(S_{i_0}) \cap \cdots \cap \mathrm{B}(S_{i_n})$ is compact. With this notation, the definition of $\operatorname{Skl} X \leq n$ can be rephrased as: $\operatorname{Skl} X \leq n$ if there exists a base $\mathcal{B}$ for the open sets such that $\kappa(\mathcal{B}) \leq n$.

As before we shall reserve $\mathcal{B}$, possibly with index, to denote a base for the open sets of X and the letter $\mathcal{F}$ to denote a base for the closed sets of X. In addition, we use $\mathcal{A}$ to denote a locally finite open collection and $\mathcal{I}$ to denote the collection of all one-point sets $\{ x \}$ formed from the isolated points x of X. It is to be noted that both $\mathcal{A}$ and $\mathcal{I}$ are at most countable and thus may be indexed by a subset of $\mathbb{N}$.

Let us remark that the use of script letters in this section has deviated from their earlier use as classes of spaces. As this section will be dealing only with the class $\mathcal{K}$ of compact spaces, there should be no confusion arising from the new convention in this section. We shall reserve the use of script letters to denote collections that are related to bases.

We begin by stating two lemmas whose easy proofs will be omitted.

5.2. Lemma. *In a space X let $\mathbf{V} = \{ V_\alpha : \alpha \in A \}$ be an open collection that admits a locally finite refinement $\{ W_\alpha : \alpha \in A \}$ by a collection of nonempty sets such that $W_\alpha \subset V_\alpha$, $\alpha \in A$. Let $\mathcal{B}$ be a base for the open sets. Then $(\mathcal{B} \setminus \mathbf{V}) \cup (\mathbf{V} \cap \mathcal{I})$ is also a base.*

5.3. Lemma. *If $\mathbf{L}$ is a locally finite collection of sets in a space X and $\mathcal{B}$ is any base for the open sets of X, then there is a subcollection $\mathcal{B}_1$ of $\mathcal{B}$ such that $\mathcal{B}_1$ is a base and the collection $\{ L : L \in \mathbf{L},\ L \cap U \neq \emptyset \}$ is finite for each U in $\mathcal{B}_1$.*

The following lemma has been implicitly established in the proof of Theorem I.12.1.

5.4. Lemma. *Let $\mathcal{B}$ be a base for the open sets of a space X. Suppose that F is a closed set and V is an open set such that $F \subset V$. Then there is a locally finite open collection $\mathcal{A}$ such that*

(a) $F \subset \bigcup \mathcal{A} \subset \bigcup \mathrm{cl}(\mathcal{A}) \subset V$,
(b) *for each U in $\mathcal{A}$ there is a finite subcollection $\boldsymbol{E}(U)$ of $\mathcal{B}$ such that $\mathrm{B}(U) \subset \bigcup \mathrm{B}(\boldsymbol{E}(U))$ and $U \cap \left(\bigcup \mathrm{B}(\boldsymbol{E}(U))\right) = \emptyset$.*

In the proof of Theorem I.12.1 we used a trick to avoid the indexing pitfall discussed in Example I.6.13. The same pitfall must be avoided in the proof of Kimura's characterization. The next lemma provides a second trick.

5.5. Lemma. *For each base $\mathcal{B}$ there are subcollections $\mathcal{B}^{(1)}$ and $\mathcal{B}^{(2)}$ of $\mathcal{B}$ such that $\mathcal{B}^{(1)}$ and $\mathcal{B}^{(2)}$ are bases and $\mathcal{B}^{(1)} \cap \mathcal{B}^{(2)} = \mathcal{I}$.*

Proof. The first two steps of the inductive proof will be indicated. Let $\mathcal{B}_1$ be the collection of all sets in $\mathcal{B}$ with diameter less than 1. The collection $\mathcal{B}_1$ is a base. Since $\mathcal{B}_1$ is a cover, it has a locally finite open refinement $\boldsymbol{W}_1$. There is a subcollection $\mathcal{V}_1$ of $\mathcal{B}_1$ such that $\boldsymbol{W}_1$ is a shrinking of $\mathcal{V}_1$. By Lemma 5.2 the collection $(\mathcal{B}_1 \setminus \mathcal{V}_1) \cup (\mathcal{V}_1 \cap \mathcal{I})$ is a base for the open sets. Let $\mathcal{B}_2$ be the collection of all elements of this base with diameter less than $\frac{1}{2}$. As above, there is a subcollection $\mathcal{V}_2$ of $\mathcal{B}_2$ that has a locally finite shrinking. The induction is completed in the obvious way. When the inductive construction has been completed, we define $\mathcal{B}^{(1)} = \mathcal{I} \cup \left(\bigcup\{\, \mathcal{V}_i : i \text{ is odd}\,\}\right)$ and $\mathcal{B}^{(2)} = \mathcal{I} \cup \left(\bigcup\{\, \mathcal{V}_i : i \text{ is even}\,\}\right)$.

The following is the key lemma. It is, in fact, the inductive step in the construction of the base that was mentioned in the introduction.

5.6. Lemma. *Let $\mathcal{A}$ be a locally finite collection of regularly open sets of a space X. Suppose that $\mathcal{B}$ is a base for the open sets of X consisting of regularly open sets such that $\kappa(\mathcal{A} \cup \mathcal{B}) \leq n$. Further suppose that $\boldsymbol{E}$ is a finite subcollection of $\mathcal{A}$, F is a closed set and V is an open set such that*

$$F \subset V \subset X \setminus \left(\bigcup \mathrm{B}(\boldsymbol{E})\right) \quad \text{and} \quad \mathrm{B}(V) \subset \bigcup \mathrm{B}(\boldsymbol{E}).$$

Then there exist a regularly open set U and a subcollection $\mathcal{B}_1$ of $\mathcal{B}$ such that

(a) $F \subset U \subset \mathrm{cl}(U) \subset V$,
(b) *$\mathcal{B}_1$ is a base for the open sets,*
(c) $\kappa(\mathcal{A} \cup \{\, U \,\} \cup \mathcal{B}_1) \leq n$.

Proof. In view of Lemma 5.2 we may assume $\mathcal{A} \cap \mathcal{B} \subset \mathcal{I}$. Let $\mathcal{B}^{(1)}$ and $\mathcal{B}^{(2)}$ be bases as in Lemma 5.5. Define $\boldsymbol{H}$ to be the collection $\{\, \bigcap \mathrm{B}(\mathcal{A}') : \mathcal{A}' \in [\mathcal{A} \setminus \boldsymbol{E}]^n \,\}$. As $\mathcal{A}$ is a locally finite collection, it will follow that $\boldsymbol{H}$ is also locally finite and therefore countable as well. Since $\boldsymbol{E}$ is a finite subcollection of $\mathcal{A}$ such that $\mathrm{B}(V) \subset \bigcup \mathrm{B}(\boldsymbol{E})$ and $\kappa(\mathcal{A}) \leq n$, we have that $H \cap \mathrm{B}(V)$ is compact for each H in $\boldsymbol{H}$. Let $\{ H_i : i \in \mathrm{N} \}$ be an indexing of the collection $\boldsymbol{H}$. If $\boldsymbol{H}$ happens to be finite, then we complete the listing with empty sets. We make two observations. First, each point of an H_i is a limit point of X. Second, there exists a locally finite open cover of X such that for each element D of this cover we have $\mathrm{cl}(D) \cap H_i \neq \emptyset$ for only finitely many i. Using these observations and an inductive construction, we have a sequence $\mathbf{V}_i$, $i \in \mathrm{N}$, of disjoint finite subcollections of $\mathcal{B}^{(1)} \setminus \mathcal{I}$ such that for each i in N

(i) $H_i \cap \mathrm{B}(V) \subset \bigcup \mathbf{V}_i$,
(ij) $\bigcup \{ \mathbf{V}_j : j \in \mathrm{N} \}$ is a locally finite collection,
(iij) $\big(\mathrm{cl}(\bigcup \mathbf{V}_i)\big) \cap F = \emptyset$,
(iv) $\big(\mathrm{cl}(\bigcup \mathbf{V}_i)\big) \cap \big(H_j \setminus (\bigcup \mathbf{V}_j)\big) = \emptyset$ for each j that is less than i.

Let $F' = F \cup \bigcup \{ \big(H_i \cap \mathrm{cl}(V)\big) \setminus \big(\bigcup \mathbf{V}_i\big) : i \in \mathrm{N} \}$. By the first observation made above we may assume that $\bigcup \{ \mathbf{V}_j : j \in \mathrm{N} \}$ and $\mathcal{I}$ are disjoint. Then by (i) and (ij) we have $F \subset F' \subset V$. Since $\boldsymbol{H}$ is a locally finite collection, it will follow that F' is closed. By (ij) and Lemma 5.2 we have that $\mathcal{B}^{(3)} = \mathcal{B}^{(1)} \setminus \bigcup \{ \mathbf{V}_j : j \in \mathrm{N} \}$ is a base. There exists by Lemma 5.4 a locally finite subcollection $\boldsymbol{W}$ of $\mathcal{B}^{(3)}$ with the property that if $W = \bigcup \boldsymbol{W}$, then

- $F' \subset W \subset \mathrm{cl}(W) \subset V$,
- $\{\mathrm{B}(W) \cap \mathrm{B}(D) : D \in \boldsymbol{W}\}$ is a locally finite collection whose union is $\mathrm{B}(W)$.

Since $\mathrm{B}(W) \cap \mathrm{B}(D) \neq \emptyset$ implies that D has more than one point, we may assume $\boldsymbol{W} \subset (\mathcal{B}^{(1)} \setminus \mathcal{I})$ because the set $F' \cap \big(\bigcup(\boldsymbol{W} \cap \mathcal{I})\big)$ is an open-and-closed set contained in F' which can be added back in at the end of the argument. Let

$$U = W \setminus \bigcup \{ \mathrm{cl}(\bigcup \mathbf{V}_i) : i \in \mathrm{N} \}.$$

From (ij) we have that U is open and $\big\{ \mathrm{B}(D) : D \in \bigcup \{ \mathbf{V}_i : i \in \mathrm{N} \} \big\}$ is a locally finite collection. Let $\mathbf{V}$ be the union of the two collections $\bigcup \{ \mathbf{V}_i : i \in \mathrm{N} \}$ and $\boldsymbol{W}$. Then $\mathbf{V} \subset \mathcal{B}^{(1)} \setminus \mathcal{I}$ and also the collection $\boldsymbol{L} = \{ \mathrm{B}(U) \cap \mathrm{B}(D) : D \in \mathbf{V} \}$ is locally finite and $\bigcup \boldsymbol{L} = \mathrm{B}(U)$. By Lemma 5.3 there is a subcollection $\mathcal{B}_1$ of $\mathcal{B}^{(2)}$ such that $\mathcal{B}_1$ is a base and the collection $\{ E \in \boldsymbol{L} : E \cap D \neq \emptyset \}$ is finite for each D in $\mathcal{B}_1$. It follows that condition (b) is satisfied. We can verify condition (a) by using (i) and (iij) and the inclusion $F \subset F'$. Condition (c) remains to be verified. Let

$$\mathcal{G} \in \{ \mathrm{B}(\mathcal{A}') : \mathcal{A}' \in [\mathcal{A} \cup \{ U \} \cup \mathcal{B}_1]^{n+1} \}.$$

Only those $\mathcal{G}$ with $\mathrm{B}(U) \in \mathcal{G}$ need to be considered. For such a $\mathcal{G}$ we have either $\mathcal{G} \cap \mathrm{B}(\mathcal{B}_1) = \emptyset$ or $\mathcal{G} \cap \mathrm{B}(\mathcal{B}_1) \neq \emptyset$.

The case $\mathcal{G} \cap \mathrm{B}(\mathcal{B}_1) = \emptyset$.
In this case, $\mathcal{G} \setminus \{ \mathrm{B}(U) \}$ is a member of $\{ \mathrm{B}(\mathcal{A}') : \mathcal{A}' \in [\mathcal{A}]^n \}$. As the inclusions $V \subset X \setminus \bigcup \mathrm{B}(\boldsymbol{E})$ and $\bigcap \mathcal{G} \subset \mathrm{B}(U) \subset V$ hold, we have that $\mathcal{G} \setminus \{ \mathrm{B}(U) \}$ is a member of $\boldsymbol{H}$. So there is an i in $\mathbb{N}$ such that $H_i = \bigcap(\mathcal{G} \setminus \{ \mathrm{B}(U) \})$ holds. Let us show $\bigcap \mathcal{G} = H_i \cap \mathrm{B}(U)$ is compact. Since for each i in $\mathbb{N}$

$$(H_i \cap \mathrm{cl}(V)) \setminus \bigcup \mathbf{V}_i \subset F' \subset W \subset \mathrm{cl}(W) \subset V,$$

we have

$$\begin{aligned} H_i \cap \mathrm{B}(W) &= H_i \cap (\mathrm{cl}(W) \setminus W) \\ &\subset H_i \cap \Big(V \setminus ((H_i \cap \mathrm{cl}(V)) \setminus \bigcup \mathbf{V}_i)\Big) \\ &= (H_i \cap V \cap (\bigcup \mathbf{V}_i)) \cup \Big((H_i \cap V) \setminus (H_i \cap \mathrm{cl}(V))\Big) \\ &\subset \bigcup \mathbf{V}_i. \end{aligned}$$

By (iv) we have $H_i \cap \mathrm{B}(\bigcup \mathbf{V}_j) \subset \bigcup \mathbf{V}_i$ for each j that is larger than i. Consequently,

$$\begin{aligned} H_i \cap \mathrm{B}(U) &\subset \Big(H_i \cap (\mathrm{B}(W) \cup \bigcup \{ \mathrm{B}(\bigcup \mathbf{V}_j) : j \in \mathbb{N} \})\Big) \setminus \bigcup \mathbf{V}_i \\ &\subset H_i \cap \bigcup \{ \mathrm{B}(\bigcup \mathbf{V}_j) : j \leq i \} \\ &\subset H_i \cap \bigcup \{ \mathrm{B}(D) : D \in \bigcup \{ \mathbf{V}_j : j \leq i \} \}. \end{aligned}$$

Since $\bigcup \{ \mathbf{V}_j : j \in \mathbb{N} \} \subset \mathcal{B}^{(1)} \setminus \mathcal{I}$ and $\kappa(\mathcal{A} \cup \mathcal{B}^{(1)}) \leq n$ hold, the compactness of $\bigcap \mathcal{G} = H_i \cap \mathrm{B}(U)$ will now follow.

The case $\mathcal{G} \cap \mathrm{B}(\mathcal{B}_1) \neq \emptyset$.
Let $B(D_0) \in \mathcal{G} \cap \mathrm{B}(\mathcal{B}_1)$. We may assume

$$\emptyset \neq \mathrm{B}(D_0) \cap \mathrm{B}(U) = \bigcup \{ \mathrm{B}(D_0) \cap \mathrm{B}(U) \cap \mathrm{B}(D) : D \in \mathbf{V} \}.$$

Note that $\mathrm{B}(U) \cap \mathrm{B}(D) \in \boldsymbol{L}$ for each D in $\mathbf{V}$. The collection

$$\mathbf{V}' = \{ D : D \in \mathbf{V}, \mathrm{B}(D_0) \cap \mathrm{B}(U) \cap \mathrm{B}(D) \neq \emptyset \}$$

is finite by the construction of $\mathcal{B}_1$. Now consider the collection $\mathcal{G}' = \mathcal{G} \setminus \{ \mathrm{B}(D_0), \mathrm{B}(U) \}$. We have that $\mathcal{G}'$ is a member of the collection $\{ \mathrm{B}(\mathcal{A}'') : \mathcal{A}'' \in [\mathcal{A} \cup \mathcal{B}_1]^{n-1} \}$. If $D \in \mathbf{V}'$, then $\mathrm{B}(D) \neq \emptyset$; so $D \in \mathcal{B}^{(1)} \setminus \mathcal{B}^{(2)}$ holds. Also $\mathrm{B}(D_0) \neq \emptyset$ yields $D_0 \in \mathcal{B}^{(2)} \setminus \mathcal{B}^{(1)}$. Finally

$$\begin{aligned} \bigcap \mathcal{G} &= (\bigcap \mathcal{G}') \cap \mathrm{B}(D_0) \cap \mathrm{B}(U) \\ &= \bigcup \{ (\bigcap \mathcal{G}') \cap \mathrm{B}(D_0) \cap \mathrm{B}(U) \cap \mathrm{B}(D) : D \in \mathbf{V}' \} \end{aligned}$$

and the compactness of $\bigcap \mathcal{G}$ follows from $\kappa(\mathcal{A} \cup \mathcal{B}) \leq n$ and the finiteness of $\mathbf{V}'$.

The lemma is completely proved.

Before embarking on the proof of Kimura's theorem we shall state another lemma whose easy proof will be omitted.

5.7. Lemma. *Let $\mathcal{A}$ and $\mathcal{B}$ be as in Lemma 5.6. Suppose that $\mathbf{E}$ is a finite subcollection of $\mathcal{A}$ and let $U = \bigcup \mathbf{E}$. Then*

$$\kappa \big((\mathcal{A} \setminus \mathbf{E}) \cup \{ U \} \cup \mathcal{B} \big) \leq n.$$

Kimura's theorem will be proved by the existence of an appropriate Wallman-type compactification. The next lemma provides the required normal base.

5.8. Lemma. *Let n be a natural number and X be a space such that $0 \leq \operatorname{Skl} X \leq n$. Then there are collections $\mathcal{F}$ and $\mathcal{G}$ of regularly closed subsets of X such that*

(a) *$\mathcal{F}$ is a countable normal base for the closed sets,*
(b) *$\mathcal{G} \subset \mathcal{F}$ holds and $\mathcal{G}$ can be represented as $\mathcal{G} = \{ G_i : i \in \mathrm{N} \} \cup \{ \mathrm{cl}(X \setminus G_i) : i \in \mathrm{N} \}$,*
(c) *for each pair of disjoint elements of $\mathcal{F}$ there is a screening by a pair $\big(G_i, \mathrm{cl}(X \setminus G_i)\big)$ of elements of $\mathcal{G}$,*
(d) *$\kappa \big(\{ X \setminus G_i : i \in \mathrm{N} \} \big) \leq n$.*

TABLE

$$\begin{array}{lllll}
(S_{00}, T_{00}) & (S_{01}, T_{01}) & (S_{02}, T_{02}) & (S_{03}, T_{03}) & \cdots \\
(S_{10}, T_{10}) & (S_{11}, T_{11}) & (S_{12}, T_{12}) \ldots\ldots\ldots\ldots & & \\
(S_{20}, T_{20}) & (S_{21}, T_{21}) \ldots\ldots\ldots\ldots\ldots\ldots & & & \\
\ldots\ldots\ldots\ldots\ldots\ldots\ldots\ldots\ldots\ldots & & & &
\end{array}$$

The proof of the lemma is not very difficult, but it will require a lot of careful bookkeeping. To make the bookkeeping easier, the above table will be used.

Proof. Let $\mathcal{B}$ be a countable base for the open sets of X such that $\kappa(\mathcal{B}) \leq n$. We may assume that each element of $\mathcal{B}$ is regularly open. Define $\mathcal{H} = \{\operatorname{cl}(U) : U \in \mathcal{B}\} \cup \{(X \setminus U) : U \in \mathcal{B}\}$. Then list the family of all ordered pairs of disjoint elements of $\mathcal{H}$ as $\{(S_{i0}, T_{i0}) : i \in \mathbb{N}\}$ and use this listing to form the 0-th column of the table. The remaining columns of the table will be inductively defined.

To aid the reader, the construction will be briefly outlined. The idea of the proof is to inductively define screenings of disjoint pairs. The elements of these screenings will be the members of the base $\mathcal{F}$ that is to be constructed. The listed pairs in the 0-th column will guarantee that $\mathcal{F}$ will be a base for the closed sets. The other columns will be used to assure that $\mathcal{F}$ is a normal base. That is, immediately after a screening has been defined, a new column is made by listing all of the newly formed disjoint pairs for which a screening must be defined at a later stage of the construction. To accomplish this, the canonical ordering of pairs of natural numbers is used. That is, the pair (S_{kl}, T_{kl}) is indexed by $j = \varphi(k, l)$, where

$$\varphi(k, l) = 0 + 1 + 2 + \cdots + (k + l) + l.$$

With this outline in mind, we continue the proof.

Inductively on $j = \varphi(k,l)$ we shall define

(i) a collection $\{ (S_{i\,j+1}, T_{i\,j+1}) : i \in \mathbb{N} \}$ (which will become the $(j+1)$-th column),
(ij) a pair of regularly closed sets F_j and $G_j = \mathrm{cl}(X \setminus F_j)$,
(iij) a locally finite collection $\mathcal{A}_j$ of regularly open sets,
(iv) a base $\mathcal{B}_j$ for the open sets consisting of regularly open sets

so that they collectively satisfy the following conditions for each j in $\mathbb{N}$:

(1) $S_{i\,j+1}$ and $T_{i\,j+1}$ are disjoint closed sets for each i in $\mathbb{N}$.
(2) If $j = \varphi(k,l)$ and $l > 1$, then the pair (F_j, G_j) is a screening of (S_{kl}, T_{kl}), so $S_{kl} \cap G_j = \emptyset$, $F_j \cap T_{kl} = \emptyset$ and $F_j \cup G_j = X$.
(3) $X \setminus G_j \in \mathcal{A}_j$.
(4) $\mathcal{A}_j \subset \mathcal{A}_{j+1}$ and $\mathcal{B}_{j+1} \subset \mathcal{B}_j$.
(5) $\kappa(\mathcal{A}_j \cup \mathcal{B}_j) \leq n$.
(6) For each i in $\mathbb{N}$ the pair $(S_{i\,j+1}, T_{i\,j+1})$ is of one of the following two types:
Type 1: There is a finite subcollection $\boldsymbol{E}$ of $\mathcal{A}_j$ such that $S_{i\,j+1} \subset X \setminus T_{i\,j+1} \subset X \setminus \bigcup \mathrm{B}(\boldsymbol{E})$ and $\mathrm{B}(T_{i\,j+1}) \subset \bigcup \mathrm{B}(\boldsymbol{E})$.
Type 2: There is a finite collection of distinct pairs (i_v, j_v), $0 \leq v \leq u$, each of whose indices $\varphi(i_v, j_v)$ are at most j and whose corresponding pairs $(S_{i_v\,j_v}, T_{i_v\,j_v})$ are of type 1 such that
$$S_{i\,j+1} = \bigcup \{ S_{i_v\,j_v} : 1 \leq v \leq u \},$$
$$T_{i\,j+1} = \bigcap \{ T_{i_v\,j_v} : 1 \leq v \leq u \}.$$
(That is, the pair $(S_{i\,j+1}, T_{i\,j+1})$ is constructed out of finitely many pairs of the first type that have been previously defined.)

Suppose $j = 0$. Then $j = \varphi(0,0)$. By Lemma 5.4 there is a locally finite open collection $\mathcal{A}'_0$ such that

- $S_{00} \subset \bigcup \mathcal{A}'_0 \subset \bigcup \mathrm{cl}(\mathcal{A}'_0) \subset X \setminus T_{00}$,
- for each U in $\mathcal{A}'_0$ there is a finite subcollection $\boldsymbol{E}(U)$ of $\mathcal{B}$ such that $\mathrm{B}(U) \subset \bigcup \mathrm{B}(\boldsymbol{E}(U))$ and $U \cap \left(\bigcup \mathrm{B}(\boldsymbol{E}(U))\right) = \emptyset$.

We may assume further that each U in $\mathcal{A}'_0$ is a regularly open set. Because $\mathcal{A}'_0$ is locally finite, it is countable and can be denoted as $\mathcal{A}'_0 = \{ U_i : i \in \mathbb{N} \}$. There is a closed shrinking $\boldsymbol{L}_0 = \{ L_i : i \in \mathbb{N} \}$ of $\mathcal{A}'_0$ that covers S_{00}. The collection of all pairs $(L_i \cap S_{00}, X \setminus U_i)$ is then listed in the 1-th column as (S_{i1}, T_{i1}). It can also be assumed that each S_{i1} is regularly closed. Note that all of these pairs are of type 1. We let $F_0 = \emptyset$, $G_0 = X$, $\mathcal{A}_0 = (\mathcal{A}'_0 \cap \mathcal{B}) \cup \{\emptyset\}$ and $\mathcal{B}_0 = (\mathcal{B} \setminus \mathcal{A}_0) \cup (\mathcal{A}_0 \cap \mathcal{I})$. By Lemma 5.2, $\mathcal{B}_0$ is a base. It is easily seen that the conditions (1) through (6) are satisfied for $j = 0$.

Suppose that $j \geq 1$ and that the constructions and definitions have been made for all natural numbers smaller than j. There are two cases to consider.

Case 1: $j = \varphi(k,l)$ *and* $l = 0$.
This case is similar to the case $j = 0$. By Lemma 5.4 there is a locally finite open collection $\mathcal{A}'_j$ such that

- $S_{k0} \subset \bigcup \mathcal{A}'_j \subset \bigcup \mathrm{cl}(\mathcal{A}'_j) \subset X \setminus T_{k0}$,
- for each U in $\mathcal{A}'_j$ there is a finite subcollection $\boldsymbol{E}(U)$ of $\mathcal{B}$ such that $\mathrm{B}(U) \subset \bigcup \mathrm{B}(\boldsymbol{E}(U))$ and $U \cap (\bigcup \mathrm{B}(\boldsymbol{E}(U))) = \emptyset$.

Write $\mathcal{A}'_j = \{U_i : i \in \mathrm{N}\}$. Let $\boldsymbol{L}_0 = \{L_i : i \in \mathrm{N}\}$ be a closed shrinking of $\mathcal{A}'_j$ that covers S_{k0}. List the pairs $(L_i \cap S_{00}, X \setminus U_i)$ in the $(j+1)$-th column as $(S_{i\,j+1}, T_{i\,j+1})$. We may also assume that these pairs consist of regularly closed sets. Note that all of these pairs are also of type 1. Define the collections $\mathcal{A}_j = \mathcal{A}_{j-1} \cup (\mathcal{A}'_j \cap \mathcal{B}_{j-1})$, $\mathcal{B}_j = (\mathcal{B}_{j-1} \setminus \mathcal{A}_j) \cup (\mathcal{A}_j \cap \mathcal{I})$, and the sets $F_j = \emptyset$, $G_j = X$. Again it is easily verified that the conditions (1) through (6) are satisfied for j.

Case 2: $j = \varphi(k,l)$ *and* $l > 0$.
Here the pair (S_{kl}, T_{kl}) has been determined at an earlier stage. If $S_{kl} = \emptyset$, then we let $(F_j, G_j) = (S_{i\,j+1}, T_{i\,j+1}) = (\emptyset, X)$ for each i, $\mathcal{A}_j = \mathcal{A}_{j-1}$ and $\mathcal{B}_j = \mathcal{B}_{j-1}$. This is the trivial case. There are two more cases to be considered.

Case 2A: *The pair* (S_{kl}, T_{kl}) *is of type 1.*
By the construction there is a finite subcollection $\boldsymbol{E}_k$ of $\mathcal{A}_{l-1}$ such that $S_{kl} \subset X \setminus T_{kl} \subset X \setminus \bigcup \mathrm{B}(\boldsymbol{E}_k)$ and $\mathrm{B}(T_{kl}) \subset \bigcup \mathrm{B}(\boldsymbol{E}_k)$. Note that $\mathcal{A}_{l-1} \subset \mathcal{A}_{j-1}$. By Lemma 5.6 there exists a regularly open set U and a subcollection $\mathcal{B}_j$ of $\mathcal{B}_{j-1}$ such that

- $S_{kl} \subset U \subset \mathrm{cl}(U) \subset X \setminus T_{kl}$,
- $\mathcal{B}_j$ is a base for the open sets,
- $\kappa(\mathcal{A}_{j-1} \cup \{U\} \cup \mathcal{B}_j) \leq n$.

We define $F_j = \mathrm{cl}(U)$, $G_j = \mathrm{cl}(X \setminus U)$ and $\mathcal{A}_j = \mathcal{A}_{j-1} \cup \{U\}$. It is easily seen that the conditions (2), (3), (4) and (5) are satisfied. Define

$$\mathcal{D}_j = \{F_i : i \leq j\} \cup \{G_i : i \leq j\}.$$

Let us consider the finite collection of all disjoint nonempty pairs of the collection $\mathcal{D}_j{}^{\vee\wedge}$. Let (S,T) be such a pair. Then T can be represented as the intersection $T = T_1 \cap \cdots \cap T_u$ for some finite subcollection T_v, $1 \le v \le u$, of $\mathcal{D}_j{}^{\vee}$. Let $\{L_1, \ldots, L_k\}$ be a closed shrinking of $\{T_1, \ldots, T_k\}$ that covers S. Then $(L_v \cap S, T_v)$ is a pair of type 1 for each v. So we have that (S,T) is a pair of type 2 for the index $j+1$ when $u > 1$. We list the collection of all such pairs $(L_v \cap S, T_v)$ formed from all disjoint nonempty pairs of $\mathcal{D}_j{}^{\vee\wedge}$ in the $(j+1)$-th column and complete the listing with the pair $(\emptyset, X)$.

Case 2B: The pair (S_{kl}, T_{kl}) is of type 2.
There is a finite collection of distinct pairs (i_v, j_v), $0 \le v \le u$, with type 1 pairs $(S_{i_v\, j_v}, T_{i_v\, j_v})$ having indices $\varphi(i_v, j_v)$ at most j such that

$$S_{i\, j+1} = \bigcup\{ S_{i_v\, j_v} : 1 \le v \le u\}$$
$$T_{i\, j+1} = \bigcap\{ T_{i_v\, j_v} : 1 \le v \le u\}.$$

By a simple inductive process similar to that of case 2A we can construct regularly open sets $U_0, \ldots, U_u$ and a subcollection $\mathcal{B}_j$ of $\mathcal{B}_{j-1}$ such that

- $S_{i_v j_v} \subset U_v \subset \mathrm{cl}(U_v) \subset X \setminus T_{i_v j_v}$, $0 \le v \le u$,
- $\mathcal{B}_j$ is a base for the open sets,
- $\kappa(\mathcal{A}_{j-1} \cup \{U_0\} \cup \cdots \cup \{U_u\} \cup \mathcal{B}_j) \le n$.

Let $U = U_0 \cup \cdots \cup U_k$. Then in view of Lemma 5.7,

- $S_{kl} \subset U \subset \mathrm{cl}(U) \subset X \setminus T_{kl}$,
- $\kappa(\mathcal{A}_{j-1} \cup \{U\} \cup \mathcal{B}_j) \le n$.

The rest of the construction, in particular the definitions of the pair (F_j, G_j), the collection $\mathcal{A}_j$, the collection $\mathcal{D}_j$ and the listing of the disjoint pairs arising from $\mathcal{D}_j{}^{\vee\wedge}$ in the $(j+1)$-th column, is the same as in case 2A. This completes the inductive construction.

Finally, we put

$$\mathcal{F} = \bigcup\{ \mathcal{D}_j{}^{\vee\wedge} : j = \varphi(k,l),\, k \in \mathbb{N},\, l \in \mathbb{N} \setminus \{0\}\}$$
$$\mathcal{G} = \{G_i : i \in \mathbb{N}\} \cup \{\mathrm{cl}(X \setminus G_i) : i \in \mathbb{N}\}.$$

The only condition that might not be clear at the first glance is that $\mathcal{F}$ is a disjunctive base. Suppose that K is a closed set and x is not a member of K. Let (S_{k0}, T_{k0}) be the pair with minimal k

such that $x \in S_{k0}$ and $K \subset X \setminus T_{k0}$ hold. Write $j = \varphi(k, 0)$. Then let $(S_{i\,j+1}, T_{i\,j+1})$ be the pair with minimal i such that $x \in S_{i\,j+1}$ and $T_{k0} \subset T_{i\,j+1}$ hold. With $l = \varphi(i, j+1)$ we have that (F_l, G_l) is a screening of $(S_{i\,j+1}, T_{i\,j+1})$ and therefore of $(\{x\}, K)$ as well. This completes the proof of Lemma 5.8.

In the proof of the last lemma we have done all the hard work for a proof of the final solution of de Groot's problem.

5.9. Theorem (Kimura [1988]). *Let $n \geq 0$. Then*

$$\operatorname{Skl} X \leq n \quad \textit{if and only if} \quad \operatorname{def} X \leq n.$$

Proof. As we have remarked in the introduction, it remains to be proved that $\operatorname{def} X \leq n$ when $\operatorname{Skl} X \leq n$. Suppose that $\mathcal{F}$ and $\mathcal{G}$ are the collections that are provided by Lemma 5.8. Let $Y = \omega(\mathcal{F}, X)$. In view of (a) and (b), we have that $\mathrm{B}\,(G)$ is in $\mathcal{F}$ for every G in $\mathcal{G}$. As $\mathrm{B}_X(G) = G \cap \mathrm{cl}_X(X \setminus G)$, by Theorem 2.2 we have

$$\mathrm{cl}_Y(\mathrm{B}_X(G)) = \mathrm{cl}_Y(G) \cap \mathrm{cl}_Y(X \setminus G) = \mathrm{B}_Y(\mathrm{cl}_Y(G)).$$

The sets $\mathrm{cl}_Y(G)$, $\mathrm{cl}_Y(X \setminus G)$, $\mathrm{ex}_Y(\mathrm{int}_X G)$ and $\mathrm{ex}_Y(X \setminus G)$ all have the same boundary, namely $\mathrm{cl}_Y(\mathrm{B}_X(G))$. From (c) of Lemma 5.8 it follows that $\mathcal{G}$ is a base for the closed sets of X and also that $\mathrm{cl}_Y(\mathcal{G})$ is a base for the closed sets of Y. In view of Theorem 2.2, we have

$$\begin{aligned}
\mathrm{B}_Y(\mathrm{ex}_Y(X \setminus G_{i_0})) \cap \cdots \cap \mathrm{B}_Y(\mathrm{ex}_Y(X \setminus G_{i_n}))& \\
&= \mathrm{cl}_Y(\mathrm{B}_X(G_{i_0})) \cap \cdots \cap \mathrm{cl}_Y(\mathrm{B}_X(G_{i_n})) \\
&= \mathrm{cl}_Y\big(\mathrm{B}_X(G_{i_0}) \cap \cdots \cap \mathrm{B}_X(G_{i_n})\big) \\
&= \mathrm{B}_X(G_{i_0}) \cap \cdots \cap \mathrm{B}_X(G_{i_n})
\end{aligned}$$

for any $n+1$ indices $i_0 < \cdots < i_n$. The last equality holds because the last set is compact. So $\{\mathrm{ex}_Y(X \setminus G_i) \cap (Y \setminus X) : i \in \mathbb{N}\}$ is a collection of open sets of $Y \setminus X$ that satisfies the requirements of Theorem V.4.5 and Corollary V.4.4 with $\mathcal{P} = \{\emptyset\}$. It follows that $\operatorname{ind}(Y \setminus X) \leq n$. The last result can also be deduced directly along the same lines as the proof of Lemma I.6.10.

There are two corollaries. The first one is an immediate consequence of Kimura's theorem and Proposition V.3.5.

5.10. Corollary. *If X is a space with $0 \leq \operatorname{def} X$, then for every open subspace Y of X*

$$\operatorname{def} Y \leq \operatorname{def} X.$$

In many examples, notably Pol's example, a space is constructed by attaching a locally compact space to a compact space. The next corollary gives an upper bound for the deficiency of the resulting space.

5.11. Corollary. *Let $X = Y \cup Z$, where Y is closed and Z is not compact. Then*

$$\operatorname{def} X \leq \operatorname{ind} Y + \operatorname{def} Z + 1.$$

Proof. The theorem is obvious for $\operatorname{ind} Y = -1$. So we may assume $\operatorname{ind} Y = n \geq 0$. Let $\operatorname{def} Z = m$. As Z is not compact, we have $m \geq 0$. By the previous corollary, $\operatorname{def}(Z \setminus Y) \leq \operatorname{def} Z \leq m$. Consequently we may assume that Y and Z are disjoint. Suppose that $\mathcal{B}$ is a countable base for the open sets. Consider the countable collection $\{ (F_i, U_i) : i \in \mathbb{N} \}$ of all pairs F_i from $\operatorname{cl}_X(\mathcal{B})$ and U_i from $\mathcal{B}$ such that $F_i \subset U_i$ and $U_i \cap Y = \emptyset$ or $F_i \cap Y \neq \emptyset$. If $F_i \cap Y \neq \emptyset$, then we replace F_i with $F_i \cap Y$. It is easily seen that if $\{ V_i : i \in \mathbb{N} \}$ is an open collection such that $F_i \subset V_i \subset \operatorname{cl}(V_i) \subset U_i$, then $\{ V_i : i \in \mathbb{N} \}$ is a base for the open sets. The indexing set $\mathbb{N}$ is the disjoint union of $N_1 = \{ i : U_i \cap Y = \emptyset \}$ and $N_2 = \{ i : Y \cap F_i \neq \emptyset \}$. Let $\mathcal{B}'$ be a base that witnesses the fact that $\operatorname{Skl} Z \leq m$. For each i in N_1 for which there exists a V in $\mathcal{B}'$ such that $F_i \subset V \subset U_i$ we select one such V and denote it as V_i. The collection formed in this way will be denoted by $\mathcal{B}_Z = \{ V_i : i \in N_3 \}$ where $N_3 \subset N_1$ holds. We shall construct inductively a collection $\{ S_i : i \in N_2 \}$ satisfying the following conditions.

$$\text{(1)} \qquad S_i \text{ is a partition between } F_i \text{ and } X \setminus U_i, \qquad i \in N_2.$$

When $i_1 < \cdots < i_k$ and $1 \leq k \leq n+1$,

$$\text{(2)} \qquad \operatorname{ind}(S_{i_1} \cap \cdots \cap S_{i_k} \cap Y) \leq n - k.$$

When $i_1 < \cdots < i_{n+2}$,

$$S_{i_1} \cap \cdots \cap S_{i_{n+2}} = \emptyset. \tag{3}$$

The construction is similar to the one in the proof of Lemma I.6.11. Note that from (2) we have

$$S_{i_1} \cap \cdots \cap S_{i_{n+1}} \cap Y = \emptyset,$$

when $i_1 < \cdots < i_{n+1}$, in other words, the sets $S_{i_1} \cap \cdots \cap S_{i_{n+1}}$ and Y are disjoint. Suppose for m that the sets S_i, $i \leq m$, have been constructed. Define

$$U'_{m+1} = U_{m+1} \setminus \bigcup\{\bigcap\{\, S_{i_1} \cap \cdots \cap S_{i_{n+1}} : i_1 < \cdots < i_{n+1} \leq m \,\}\}$$

and note that $F_{m+1} \subset U'_{m+1}$. Applying Lemma I.4.11, we get a partition S_{m+1} between F_{m+1} and $X \setminus U'_{m+1}$ such that (2) holds whenever $i_1 < \cdots < i_k \leq m+1$ holds. From the construction it follows that (3) holds whenever $i_1 < \cdots < i_{n+2} \leq m+1$ holds. In view of (1), for each i in N_2 there is an open set V_i such that

- $F_i \subset V_i \subset \operatorname{cl}(V_i) \subset U'_i$,
- $\mathrm{B}_X(V_i) \subset S_i$.

Finally let $\mathcal{B}^* = \{\, V_i : i \in N_2 \cup N_3 \,\}$. Now suppose that a collection

$$\boldsymbol{D} = \{\, \mathrm{B}\,(V_{i_k}) : i_1 < \cdots < i_{n+m+2} \,\}$$

is given. Then either $n+2$ indices out of $\{\, i_1, \ldots, i_{n+m+2} \,\}$ are in N_2 or $m+1$ indices are in N_3. In the first case, the intersection of $\boldsymbol{D}$ is empty by formula (3). In the second case the intersection of $\boldsymbol{D}$ is compact by the construction of $\mathcal{B}^*$. It follows that $\mathcal{B}^*$ witnesses the fact that $\operatorname{Skl} X \leq n+m+1$. By Kimura's theorem, $\operatorname{def} X \leq n+m+1$.

The inductive dimension ind in the formula of the last theorem cannot be replaced by def as the next example will show.

5.12. Example. In Example I.5.10.f we have studied the subspace X of $\mathbb{R}^2$ given by $X = \mathrm{S}_1\big((0,0)\big) \cup \{\,(1,0)\,\}$. It was proved there that $\operatorname{def} X = 1$. For this example, consider $Y = \{\,(1,0)\,\}$ and $Z = \mathrm{S}_1\big((0,0)\big)$. Obviously $\operatorname{def} Y = -1$ and $\operatorname{ind} Y = 0$ hold. As Z is locally compact, we have $\operatorname{def} Z = 0$.

6. The inequality $\mathcal{K}$-dim $\geq$ $\mathcal{K}$-def

Smirnov characterized the compactness deficiency in the series of papers [1965], [1966] and [1966a]. In these papers the notion of border covers made its first appearance. The starting point of Smirnov's work is the dimension theory of proximity spaces and its relationship to compactifications. A brief presentation of this approach to the compactness deficiency will be made. The inequality in the title of this section will come out of this endeavor.

The first part of the section will be devoted to a brief outline of the more important facts about proximity spaces. Proximity spaces were introduced by Efremovich in [1952] and extensively studied by Smirnov in [1952] and [1954]. For more details the reader is referred to Engelking [1977] and Naimpally and Warrack [1970]. The reader is referred to Isbell [1964] for more details on proximity dimensions.

6.1. Definition. A *proximity relation* on a nonempty set X is a relation δ between subsets of X such that the following conditions are satisfied: For subsets A, B and C of X and for elements x and y of X

(P1) if $(A, B) \in \delta$, then $(B, A) \in \delta$,
(P2) $(A, B \cup C) \in \delta$ if and only if $(A, B) \in \delta$ or $(A, C) \in \delta$,
(P3) $(\{x\}, \{y\}) \in \delta$ if and only if $x = y$,
(P4) $(X, \emptyset) \notin \delta$,
(P5) if $(A, B) \notin \delta$, then there are subsets E and F of X such that $E \cup F = X$, $(E, B) \notin \delta$ and $(A, F) \notin \delta$.

If δ is a proximity relation, then the pair (X, δ) is called a *proximity space*. The sets A and B are *near* or *proximal* if $(A, B) \in \delta$. If $(A, B) \notin \delta$, then A and B are said to be *far*.

We shall frequently use the following proposition: *If A and B are far and if $C \subset A$ and $D \subset B$, then C and D are far.*

The natural proximity relation on a metric space (X, d) is defined by $(A, B) \in \delta$ if and only if $d(A, B) = 0$.

Suppose that (X, δ) is a proximity space. We shall say that a subset A is *deep inside* B — denoted by $A \Subset B$ — if A and $X \setminus B$ are far. The following properties are easily verified:

(1) If $A \Subset B$, then $X \setminus B \Subset X \setminus A$.
(2) If $A \Subset B$, then $A \subset B$.

(3) If $A_1 \subset A \Subset B \subset B_1$, then $A_1 \Subset B_1$.
(4) If $A_i \Subset B_i$ for $i = 1, 2$, then $A_1 \cup A_2 \Subset B_1 \cup B_2$.
(5) If $A \Subset B$, then there exists a C such that $A \Subset C \Subset B$.
(6) $\emptyset \Subset \emptyset$.

6.2. Definition. Let (X_1, δ_1) and (X_2, δ_2) be proximity spaces. A function $f\colon X_1 \to X_2$ is called a *δ-function* provided that f satisfies the condition $(f[A], f[B]) \in \delta_2$, whenever $(A, B) \in \delta_1$. A δ-function is a *δ-homeomorphism* (or *equimorphism*) if f is bijective and f^{-1} is also a δ-function.

Let (X, δ) be a proximity space and let Z be a subset of X. Then there is a natural proximity δ_Z defined on Z. Namely, for subsets A and B of Z the relation δ_Z given by

$$(A, B) \in \delta_Z \quad \text{if and only if} \quad (A, B) \in \delta$$

is a proximity relation on Z. The proximity space (Z, δ_Z) is called a *subspace* of (X, δ).

6.3. Definition. The *topology* of the proximity space (X, δ) is defined by its closure operator: For a subset A of X,

$$\mathrm{cl}_X(A) = \{\, x : (\{\, x \,\}, A) \in \delta \,\}.$$

Equivalently, U is a neighborhood of x if and only if $\{\, x \,\} \Subset U$.

For a metric space (X, d) the topologies induced by the metric and the natural proximity coincide.

Clearly, every δ-function is continuous and the topology of a proximity subspace (Z, δ_Z) of (X, δ) coincides with the relative topology of Z in X.

For subsets A and B of (X, δ) we have

$$(A, B) \in \delta \quad \text{if and only if} \quad \big(\mathrm{cl}_X(A), \mathrm{cl}_X(B)\big) \in \delta.$$

It then follows that

$$A \Subset B \quad \text{if and only if} \quad \mathrm{cl}_X(A) \Subset \mathrm{int}_X B.$$

The topology of a proximity space (X, δ) is completely regular. Conversely, if X is a completely regular space, then there is a proximity relation δ on X such that the topology of (X, δ) agrees with the topology of X. Such a δ is called a proximity relation on the topological space X. These facts will follow easily from the relation between proximity spaces and compact Hausdorff spaces which will be discussed next.

If X is a compact Hausdorff space, then there is one and only one proximity relation δ on X such that the topologies of X and (X, δ) agree, namely

$$(A, B) \in \delta \quad \text{if and only if} \quad \mathrm{cl}_X(A) \cap \mathrm{cl}_X(B) \neq \emptyset.$$

Furthermore, a continuous mapping of a compact Hausdorff space into a proximity space is a δ-function. Two compact Hausdorff spaces are homeomorphic if and only if they are equimorphic. Therefore a proximity relation need not be specified for a compact Hausdorff space.

6.4. Definition. A proximity space (Y, δ_1) is called a δ-*extension* of the proximity space (X, δ) if (X, δ) is a subspace of (Y, δ_1) and X is dense in Y.

The fundamental results about compact extensions of proximity spaces, which were first proved by Smirnov in [1952], will be summarized next.

6.5. Theorem (Uniqueness theorem). *Each proximity space has a unique compact δ-extension.*

Here uniqueness means the following: If Y_1 and Y_2 are compact δ-extensions of the proximity space (X, δ), then there is an equimorphism of Y_1 onto Y_2 whose restriction to X is the identity map on (X, δ). The unique compact Hausdorff extension of a proximity space (X, δ) provided by the above theorem will be denoted by δX. The set $\delta X \setminus X$ is called the *remainder* of X in δX.

For a fixed completely regular topological space X there is a natural one-to-one correspondence between the collection of all proximity relations on X and the collection of all classes of topologically equivalent Hausdorff compactifications of X.

Each δ-function f from (X_1, δ_1) into (X_2, δ_2) can be uniquely extended to a continuous map δf of δX_1 into δX_2.

Let us now turn to proximity dimension. The natural generalization of covering dimension to proximity spaces was defined by Smirnov in [1956].

6.6. Definition. Let (X, δ) be a proximity space. A finite cover $\{V_1, \ldots, V_k\}$ of X is called a *δ-cover* of X if there exists a cover $\{A_1, \ldots, A_k\}$ of X such that $A_i \Subset V_i$ for each i.

The *proximity dimension* of X is minus one, $\delta\dim X = -1$, if and only if $X = \emptyset$. For n in $\mathbb{N}$, $\delta\dim X \leq n$ if for each δ-cover $\boldsymbol{V}$ of X there exists a δ-cover $\boldsymbol{W}$ of X such that $\boldsymbol{W}$ is a refinement of $\boldsymbol{V}$ and $\operatorname{ord} \boldsymbol{W} \leq n + 1$. As usual, $\delta\dim X = n$ means $\delta\dim X \leq n$ holds but $\delta\dim X \leq n - 1$ fails.

Observe that if $\{V_1, \ldots, V_n\}$ is a δ-cover, then so is the collection $\{\operatorname{int} V_1, \ldots, \operatorname{int} V_n\}$. It will follow from Lemma I.8.8 that every finite open cover of a compact space is a δ-cover. Consequently we have $\delta\dim X = \dim X$ for a compact spaces X.

6.7. Theorem. *Let (X, δ) be a proximity space. Then*

$$\delta\dim X = \dim \delta X$$

Proof. Suppose $\dim \delta X \leq n$. Let $\boldsymbol{V} = \{V_1, \ldots, V_k\}$ be a δ-cover of the space X. Because $\boldsymbol{V}$ is a δ-cover and not just a cover, we may assume that each V_i is an open set. Then the open collection $\{\operatorname{ex}_{\delta X}(V_1), \ldots, \operatorname{ex}_{\delta X}(V_k)\}$ is also a cover of δX. This cover has an open refinement $\boldsymbol{W}$ of order at most $n+1$. As $\boldsymbol{W}$ is a δ-cover, its trace on X is also a δ-cover of order at most $n+1$. It follows that $\delta\dim X \leq n$.

Suppose $\delta\dim X \leq n$. Let $\boldsymbol{U} = \{U_1, \ldots, U_k\}$ be an open cover of the space δX. We may assume further that the sets U_i are regularly open. The trace of $\boldsymbol{U}$ on X is a δ-cover. Let $\{W_1, \ldots, W_l\}$ be a δ-cover of order at most $n+1$ that refines this trace. There is a cover $\boldsymbol{A} = \{A_1, \ldots, A_l\}$ such that $A_i \Subset W_i$, $i = 1, \ldots, l$. For each i the open set $\operatorname{ex}(W_i)$ is contained in some U_j because each member of $\boldsymbol{U}$ is regularly open and Lemma V.2.10 (a) applies. As $\boldsymbol{A}$ covers X, it will follow that $\{\operatorname{ex}(W_1), \ldots, \operatorname{ex}(W_l)\}$ is a cover of δX that refines $\boldsymbol{U}$. That the order of this cover is at most $n+1$ will follow from Lemma V.2.10 (c). Hence, $\dim \delta X \leq n$.

We have as a corollary the following subspace theorem for $\delta\dim$.

6.8. Corollary (Subspace theorem). *For any subspace Z of a proximity space (X, δ),*

$$\delta\dim Z \leq \delta\dim X.$$

Proof. We have by the uniqueness theorem that $\mathrm{cl}_{\delta X}(Z) = \delta Z$. The closed subspace theorem for dim together with Theorem 6.7 yield $\delta\dim Z = \dim \delta Z \leq \dim \delta X = \delta\dim X$.

The following addition theorem can be proved in a similar fashion. Notice that the summands need not be closed sets.

6.9. Corollary (Addition theorem). *Suppose $X = Y \cup Z$ for a proximity space (X, δ). Then*

$$\delta\dim X = \max\{\, \delta\dim Y, \delta\dim Z \,\}.$$

With this brief introduction to proximity dimension we can now turn towards the characterization of the compactness deficiency. Following Smirnov [1966], we shall present first his characterization using proximity spaces. The key notion is that of an extendable $\mathcal{K}$-border cover.

6.10. Definition. A family $\{\, U_1, \ldots, U_k \,\}$ of open sets of a proximity space (X, δ) is called an *extendable $\mathcal{K}$-border cover* of X if its enclosure $K = X \setminus \bigcup\{\, U_i : i = 1, \ldots, k \,\}$ is compact and if for each neighborhood U of K the collection $\{\, U, U_1, \ldots, U_k \,\}$ is a δ-cover of X.

The name "extendable $\mathcal{K}$-border cover" is explained by the following lemma.

6.11. Lemma. *A family $\{\, U_1, \ldots, U_k \,\}$ of open sets of a proximity space (X, δ) is an extendable $\mathcal{K}$-border cover if and only if $\delta X \setminus X \subset \mathrm{ex}_{\delta X}(U_1) \cup \cdots \cup \mathrm{ex}_{\delta X}(U_k)$.*

Proof. Suppose that $\delta X \setminus X \subset \mathrm{ex}_{\delta X}(U_1) \cup \cdots \cup \mathrm{ex}_{\delta X}(U_k)$ holds and define

$$K = \delta X \setminus \big(\mathrm{ex}_{\delta X}(U_1) \cup \cdots \cup \mathrm{ex}_{\delta X}(U_k)\big).$$

The set K is compact. As $K = X \setminus (U_1 \cup \cdots \cup U_k)$ holds, K is also a subset of X. Select an open neighborhood U of K in X. Then $\boldsymbol{U} = \{\, \mathrm{ex}_{\delta X}(U), \mathrm{ex}_{\delta X}(U_1), \ldots, \mathrm{ex}_{\delta X}(U_k) \,\}$ is an open cover of the compact space δX. Since δX is compact, the collection $\boldsymbol{U}$ is also a δ-cover. Thereby the trace of $\boldsymbol{U}$ on X is a δ-cover of X. It follows that $\{\, U_1, \ldots, U_k \,\}$ is an extendable $\mathcal{K}$-border cover.

Conversely, suppose that $\{U_1, \ldots, U_k\}$ is an extendable $\mathcal{K}$-border cover of (X, δ). Define $K = X \setminus (U_1 \cup \cdots \cup U_k)$. Let y be in $\delta X \setminus X$ and let V and U be disjoint open neighborhoods of y and K respectively. As the collection $\{U \cap X, U_1, \ldots, U_k\}$ is a δ-cover of X, there is a shrinking $\{H, H_1, \ldots, H_k\}$ such that $H \Subset U$ and $H_i \Subset U_i$, for $i = 1, \ldots, k$. Thus, $\delta X = \mathrm{cl}_{\delta X}(H) \cup \mathrm{cl}_{\delta X}(H_1) \cup \cdots \cup \mathrm{cl}_{\delta X}(H_k)$. As $\mathrm{cl}_{\delta X}(H) \subset \mathrm{cl}_{\delta X}(U)$ holds, we have that y is not in $\mathrm{cl}_{\delta X}(H)$. It follows that $y \in \mathrm{cl}_{\delta X}(H_i) \Subset \mathrm{ex}_{\delta X}(U_i)$ must hold for some i. Consequently, $\delta X \setminus X \subset \mathrm{ex}_{\delta X}(U_1) \cup \cdots \cup \mathrm{ex}_{\delta X}(U_k)$.

6.12. Definition. The *boundary dimension* $\dim^\infty X$ of a proximity space (X, δ) is defined as follows. $\dim^\infty X = -1$ if and only if X is compact. For n in $\mathbf{N}$, $\dim^\infty X \leq n$ if for each extendable $\mathcal{K}$-border cover $\boldsymbol{U}$ of X there exists an extendable $\mathcal{K}$-border cover $\boldsymbol{V}$ of X such that $\boldsymbol{V}$ is a refinement of $\boldsymbol{U}$ and $\mathrm{ord}\, \boldsymbol{V} \leq n + 1$.

The next theorem concerns the topological notion of compactness deficiency in the setting of proximity spaces. It should be no surprise that a topological condition such as Lindelöf at infinity (see Definition 3.10) has been included in its hypothesis. The need for this hypothesis will become apparent in the proof of the theorem.

6.13. Theorem. *Let (X, δ) be a proximity space. Suppose that X is Lindelöf at infinity. Then*

$$\dim^\infty X = \dim(\delta X \setminus X).$$

Consequently,

$$\dim^\infty X \geq \mathcal{K}\text{-def}\, X.$$

Proof. Notice first that $\delta X \setminus X$ is a normal space because it is Lindelöf and thus $\dim(\delta X \setminus X)$ is well defined.

Suppose $\dim(\delta X \setminus X) \leq n$. Let $\{U_1, \ldots, U_k\}$ be an extendable $\mathcal{K}$-border cover of X. By Lemma 6.11 we have

$$\delta X \setminus X \subset \mathrm{ex}_{\delta X}(U_1) \cup \cdots \cup \mathrm{ex}_{\delta X}(U_k).$$

Since $\delta X \setminus X$ is normal and $\dim(\delta X \setminus X) \leq n$, the cover of $\delta X \setminus X$ consisting of the sets $(\delta X \setminus X) \cap \mathrm{ex}_{\delta X}(U_i)$, $i = 1, \ldots, k$, has a closed shrinking $\boldsymbol{H} = \{H_1, \ldots, H_k\}$ in $\delta X \setminus X$ such that $\mathrm{ord}\, \boldsymbol{H} \leq n + 1$. Let us observe that disjoint closed subsets of $\delta X \setminus X$ will have disjoint open neighborhoods in δX because X is Lindelöf at infinity.

To see this, we consider disjoint sets G and H that are closed in the space $\delta X \setminus X$. Then $F = \mathrm{cl}_{\delta X}(G) \cap \mathrm{cl}_{\delta X}(H)$ is a compact subset X. The condition that X is Lindelöf at infinity will yield an open neighborhood V of $\delta X \setminus X$ such that $V \cap F = \emptyset$ and V is a normal space. (See the proof of the converse in Lemma 3.11 where it is shown that V is an F_σ-set of δX.) The existence of the disjoint open neighborhoods of G and H will follow from the normality of V. We can now modify the proof of Proposition II.4.4 to get an open collection $\boldsymbol{W} = \{ W_1, \ldots, W_k \}$ in δX such that $H_i \subset W_i \subset \mathrm{ex}_{\delta X}(U_i)$ for each i and such that $\boldsymbol{W}$ is combinatorially equivalent to $\boldsymbol{H}$. In particular $\mathrm{ord}\, \boldsymbol{W} \leq n+1$. Obviously, $(\delta X \setminus X) \subset W_1 \cup \cdots \cup W_k$. It will follow from Lemma 6.11 that $\{ W_1 \cap X, \ldots, W_k \cap X \}$ is an extendable $\mathcal{K}$-border cover of order at most $n+1$. Hence $\dim^\infty X \leq n$.

Suppose that $\dim^\infty X \leq n$ and let $\{ V_1, \ldots, V_k \}$ be an open cover of $\delta X \setminus X$. By Lemma I.8.8 there is a closed cover $\{ G_1, \ldots, G_k \}$ of $\delta X \setminus X$ such that $G_i \subset V_i$ for each i. From the observation made in the last paragraph there is an open collection $\{ U_1, \ldots, U_k \}$ in δX such that $G_i \subset U_i$ and $\mathrm{cl}_{\delta X}(U_i) \cap (\delta X \setminus X) \subset V_i$ for each i. The collection $\{ U_1 \cap X, \ldots, U_k \cap X \}$ is an extendable $\mathcal{K}$-border cover of X by Lemma 6.11. From $\dim^\infty X \leq n$ we have an extendable $\mathcal{K}$-border cover $\{ W_1, \ldots, W_l \}$ that refines this cover and has order at most $n+1$. By Lemma V.2.10, $\{ \mathrm{ex}_{\delta X}(W_1), \ldots, \mathrm{ex}_{\delta X}(W_l) \}$ is a refinement of $\{ \mathrm{cl}_{\delta X}(U_1), \ldots, \mathrm{cl}_{\delta X}(U_k) \}$ with order at most $n+1$. So $\{ (\mathrm{ex}_{\delta X}(W_1)) \cap (\delta X \setminus X), \ldots, (\mathrm{ex}_{\delta X}(W_l)) \cap (\delta X \setminus X) \}$ has order at most $n+1$ and refines $\{ V_1, \ldots, V_k \}$. Thereby $\dim(\delta X \setminus X) \leq n$ follows.

We have need for two more lemmas. The first lemma has been included for our convenience because it is not readily available in the stated form.

6.14. Lemma. *Every finite open cover of a normal space has a finite open star refinement.*

Proof. Let $\mathbf{V} = \{ V_1, \ldots, V_k \}$ be a finite open cover of the normal space X. Let $\{ F_1, \ldots, F_k \}$ be a closed shrinking of $\mathbf{V}$. Then there are continuous functions $f_i \colon X \to \mathbb{I}$, $i = 1, \ldots, k$, such that $f_i[F_i] = -1$ and $f_i[X \setminus V_i] = 1$ for each i. It is clear that

$$\bigwedge \{ \{ f_i^{-1}[(-\infty, 1)], f_i^{-1}[(-1, \infty)] \} : i = 1, \ldots, k \}$$

is a cover that refines $\boldsymbol{V}$. As the collection

$$\{(-\infty,-\tfrac{1}{3}),(-1,\tfrac{1}{3}),(-\tfrac{1}{3},1),(\tfrac{1}{3},\infty)\}$$

is a star refinement of $\{(-\infty,1),(-1,\infty)\}$, it will follow that the meet of the collections

$$\{f_i^{-1}[(-\infty,-\tfrac{1}{3})],f_i^{-1}[(-1,\tfrac{1}{3})],f_i^{-1}[(-\tfrac{1}{3},1)],f_i^{-1}[(\tfrac{1}{3},\infty)]\},$$

$i=1,\ldots,k$, is a finite open star refinement of $\boldsymbol{V}$ and the lemma is proved.

The second lemma will be proved by an argument that was used in the first part of the proof of Theorem 6.13.

6.15. Lemma. *Let (X,δ) be a proximity space and suppose that X is Lindelöf at infinity. Then every extendable $\mathcal{K}$-border cover of X has a star refinement which is an extendable $\mathcal{K}$-border cover.*

Proof. Let $\boldsymbol{U}=\{U_1,\ldots,U_l\}$ be an extendable $\mathcal{K}$-border cover. By Lemma 6.11, the collection $\{\operatorname{ex}_{\delta X}(U_1),\ldots,\operatorname{ex}_{\delta X}(U_l)\}$ is a cover of $\delta X\setminus X$. As the space $\delta X\setminus X$ is normal, the open collection $\{\operatorname{ex}_{\delta X}(U_i)\cap(\delta X\setminus X):i=1,\ldots,l\}$ has a finite open star refinement $\boldsymbol{O}$ by the last lemma. The cover $\boldsymbol{O}$ of $\delta X\setminus X$ has a closed shrinking $\boldsymbol{H}=\{H_1,\ldots,H_k\}$ in $\delta X\setminus X$. By the argument employed in the first part of the proof of Theorem 6.13, there is an open collection $\boldsymbol{W}=\{W_1,\ldots,W_k\}$ in δX that is combinatorially equivalent to $\boldsymbol{H}$ and satisfies the condition $H_i\subset W_i$ for each i. For each W_j in $\boldsymbol{W}$ we define V_j by

$$V_j=W_j\cap\bigcap\{\operatorname{ex}_{\delta X}(U_i):H_j\subset\operatorname{ex}_{\delta X}(U_i)\}.$$

Let $\boldsymbol{V}=\{V_j\cap X:j=1,\ldots,k\}$. Suppose that j is in $\{1,\ldots,k\}$. Then there exists an m such that the star of H_j in $\boldsymbol{H}$ is contained in $\operatorname{ex}_{\delta X}(U_m)$. It follows that $V_j\cap X\subset U_m$ holds. Let us show that the star of V_j in $\boldsymbol{V}$ is contained in U_m. If $V_j\cap V_q\neq\emptyset$ for some q, then $H_j\cap H_q\neq\emptyset$, whence $H_q\subset U_m$. Thus it follows that $V_q\subset U_m$.

The compactness deficiency is characterized by the existence of a special collection of $\mathcal{K}$-border covers that will now be defined. Observe that the notion being defined is a topological one.

6.16. Definition. Let X be a topological space. A collection Σ of finite $\mathcal{K}$-border covers is called a *border structure of order* n for X if the following conditions are satisfied.

($\Sigma 1$) If $\boldsymbol{U}$ and $\boldsymbol{V}$ are in Σ, then there is a $\boldsymbol{W}$ in Σ such that $\boldsymbol{W}$ is a star refinement of both $\boldsymbol{U}$ and $\boldsymbol{V}$.

($\Sigma 2$) For every point x of X and for every neighborhood U of x there exists a neighborhood V of x and an element $\boldsymbol{W}$ of Σ such that the star of V in $\boldsymbol{W}$ is contained in U, that is, $\mathrm{St}_{\boldsymbol{W}}(V) \subset U$.

($\Sigma 3$) Every $\boldsymbol{U}$ in Σ has order at most n.

Smirnov's topological characterization reads as follows.

6.17. Theorem. *Suppose that the space X is Lindelöf at infinity. Then $\mathcal{K}$-def $X \leq n$ if and only if there is a border structure for X of order at most $n+1$.*

Proof. Observe that the space $Y \setminus X$ is normal for every compactification Y of X. We shall prove the necessity of the conditon first. Suppose $\mathcal{K}$-def $X \leq n$. Then there exists a compactification Y of X such that $\dim(Y \setminus X) \leq n$. Let δ be the unique proximity on X such that $Y = \delta X$. By Theorem 6.13 we have $\dim^{\infty} X \leq n$ for this proximity. Let Σ be the collection of all extendable $\mathcal{K}$-border covers of order at most $n+1$. We shall show that Σ is a border structure of order at most $n+1$. Obviously the elements of Σ are $\mathcal{K}$-border covers. Let $\boldsymbol{U}$ and $\boldsymbol{V}$ be in Σ. The meet $\boldsymbol{U} \wedge \boldsymbol{V}$ is an extendable border cover of X as is easily seen with the help of Lemma 6.11. By Lemma 6.15 there is a star refinement $\boldsymbol{O}$ that is an extendable $\mathcal{K}$-border cover. As $\dim^{\infty} X \leq n$, there is a $\mathcal{K}$-border cover refinement $\boldsymbol{W}$ of $\boldsymbol{O}$ of order at most $n+1$. This shows that condition ($\Sigma 1$) is satisfied.

Let x be in X and U be an open neighborhood of x. In view of axiom (P5) there is a closed neighborhood V of x such that V is deep inside U. The binary cover $\{U, X \setminus V\}$ is a δ-cover of X and hence an extendable $\mathcal{K}$-border cover of X. By Lemma 6.15 there is a finite extendable $\mathcal{K}$-border cover $\boldsymbol{V}$ that star refines the binary cover. Let $\boldsymbol{W}$ be an extendable $\mathcal{K}$-border cover of order at most $n+1$ that refines $\boldsymbol{V}$. It is easily seen that the condition ($\Sigma 2$) is satisfied.

Obviously condition ($\Sigma 3$) is satisfied. Thereby the necessity part is proved.

The proof of the sufficiency of the condition will be given in three steps. First, using the border structure, we shall define a proximity relation δ on X that is compatible with the topology. Then we shall show that every element of Σ is an extendable $\mathcal{K}$-border cover of (X, δ). Finally $\dim^{\infty} X \leq n$ will be proved. The theorem will then follow from Theorem 6.13.

Suppose that Σ is a border structure for X of order at most $n+1$. We shall use the structure Σ to define a proximity relation δ on X that is compatible with the topology of X.

Definition. Let A and B be subsets of X. Then δ is defined as follows. $(A, B) \notin \delta$ if $\mathrm{cl}_X(A) \cap \mathrm{cl}_X(B) = \emptyset$ and there exists a $\boldsymbol{U}$ in Σ such that $B \cap \mathrm{St}_{\boldsymbol{U}}(A) = \emptyset$. Otherwise $(A, B) \in \delta$.

We shall verify that the axioms of Definition 6.1 are satisfied and that δ is compatible with the topology of X. The axioms (P1), (P4) and a part of (P3), namely that $(\{x\}, \{y\}) \in \delta$ holds whenever $x = y$ holds, are obviously satisfied.

We shall verify a part of (P2) and leave the remainder of its proof to the reader. Let us prove that if $(A, B) \notin \delta$ and $(A, C) \notin \delta$, then $(A, B \cup C) \notin \delta$. Suppose that $(A, B) \notin \delta$ and $(A, C) \notin \delta$. It is evident that $\mathrm{cl}(A) \cap (\mathrm{cl}(B) \cup \mathrm{cl}(C)) = \emptyset$. As $(A, B) \notin \delta$, there is a $\boldsymbol{U}$ in Σ such that $B \cap \mathrm{St}_{\boldsymbol{U}}(A) = \emptyset$. Similarly, there is a $\boldsymbol{V}$ in Σ such that $C \cap \mathrm{St}_{\boldsymbol{V}}(A) = \emptyset$. By ($\Sigma$1) there is a $\boldsymbol{W}$ in Σ that is a common refinement of both $\boldsymbol{U}$ and $\boldsymbol{V}$. Then we have $(B \cup C) \cap \mathrm{St}_{\boldsymbol{W}}(A) = \emptyset$. Thus it follows that $(A, B \cup C) \notin \delta$.

Before we continue with the verification of the axioms, let us first show that the topology of X is compatible with δ. If $x \in \mathrm{cl}(A)$, then obviously $(\{x\}, A) \in \delta$. Suppose that $x \notin \mathrm{cl}(A)$. Then we have $\mathrm{cl}(\{x\}) \cap \mathrm{cl}(A) = \emptyset$ and, in view of (Σ2), there is a $\boldsymbol{W}$ in Σ such that $\mathrm{St}_{\boldsymbol{W}}(x) \subset X \setminus \mathrm{cl}(A)$. Consequently $\mathrm{cl}(A) \cap \mathrm{St}_{\boldsymbol{W}}(x) = \emptyset$. By the definition of δ we have $(A, \{x\}) \notin \delta$. This completes the proof of the compatibility.

Now the other part of (P3) will easily follow. That is to say, we have $(\{x\}, \{y\}) \notin \delta$ whenever $x \neq y$ because points are closed.

Only axiom (P5) remains to be verified. First let us observe that if S is a compact set and T is a closed set such that $S \cap T = \emptyset$, then there is a neighborhood U of S such that U and T are far. This can be shown in the following way. From (Σ2) it follows that for each point x of S there is an open neighborhood U_x such that U_x and T

are far, or $(U_x, T) \notin \delta$. Using the compactness of S, we select a finite collection of such neighborhoods whose union is an open set U that contains S. By (P2), which has already been verified, we have $(U, T) \notin \delta$. Turning to the verification of (P5), we assume that A and B are far. Then $\operatorname{cl}(A) \cap \operatorname{cl}(B) = \emptyset$ and there exists a $\boldsymbol{V}$ in Σ such that $B \cap \operatorname{St}_{\boldsymbol{V}}(A) = \emptyset$. As every set in $\boldsymbol{V}$ is open, we have the equality $\operatorname{cl}(B) \cap \operatorname{St}_{\boldsymbol{V}}(\operatorname{cl}(A)) = \emptyset$. It follows that we may assume A and B are also closed. There is a $\boldsymbol{W}$ in Σ that is a star refinement of $\boldsymbol{V}$. Then $\operatorname{St}_{\boldsymbol{W}}(A) \cap \operatorname{St}_{\boldsymbol{W}}(B) = \emptyset$. Let $C = X \setminus \bigcup \boldsymbol{W}$. By our observation there is a neighborhood U of $C \cap A$ such that U and B are far. Similarly there is a neighborhood V of $C \cap B$ such that V and A are far. Define $F = X \setminus \left(U \cup \operatorname{St}_{\boldsymbol{W}}(A)\right)$ and $E = X \setminus \left(V \cup \operatorname{St}_{\boldsymbol{W}}(B)\right)$. As $\operatorname{St}_{\boldsymbol{W}}(A) \cap F = \emptyset$, it follows by (P2) that F and A are far. Similarly E and B are far. Finally let $F' = F \cup V$ and $E' = E \cup U$. Then $(A, F') \notin \delta$ and $(B, E') \notin \delta$. An easy computation will show $E' \cup F' = X$. It follows that axiom (P5) is satisfied. Thus we have completed the first step of the proof of the sufficiency.

In the second step of the proof we shall show that every element $\boldsymbol{U}$ of Σ is an extendable $\mathcal{K}$-border cover of the proximity space (X, δ) that has just been defined. Suppose that $\boldsymbol{U} = \{ U_1, \ldots, U_k \}$ is an element of Σ. Let $\boldsymbol{V}$ be an element of Σ that is a star refinement of $\boldsymbol{U}$. Denote the enclosure of $\boldsymbol{V}$ by B. Let V_i be an element of $\boldsymbol{V}$. Then there is an U_j in $\boldsymbol{U}$ such that $\operatorname{St}_{\boldsymbol{V}}(V_i) \subset U_j$. So it follows that $V_i \Subset U_j$. For each j in $\{ 1, \ldots, k \}$ we define

$$W_j = \bigcup \{ V_i : V_i \Subset U_j,\ V_i \not\Subset U_l \text{ for } l < j \}.$$

Then $W_j \Subset U_j$ and $\bigcup \{ W_j : j = 1, \ldots, k \} = X \setminus B$. For each neighborhood W of B the sets $X \setminus W$ and B are far and consequently $\{ W, W_1, \ldots, W_k \}$ is a δ-cover. It follows that $\boldsymbol{U}$ is extendable.

For the final part of the proof we shall show $\dim^{\infty} X \leq n$ holds. Suppose that $\boldsymbol{U} = \{ U_1, \ldots, U_k \}$ is an extendable $\mathcal{K}$-border cover. Let $B = X \setminus \boldsymbol{U}$ and let U_0 be a neighborhood of B. There are closed sets H_0, H_1, ..., H_k such that $X = H_0 \cup H_1 \cup \cdots \cup H_k$ holds and $H_i \Subset U_i$ hold for $i = 0, 1, \ldots, k$. There are $\boldsymbol{V}_0$, $\boldsymbol{V}_1$, ..., $\boldsymbol{V}_k$ in Σ such that $\operatorname{St}_{\boldsymbol{V}_i}(H_i) \subset U_i$ for $i = 0, 1, \ldots, k$. Choose a member $\boldsymbol{W}$ of Σ that is a common refinement of all the $\boldsymbol{V}_i$'s. This $\boldsymbol{W}$ refines $\boldsymbol{U}$ and has order at most $n + 1$. It follows that $\dim^{\infty} X \leq n$.

As a corollary to Smirnov's theorem we get the inequality in the title of this section.

6.18. Corollary. *Let X be a hereditarily normal space that is Lindelöf at infinity. Then*

$$\mathcal{K}\text{-dim}\, X \geq \mathcal{K}\text{-def}\, X.$$

Proof. Suppose $\mathcal{K}$-dim $X \leq n$. As X is a hereditarily normal space, we infer from Lemma 6.14 that the family of all finite $\mathcal{K}$-border covers of order at most $n+1$ is a border structure of order at most $n+1$ for X. By the theorem, $\mathcal{K}$-def $X \leq n$.

There are weight preserving compactifications with remainders of minimal dimension.

6.19. Corollary. *Suppose that X is a space that is Lindelöf at infinity and has infinite weight $w(X) = \tau$. Then the following conditions are equivalent.*

(a) *$\mathcal{K}$-def $X \leq n$.*
(b) *A border structure Σ of order $n+1$ with $|\Sigma| \leq \tau$ exists.*
(c) *A compactification Y of X with both $\dim (Y \setminus X) \leq n$ and $w(Y) = \tau$ exists.*

Proof. We shall only sketch the proof. Providing the details will be left to the reader. That (c) implies (a) is obvious. For the proof that (a) implies (b) we let $\mathcal{B}$ be a base for the open sets of X with $|\mathcal{B}| = \tau$. By Theorem 6.17, there is a border structure Σ of order at most $n+1$. We shall inductively define subfamilies Σ_i of Σ. For each pair (U_α, U_β) of elements of $\mathcal{B}$ for which there exists a $\boldsymbol{V}$ in Σ with $\operatorname{St}_{\boldsymbol{V}}(U_\alpha) \subset U_\beta$ we select precisely one such element of Σ. The family of all collections obtained in this way is denoted by Σ_0. It can be verified that Σ_0 satisfies conditions (Σ2) and (Σ3). Suppose for $i \geq 1$ that $\Sigma_0, \ldots, \Sigma_{i-1}$ have been defined. For each finite subfamily of Σ_{i-1} we select a common star refinement $\boldsymbol{W}$ from Σ. The family Σ_i is defined to be the union of the family Σ_{i-1} and the newly formed family of the common star refinements that were selected. Finally we let $\Sigma_\omega = \bigcup\{\, \Sigma_i : i \in \mathrm{N}\,\}$. It is clear that Σ_ω is a border structure.

For the proof that (b) implies (c) we use the sufficiency part of proof of Theorem 6.17 to obtain the compactification $Y = \delta X$. Let $\mathcal{B}$ be a base witnessing the fact that $w(X) \leq \tau$. One easily verifies that $\mathcal{B}^* = \{\operatorname{ex}_{\delta X}(U) : U \in \mathcal{B}\} \cup \{\operatorname{ex}_{\delta X}(V) : V \in \boldsymbol{V} \in \Sigma_\omega\}$ is a base for the open sets of δX with cardinality at most τ.

Remark. The outline of the proof gives the following stronger result. If δ_1 denotes the proximity relation induced by Σ and δ denotes the proximity relation induced by Σ_ω, then the identity mapping of X can be extended to a continuous function of the compactification induced by δ_1 to the compactification induced by δ.

7. Historical comments and unsolved problems

The theory of the Wallman compactification was developed by Wallman in [1938]. The fact that the Wallman compactification preserves the covering dimension is in Wallman's paper. The corresponding result for the large inductive dimension was first published in [1939] by Vedenisoff.

Construction of compactifications of a space via bases for the closed sets have been discussed by Shanin in his papers [1943], [1943a] and [1943b]. The theorems in the first part of Section 2 are from the [1964] paper by Frink. Frink posed the question of whether every compactification was a Wallman compactification. A lot of effort was put into investigating this question; Ul′janov in [1977] finally answered the question in the negative. Sklyarenko [1958a] and Engelking and Sklyarenko [1963] present extensive studies of compactifications that allow extensions of mappings in the general situation as well as in the special case of rim-compact spaces. The thesis of de Vries [1962] also has many results along these lines. The first results on the extensions of families of mappings to compactifications were given by de Groot and McDowell [1960] and Engelking [1960]. See also de Groot [1961].

Almost all of the results in Section 3 originate from the work of Freudenthal in [1942] and [1951]. Freudenthal's work is related to the problem of deciding whether a topological space is the underlying space of a topological group. In [1931] Freudenthal showed, for example, that a path connected topological group G has at most two ends, i.e., $\digamma X \setminus X$ consists of at most two points. This aspect of the theory is discussed by Peschke in [1990]. The publication of Freudenthal [1952] was unduly delayed because of the second world war; the manuscript was submitted in [1942]. That the Freudenthal compactification is a Wallman compactification was first proved by Njåstad in [1966]. An extensive discussion of the Freudenthal compactification can be found in Dickman and McCoy [1988]. Theorem 3.6 is due to Alexandroff and Ponomarev [1959]. The existence of maximal compactifications of rim-compact spaces, Theorem 3.7, is due to Morita [1952]. The notion of a perfect compactification was introduced by Sklyarenko in [1962], but it was implicitly introduced in Freudenthal [1942]. In [1957] Henriksen and Isbell introduced the "properties at infinity". Theorem 3.12 is due to Sklyarenko [1962]. An analysis of the fine distinction between zero-dimensional subsets and zero-dimensionally embedded subsets can be found in Diamond [1987] and Diamond, Hatzenbuhler and Mattson [1988]. The first examples like the ones presented in Example 3.5 are in Smirnov [1958]. Theorem 3.13 was first proved by Sklyarenko in [1958]. The Theorems 3.14 and 3.15 which generalize results of Freudenthal [1942] were proved by de Vries [1962]. Theorems 3.17 and 3.18 are due to Morita [1956], [1957] and [1959]. Example 3.19 was first published by Nishiura in [1964]. A metrizable Wallman compactification of a complete rim-compact metric space was constructed in 3.20; the construction also provided a proof of Zippin's Theorem I.5.2. The fact that Zippin's compactification can be obtained as a Wallman compactification was first proved by Steiner in [1969]. In [1977] van Mill presented a more general result along these lines. Zippin's result has also triggered research on the problem of characterizing spaces that can be compactified by adding at most countably many points. A characterization of such spaces is found in Charalambous [1980]. See also the papers Henriksen [1977], Hoshina [1977], [1979] and Terada [1977].

Most of the results in Section 4 were first published by de Vries in [1962]. But the notion of $(\leq n)$-Inductionally embedded sets that is used in this section is slightly different from the one introduced by de Vries. Theorem 4.3 appears here for the first time.

The main result of Section 5, Theorem 5.9, is due to Kimura [1988]. The lemmas leading up to Kimura's theorem are essentially due to him. The proof of the existence of the required compactification by means of a Wallman-type constructiom appears here for the first time. Corollary 5.11 is an improvement of a result of Aarts, Bruijning and van Mill [1982] and appears here for the first time.

The most important results from the papers [1956], [1965], [1965a], [1966] and [1966a] of Smirnov have been collected in Section 6.

Unsolved problems

1. Related to Corollary 3.16 is the following problem posed in Isbell [1964].

If X is connected, metrizable and rim-compact, must X be separable?

2. Another question from Isbell [1964] that still remains open is:

Does every rim-compact space X have a compactification Y such that $\dim(Y \setminus X) \leq 0$?

From the proof of Theorem 3.12 it will follow that the answer is yes for spaces X that are Lindelöf at infinity.

3. In Theorem 4.10 we have presented a very limited result about the inequality $\mathcal{K}$-Def $\leq \mathcal{K}$-Ind.

Prove or disprove the inequality $\mathcal{K}\text{-Def}\, X \leq \mathcal{K}\text{-Ind}\, X$ for every X in some universe that is contained in $\mathcal{N}$ and properly contains $\mathcal{M}_0$.

4. In Section 3 we have studied the property that a subset Z is zero-dimensionally embedded in a space X. We have introduced in Section 4 the property $\operatorname{Ind}[Z; X] \leq n$ for subsets Z of a space X. One can also define $\operatorname{ind}[Z; X] \leq n$ in a similar way.

Is there a meaningful theory for $\operatorname{ind}[Z; X]$?

5. Does $\operatorname{Ind}[Z, X] \leq 0$ imply that Z is zero-dimensionally embedded in X?

It will follow from Theorem 4.8 and the proof of Theorem 3.12 that the answer is yes when X is compact and Z is Lindelöf.

BIBLIOGRAPHY

AARTS, J.M.

[1966] *Dimension and deficiency in general topology*, PhD thesis (Amsterdam 1966).

[1968] Completeness degree. A generalization of dimension, Fundamenta Mathematicae **63** (1968), 27–41.

[1971] A characterization of strong inductive dimension, Fundamenta Mathematicae **70** (1971), 147–155.

[1971a] Complete extensions of metrizable spaces, preserving dimension and mappings, General Topology and its Applications **1** (1971), 349–355.

[1972] Dimension modulo a class of spaces, Nieuw Archief voor Wiskunde **20** (1972), 191–215.

AARTS, J.M., BRUIJNING, J. and MILL, J. VAN

[1985] A compactification problem of J. de Groot, Topology and its Applications **21** (1985), 217–222.

AARTS, J.M. and NISHIURA, T.

[1972] The Eilenberg-Borsuk duality theorem, Indagationes Mathematicae **34** (1972), 68–72.

[1973] Kernels in dimension theory, Transactions of the American Mathematical Society **178** (1973), 227–240.

[1973a] Covering dimension modulo a class of spaces, Fundamenta Mathematicae **78** (1973), 75–97.

AKASAKI, T.

[1965] The Eilenberg-Borsuk duality theorem for metric spaces, Duke Mathematical Journal **32** (1965), 653–660.

ALEXANDROFF, P.

[1932] Dimensionstheorie. Ein Beitrag zur Geometrie der Abgeschlossenen Mengen, Mathematische Annalen **106** (1932), 161–238.

ALEXANDROFF, P. and PASYNKOV, B.A.

[1973] *Introduction to dimension theory*, Moscow, 1973 (Russian).

ALEXANDROFF, P. and PONOMAREV, V.I.

[1959] On bicompact extensions of topological spaces, Vestnik Moskovskogo Universiteta **5** (1959), Seriya I, 93–108 (Russian).

BALADZE, V.H.

[1982] On functions of dimensional type, Trudy Tbilisskogo Matematicheskogo Instituta **68** (1982), 5–41 (Russian).

BIRKHOFF, G.

[1973] *Lattice theory*, 3rd edition, American Mathematical Society Colloquium Publications XXV, Providence (R.I.), 1973.

Brouwer, L.E.J.

[1913] Über den natürlichen Dimensionsbegriff, Journal für die reine und angewandte Mathematik **142** (1913), 146–152.

Borsuk, K.

[1937] Un théorème sur les prolongements des transformations, Fundamenta Mathematicae **29** (1937), 161–166.

Charalambous, M.G.

[1976] Spaces with increment of dimension n, Fundamenta Mathematicae **93** (1976), 97–107

[1980] Compactifications with countable remainder, Proceedings of the American Mathematical Society **78** (1980), 127–131.

Čech, E.

[1931] Sur la théorie de la dimension, Comptes Rendus hebdomadaires des séances de l'Académie Scientifique, Paris **193** (1931), 976–977.

[1932] Sur la dimension des espaces parfaitement normaux, Bulletin International Académie Tchèque Scientifique **33** (1932), 38–55.

[1933] Contribution to dimension theory, Časopis pro Pěstování Matematiky a Fysiky **62** (1933), 277–291 (Czech); French translation in *E. Čech, Topological papers*, Prague 1968, 129–142.

[1937] On bicompact spaces, Annals of Mathematics **38** (1937), 823–844.

Chigogidze, A.Ch

[1981] On the dimension of increments of Tychonoff spaces, Fundamenta Mathematicae **111** (1981), 25–36.

Choban, M.M and Attia, H.

[1990] On the dimension of remainders of extensions of normal spaces, Topology and its Applications **36** (1990), 97–109.

Deák, E.

[1976] Über gewisse Verschärfungen der beiden klassischen induktiven topologischen Dimensionsbegriffe I, Räume mit "hübschen" basen, Studia Scientiarum Mathematicarum Hungarica **11** (1976), 229–246.

Diamond, B.

[1987] Almost rimcompact spaces, Topology and its applications **25** (1987), 81–91.

Diamond, B., Hatzenbuhler, J. and Mattson, D.

[1988] On when a 0-space is rimcompact, Topology Proceedings 13 (1988), 189–202.

Dickman, R.F. and McCoy, R.A.

[1988] *The Freudenthal compactification*, Dissertationes Mathematicae (Rozprawy Matematyczne) CCLXII (Warszawa 1988).

Douwen, E.K. van

[1973] The small inductive dimension can be raised by the adjunction of a single point, Indagationes Mathematicae **35** (1973), 434–442.

Dowker, C.H.

[1953] Inductive dimension of completely normal spaces, Quarterly Journal of Mathematics, Oxford **4** (1953), 267–281.

[1955] Local dimension of normal spaces, Quarterly Journal of Mathematics, Oxford **6** (1955), 101–120.

Duda, R.

[1979] The origins of the concept of dimension, Colloquium Mathematicum **42** (1979), 95–110.

Efremovich, V.A.

[1952] Geometry of proximity I, Matematicheskiĭ Sbornik **31** (1952), 189–200 (Russian).

Eilenberg, S.

[1936] Un théorème de dualité, Fundamenta Mathematicae **26** (1936), 280–282.

Eilenberg, S. and Otto, E.

[1938] Quelques propriétés caractéristiques de la théorie de la dimension, Fundamenta Mathematicae **31** (1938), 149–153.

Engelking, R.

[1960] Sur la compactification des espaces métriques, Fundamenta Mathematicae **48** (1960), 321–324.

[1961] On the Freudenthal compactification, Bulletin de l'Académie Polonaise des Sciences **9** (1961), 379–383.

[1967] On Borel sets and B-measurable functions in metric spaces, Prace Matematyczne **10** (1967), 145–149.

[1977] *General topology* (Warszawa 1977).

[1978] *Dimension theory* (Warszawa 1978).

Engelking, R. and Sklyarenko, E.G.

[1963] On compactifications allowing extensions of mappings, Fundamenta Mathematicae **53** (1963), 65–79.

FEDORCHUK, V.V.

[1988] *Principles of dimension theory*, Current problems in mathematics, Vol. 17, Edited by R.V. Gamkrelidze, pp. 111–224, VINITI, Moscow, 1988 (Russian); *The fundamentals of dimension theory*, Encyclopaedia of Mathematical Sciences, Vol. 17 (General Topology I), 91–202, Berlin 1990.

FILIPPOV, V.V.

[1973] On the dimension of normal spaces, Doklady Akademii Nauk SSSR **209** (1973), 805–807 (Russian); Soviet Mathematical Doklady **14** (1973), 547–550.

FORGE, A.B.

[1961] Dimension preserving compactifications allowing extensions of continuous functions, Duke Mathematical Journal **28** (1961), 625–627

FREUDENTHAL, H.

[1931] Über die Enden topologischer Räume und Gruppen, Mathematische Zeitschrift **33** (1931), 692–713.

[1942] Neuaufbau der Endentheorie, Annals of Mathematics **43** (1942), 261–279.

[1951] Kompaktisierungen and bikompaktisierungen, Indagationes Mathematicae **13** (1951), 184–192.

[1952] Enden und Primenden, Fundamenta Mathematicae **39** (1952), 189–210

FRINK, O.

[1964] Compactifications and semi-normal spaces, American Journal of Mathematics **86** (1964), 602–607.

GILLMAN, L. and JERISON, M.

[1960] *Rings of continuous functions* (Princeton (N.J.) 1960)

GROOT, J. DE

[1942] *Topologische studiën*, thesis (Groningen 1942).

[1961] Linearization of mappings, *General topology and its relations to modern analysis and algebra* (Proceedings Prague Topological Symposium, Prague 1961), Prague, 1962, 191–193.

[1969] *Seminar on compactification and dimension in metric spaces*, mimeographed lecture notes University of Florida, Gainesville (FL).

GROOT, J. DE and MCDOWELL, R.H.

[1960] Extension of mappings on metric spaces, Fundamenta Mathematicae **48** (1960), 251–263.

GROOT, J. DE and NISHIURA, T.
[1966] Inductive compactness as a generalization of semicompactness, Fundamenta Mathematicae **58** (1966), 201–218.

HART, K.P.
[1985] A space X with $\operatorname{cmp} X = 1$ and $\operatorname{def} X = \infty$, preliminary report.

HART, K.P., MILL, J. VAN and VERMEER, H.
[1982] Non-equality of inductive completeness degrees, VU Report 223, 9 p, Free University, Amsterdam.

HAYASHI, Y.
[1990] An axiomatic characterization of the dimension of subsets of Euclidean spaces, Topology and its Applications **37** (1990), 83–92.

HENDERSON, D.W.
[1967] An infinite-dimensional compactum with no positive-dimensional compact subsets–a simpler construction, American Journal of Mathematics **89** (1967), 105–121.

HENRIKSEN, M.
[1977] Tychonoff spaces that have a compactification with countable remainder, *General topology and its relations to modern analysis and algebra, IV* (Proceedings Fourth Prague Topological Symposium, Prague 1976), Part B, Prague, 1977, 164–167.

HENRIKSEN, M. and ISBELL, J.R.
[1957] Some properties of compactifications, Duke Mathematical Journal **25** (1957), 83–105.

HOSHINA, T.
[1977] Compactifications by adding a countable number of points, *General topology and its relations to modern analysis and algebra, IV* (Proceedings Fourth Prague Topological Symposium, Prague 1976), Part B, Prague, 1977, 168–169.
[1979] Countable-point compactifications for metric spaces, Fundamenta Mathematicae **103** (1979), 123–132.

HUREWICZ, W.
[1927] Normalbereiche und Dimensionstheorie, Mathematische Annalen **96** (1927), 736–764.

HUREWICZ, W. and WALLMAN, H.
[1941] *Dimension theory* (Princeton (N.J.) 1941).

ISBELL, J.R.
[1964] *Uniform spaces* (Providence (R.I.) 1964).

KATĚTOV, M.

[1952] On the dimension of non-separable spaces. I, Czechoslovak Mathematical Journal **2** (1952), 333–368.

KIMURA, K.

[1986] Gaps between compactness degree and compactness deficiency for Tychonoff spaces, Tsukuba Journal of Mathematics **10** (1986), 263–268.

[1988] Solution to a compactification problem of Sklyarenko, Fundamenta Mathematicae **131** (1988) 25–33.

[1988a] The gap between $\operatorname{cmp} X$ and $\operatorname{def} X$ can be arbitrarily large, Proceedings of the American Mathematical Society **102** (1988), 1077–1080.

[1989] A separable metrizable space X for which $\operatorname{Cmp} X \neq \operatorname{def} X$, Bulletin de l'Académie Polonaise des Sciences **37** (1989), 487–495.

KOTKIN, S.V.

[1990] On dimension of non-totally normal spaces, Vestnik Moskovskogo Universiteta (1990) (Russian); Moscow University Bulletin **45** (1990), 36–39 (English translation).

KULESZA, J.

[1990] An example in the dimension theory of metrizable spaces, Topology and its Applications **35** (1990), 109–120.

KURATOWSKI, C.

[1958] *Topologie I* (Warszawa 1958) (4th edition).

[1961] *Topologie II* (Warszawa 1961) (3rd edition).

KUROVSKIĬ, A.B. and SMIRNOV, YU.M.

[1976] On the dimension Ird defined by recursion, Czechoslovak Mathematical Journal **26** (1976), 30–36 (Russian)

KUZ'MINOV, V.I.

[1968] Homological Dimension Theorie, Uspekhi Matematicheskikh Nauk **23** (1968), 3–50 (Russian); Russian Mathematical Surveys 23 (1968), 1–45 (English translation)

LAVRENTIEFF, M.

[1924] Contribution à la théorie des ensembles homéomorphes, Fundamenta Mathematicae **6** (1924), 149–160.

LEBESGUE, H.

[1911] Sur la non applicabilité de deux domaines appartenant respectivement à des espaces, de n et $n+p$ dimensions, Mathematische Annalen **70** (1911), 166–168.

[1921] Sur les correspondences entre les points de deux espaces, Fundamenta Mathematicae **2** (1921), 256–285.

LELEK, A.

[1964] Dimension and mappings of spaces with finite deficiency, Colloquium Mathematicum **12** (1964), 221–227.

[1965] On the dimension of remainders in compact extensions, Doklady Akademii Nauk SSSR **160** (1965), 534–537 (Russian); Soviet Mathematical Doklady **6** (1965), 136–140.

LIFANOV, I.K. and PASYNKOV, B.A.

[1970] Two classes of spaces and two forms of dimension, Vestnik Moskovskogo Universiteta **25** (1970), 33–37 (Russian); Moscow University Bulletin **23** (1970), 26–29 (English translation).

LOKUCIEVSKIĬ, O.V.

[1949] On the dimension of bicompacta, Doklady Akademii Nauk SSSR **67** (1949), 217–219 (Russian).

[1973] On axiomatic definition of the dimension of bicompacta, Doklady Akademii Nauk SSSR **212** (1973), 813–815 (Russian); Soviet Mathematical Doklady **14** (1973), 1477–1480.

LUXEMBURG, L.A.

[1973] On compact spaces with noncoinciding transfinite dimensions, Doklady Akademii Nauk SSSR **212** (1973), 1297–1300 (Russian); Soviet Mathematical Doklady **14** (1973), 1593–1597.

[1981] On compact metric spaces with noncoinciding transfinite dimensions, Pacific Journal of Mathematics **93** (1981), 339–386.

MARDEŠIĆ, S.

[1966] Continuous images of ordered compacta and a new dimension which neglects metric subcontinua, Transactions of the American Mathematical Society **121** (1966), 424–433.

MENGER, K.

[1923] Über die Dimensionalität von Punktmengen I, Monatsheft für Mathematik und Physik **33** (1923), 148–160.

[1928] *Dimensionstheorie* (Leipzig 1928).

[1929] Über die Dimension von Punkmengen III. Zur Bergründung einer axiomatischen Theorie der Dimension, Monatsheft für Mathematik und Physik **36** (1929), 193–218.

[1943] What is dimension? American Mathematical Monthly **50** (1943), 2–7.

MILL, J. VAN

[1977] *Supercompactness and Wallman spaces*, PhD thesis (Amsterdam 1977)

[1982] Inductive Čech completeness and dimension, Compositio Mathematica **45** (1982), 145–153.

[1989] *Infinite-dimensional topology, Prerequisites and introduction* (Amsterdam 1989).

MILL, J. VAN and REED, G.M.

[1990] *Open problems in topology* (Amsterdam 1990).

MORITA, K.

[1950] On the dimension of normal spaces I, Japanese Journal of Mathematics **20** (1950), 5–36.

[1950a] On the dimension of normal spaces II, Journal of the Mathematical Society of Japan, 16–33.

[1952] On bicompactification of semibicompact spaces, Science Reports Tokyo Bunrika Daigaku, Section A4, **92** (1952), 200–207.

[1953] On the dimension of product spaces, American Journal of Mathematics **75** (1953), 205–223.

[1954] Normal families and dimension theory for metric spaces, Mathematische Annalen **128** (1954), 350–362.

[1956] On closed mappings, Proceedings of the Japan Academy **32** (1956), 539–543.

[1957] On closed mappings II, Proceedings of the Japan Academy **33** (1957), 325–327.

[1959] On images of an open interval under closed continuous mappings, Proceedings of the Japan Academy **35** (1959), 15–19.

MRÓWKA, S.

[1985] N-compactness, metrizability and covering dimension, *Lecture notes in pure and applied mathematics, Vol. 95* (New York 1985), 247–275

NAGAMI, K.

[1965] A Nagata's metric which characterizes dimension and enlarges distance, Duke Mathematical Journal **32** (1965), 557-562.

[1970] *Dimension theory* (New York 1970).

NAGATA, J.

[1965] *Modern dimension theory* (Groningen 1965).

[1967] Some aspects of extension theory in general topology, International symposium on extension theory, Berlin **1967**, 157–161.

NAIMPALLY, S.A. and WARRACK, B.D.

[1970] *Proximity spaces* (Cambridge 1970).

NISHIURA, T.

[1961] Semi-compact spaces and dimension, Proceedings of the American Mathematical Society **12** (1961), 922–924.

[1964] On the dimension of semicompact spaces and their quasicomponents, Colloquium Mathematicum **12** (1964), 7–10.

[1966] Inductive invariants and dimension theory, Fundamenta Mathematicae **59** (1966), 243–262.

[1972] Inductive invariants of closed extensions of mappings, Colloquium Mathematicum **25** (1972), 73-78.

[1972a] An axiomatic characterization of covering dimension in metrizable spaces, in *TOPO 72, General topology and its applications*, Proceedings Second Pittsburgh International Conference 1972, (Berlin 1974)

[1977] A subset theorem in dimension theory, Fundamenta Mathematicae **95** (1977), 105–109.

NJÅSTAD, O.

[1966] On Wallman-type compactifications, Mathematische Zeitschrift **91** (1966), 267–276

OSTASZEWSKI, A.

[1990] A note on the Prabir Roy space, Topology and its Applications **35** (1990), 95–107

PASYNKOV, B.A.

[1967] On open mappings, Doklady Akademii Nauk SSSR **175** (1967), 292–295 (Russian); Soviet Mathematics Doklady **8** (1967), 853–856 (English translation).

PEARS, A.R.

[1975] *Dimension theory of general spaces* (Cambridge 1975).

PESCHKE, G.

[1990] The theory of ends, Nieuw Archief voor Wiskunde **8** (1990), 1–12.

POINCARÉ, H.

[1912] Pourquoi l'espace a trois dimensions, Revue de métaphysique et de morale **20** (1912), 483–504.

POL, E. and POL, R.

[1979] A hereditarily normal strongly zero-dimensional space containing subspaces of arbitrarily large dimension, Fundamenta Mathematicae **102** (1979), 137–142.

POL, R.

[1982] A counterexample to J. de Groot's problem cmp = def, Bulletin de l'Académie Polonaise des Sciences **30** (1982), 461–464.

[1987] On transfinite inductive compactness degree, Colloquium Mathematicum **53** (1987), 57–61.

PRZYMUSIŃSKI, T.

[1974] A note on dimension theory of metric spaces, Fundamenta Mathematicae **85** (1974), 277–284.

REICHAW, M.
[1972] On the theorem of Hurewicz on mappings which lower dimension, Colloquium Mathematicum **26** (1972), 323–329.

ROY, P.
[1962] Failure of equivalence of dimension concepts for metric spaces, Bulletin American Mathematical Society **68** (1962), 609–613.
[1968] Nonequality of dimension for metric spaces, Transactions American Mathematical Society **134** (1968), 117-132.

SAKAI, S.
[1968] An axiomatic characterization of the large inductive dimension for metric spaces, Proceedings Japan Academy **44** (1968), 782–785.

SHANIN, N.A.
[1943] On special extensions of topological spaces, Doklady Akademii Nauk SSSR **38** (1943), 6–9.
[1943a] On special extensions of topological spaces, Doklady Akademii Nauk SSSR **38** (1943), 110–113.
[1943b] On the theory of bicompact extensions of topological spaces, Doklady Akademii Nauk SSSR **38** (1943), 154–156.

SHCHEPIN, E.
[1972] An axiomatization of the dimension of metric spaces, Doklady Akademii Nauk SSSR **206** (1972), 31–32 (Russian); Soviet Mathematics Doklady **13** (1972), 1177–1179 (English translation).

SKLYARENKO, E.G.
[1958] Bicompact extensions of semibicompact spaces, Doklady Akademii Nauk SSSR **120** (1958), 1200–1203
[1958a] On the embedding of normal spaces in bicompact spaces of the same weight and of the same dimension, Doklady Akademii Nauk SSSR **123** (1958), 36–39 (Russian).
[1960] Bicompact extensions and dimension, Trudy Tbilisskogo Matematicheskogo Instituta **27** (1960), 113–114 (Russian).
[1962] Some questions in the theory of bicompactifications, Izvestiya Akademii Nauk SSSR **26** (1962), 427–452 (Russian); American Mathematical Society Translations **58** (1966), 216–244.

SMIRNOV, YU.M.
[1952] On proximity spaces, Matematicheskiĭ Sbornik **31** (1952), 543–574 (Russian); American Mathematical Society Translations, (2), **38** (1964), 5–35 (English translation).
[1954] On completeness of proximity spaces, Trudy Moskovskogo Matematicheskogo Obshchestva **3** (1954), 271–308 (Russian);

American Mathematical Society Translations (2) **38** (1964), 37–73 (English translation).

[1956] On the dimension of proximity spaces, Matematicheskiĭ Sbornik **41** (1956), 283–302 (Russian); American Mathematical Society Translations, (2), **21**, 1–20 (English translation).

[1958] Example of a non-semibicompact completely regular space with a zero-dimensional complement in its Čech compactification, Doklady Akademii Nauk SSSR **120** (1958), 1204–1206 (Russian).

[1965] Über die Dimension der Adjunkten bei Kompaktifizierungen, Monatsberichte der Deutsche Adademie der Wissenschaften zu Berlin **7** (1965), 230–232.

[1965a] Einige Bemerkungen zu meinem Bericht "Über die Dimension der Adjunkten bei Kompaktifizierungen", Monatsberichte der Deutsche Akademie der Wissenschaften zu Berlin **7** (1965), 750–753.

[1966] On the dimensions of remainders of compactifications of proximity and topological spaces, Matematicheskiĭ Sbornik **69** (1966), 141–160 (Russian); see next paper for the English translation.

[1966a] On the dimensions of remainders of compactifications of proximity and topological spaces II, Matematicheskiĭ Sbornik **71** (1966), 454–482 (Russian); American Mathematical Society Translations, Series 2, **84** (1969), 197–251 (English translation).

STEINER, A.K. and STEINER, E.F.

[1969] On countable multiple point compactifications, Fundamenta Mathematicae **65** (1969), 133–137.

STONE, A.H.

[1962] Absolute F_σ spaces, Proceedings of the American Mathematical Society **13** (1962), 495–499.

STONE, M.H.

[1937] Applications of the theory of Boolean rings to general toplogy, Transactions of the American Mathematical Society **41** (1937), 375–481.

TERADA, T.

[1977] On countable discrete compactifications, General Topology and its Applications **7** (1977), 321–327.

TUMARKIN, L.A.

[1926] Beitrag zur allgemeinen Dimensionstheorie, Matematicheskiĭ Sbornik **33** (1926), 57–86.

UL'JANOV, V.M.

[1977] Solution of a basic problem on compactifications of Wallman type, Doklady Akademii Nauk SSSR **233** (1977), 1056–1059 (Russian); Soviet Mathematical Doklady **18** (1977), 567–571.

URYSOHN, P.

[1922] Les multiplicités Cantoriennes, Comptes rendus hebdomadaires des séances de l'Academie des Sciences, Paris **175** (1922), 440–442.

VAĬNSHTEĬN, I.A.

[1949] On dimension raising mappings, Doklady Akademii Nauk SSSR **67** (1949), 9–12 (Russian).

VEDENISOFF, N.

[1939] Remarks on the dimension of topological spaces, Uchenye Zapiski Moskovskogo Universiteta **30** (1939), 131–140.

[1941] On the dimension in the sense of E. Čech, Izvestiya Akademii Nauk SSSR, Series Matematika **5** (1941), 211–216.

VRIES, H. DE

[1962] *Compact spaces and compactifications*, PhD thesis (Amsterdam 1962).

WALLMAN, H.

[1938] Lattices and topological spaces, Annals of Mathematics, **39** (1938), 112–126.

WENNER, B.R.

[1969] Remetrization on strongly countable-dimensional spaces, Canadian Journal of Mathematics **21** (1969), 748-750.

[1970] Extending maps and dimension theory, Duke Mathematical Journal **37** (1970), 627-631.

WHYBURN, G.T.

[1942] *Analytic topology*, A.M.S. Colloquium Publications **28**, American Mathematical Society (Providence,RI 1942).

ZIPPIN, L.

[1935] On semicompact spaces, American Journal of Mathematics **57** (1935), 327–341.

LIST OF SYMBOLS

List of generalized dimension functions

The proper dimension functions

Bind	basic inductive dimension, 218
Dim	large covering dimension, 94
dim	covering dimension, 43
$\dim^{\infty}$	boundary dimension, 305
ind	small inductive dimension, 3
Ind	large inductive dimension, 9
Odim	order dimension, 47

The generalized dimension functions

$\mathcal{P}$-Bind	basic inductive dimension modulo $\mathcal{P}$, 218
$\mathcal{P}$-def	small $\mathcal{P}$-deficiency, 108
$\mathcal{P}$-Def	large $\mathcal{P}$-deficiency, 108
$\mathcal{P}$-dim	small covering dimension modulo $\mathcal{P}$, 94
$\mathcal{P}$-$\dim_{\boldsymbol{G}}$	relative dimension with respect to $\boldsymbol{G}$, 199
$\mathcal{P}$-Dim	large covering dimension modulo $\mathcal{P}$, 94
$\mathcal{P}$-ind	small inductive dimension modulo $\mathcal{P}$, 76
$\mathcal{P}$-Ind	large inductive dimension modulo $\mathcal{P}$, 80
$\mathcal{P}$-Mind	mixed inductive dimension modulo $\mathcal{P}$, 241
$\mathcal{P}$-Odim	order dimension modulo $\mathcal{P}$, 234
$\mathcal{P}$-sur	$\mathcal{P}$-surplus, 107
$\mathcal{P}$-Sur	$\mathcal{P}$-Surplus, 85
ccd	covering completeness degree, 49
cmp	small inductive compactness degree, 15
Cmp	large inductive compactness degree, 21
Comp	22
def	compactness deficiency, 16
icd	small inductive completeness degree, 33
Icd	large inductive completeness degree, 33
loccom	210
Skl	23

List of other symbols

Special compactifications of a space X are denoted by small Greek letters.

αX	one point compactification
βX	Čech-Stone compactification
$\mathcal{F}X$	Freudenthal compactification, 267
δX	compact δ-extension, 302
ωX	Wallman compactification, 254

$\omega(\mathcal{F}, X)$ Wallman compactification with respect to the base $\mathcal{F}$, 260

Special subsets of Euclidean spaces are denoted by blackboard bold letters.

$\mathbb{N}$ natural numbers (starting with 0)
$\mathbb{I}$ interval $[-1, 1]$
$\mathbb{P}$ irrational numbers
$\mathbb{Q}$ rational numbers
$\mathbb{R}$ real numbers
$\mathbb{R}^n$ n-dimensional Euclidean space
$\mathbb{S}^n$ n-dimensional sphere

With four exceptions, namely $\mathcal{B}$, $\mathcal{F}$, $\mathcal{G}$ and $\mathcal{Z}$, capital script letters denote classes of spaces. This convention is not in effect in Section VI.5

$\mathcal{B}$ base for the open sets
$\mathcal{C}$ class of complete metrizable spaces, 35
$\mathcal{D}$ Dowker universe, 156
$\mathcal{E}$ class of strongly hereditarily normal spaces, 150
$\mathcal{F}$ base for the closed sets
$\mathcal{G}$ base for the closed sets
$\mathcal{K}$ class of compact spaces
$\mathcal{K}(X)$ set of metrizable compactifications of X, 16
$\mathcal{L}$ class of locally compact spaces, 86
$\mathcal{M}$ class of metrizable spaces
$\mathcal{M}_0$ class of separable metrizable spaces
$\mathcal{N}$ class of normal spaces
$\mathcal{N}_H$ class of hereditarily normal spaces, 87
$\mathcal{N}_P$ class of perfectly normal spaces
$\mathcal{N}_T$ class of totally normal spaces, 147
$\mathcal{P}$ a general class of spaces
$\mathcal{Q}$ a general class of spaces
$\mathcal{R}$ class of regular spaces
$\mathcal{R}_c$ class of completely regular spaces
$\mathcal{S}$ class of σ-compact spaces, 54
$\mathcal{T}$ class of topological spaces
$\mathcal{U}$ universe of discourse, 74
$\mathcal{U}_n$ class of subspaces of $\mathbb{R}^n$
$\mathcal{U}_\omega$ class of subspaces of Euclidean spaces
$\mathcal{Z}$ family of zero-sets of a space, 255

INDEX